RENEWING DIALOGUES IN MARXISM AND EDUCATION

Marxism and Education

This series assumes the ongoing relevance of Marx's contributions to critical social analysis and aims to encourage continuation of the development of the legacy of Marxist traditions in and for education. The remit for the substantive focus of scholarship and analysis appearing in the series extends from the global to the local in relation to dynamics of capitalism and encompasses historical and contemporary developments in political economy of education as well as forms of critique and resistances to capitalist social relations. The series announces a new beginning and proceeds in a spirit of openness and dialogue within and between Marxism and education, and between Marxism and its various critics. The essential feature of the work of the series is that Marxism and Marxist frameworks are to be taken seriously, not as formulaic knowledge and unassailable methodology but critically as inspirational resources for renewal of research and understanding, and as support for action in and upon structures and processes of education and their relations to society. The series is dedicated to the realization of positive human potentialities as education and thus, with Marx, to our education as educators.

Renewing Dialogues in Marxism and Education: Openings
Edited by Anthony Green, Glenn Rikowski, and Helen Raduntz

Renewing Dialogues in Marxism and Education

Openings

Edited by

Anthony Green, Glenn Rikowski, and Helen Raduntz

RENEWING DIALOGUES IN MARXISM AND EDUCATION
Copyright © Anthony Green, Glenn Rikowski, and Helen Raduntz, 2007.

Softcover reprint of the hardcover 1st edition 2007 978-1-4039-7496-9

All rights reserved. No part of this book may be used or reproduced in any manner whatsoever without written permission except in the case of brief quotations embodied in critical articles or reviews.

First published in 2007 by
PALGRAVE MACMILLAN™
175 Fifth Avenue, New York, N.Y. 10010 and
Houndmills, Basingstoke, Hampshire, England RG21 6XS
Companies and representatives throughout the world.

PALGRAVE MACMILLAN is the global academic imprint of the Palgrave Macmillan division of St. Martin's Press, LLC and of Palgrave Macmillan Ltd. Macmillan® is a registered trademark in the United States, United Kingdom and other countries. Palgrave is a registered trademark in the European Union and other countries.

ISBN 978-1-349-53565-1   ISBN 978-0-230-60967-9 (eBook)
DOI 10.1057/9780230609679

Library of Congress Cataloging-in-Publication Data is available from the Library of Congress.

A catalogue record for this book is available from the British Library.

Design by Newgen Imaging Systems (P) Ltd., Chennai, India.

First edition: December 2007

10 9 8 7 6 5 4 3 2 1

Transferred to Digital Printing 2011

After Marx, to the Educators... *and to our education*...

Contents

Acknowledgments   ix

Part I   Introduction and Overview: Marxism and Education—Renewing the Dialogues

Introduction and Overview: Marxism and
Education—Renewing the Dialogues   3
Anthony Green and Glenn Rikowski

1. Marxism, Education, and Dialogue   11
   Anthony Green

Part II   Neoliberalism, Globalization, Crises, and Education in the Transitional Epochs

2. Toward a Political Economy of Education in the Transitional Period   35
   Geraldine Thorpe and Pat Brady

3. The Role of Education in Capital Crisis Resolution   57
   Helen Raduntz

4. What Neoliberal Global and National Capitals Are Doing to Education Workers and to Equality—Some Implications for Social Class Analysis   71
   Dave Hill

5. Neoliberalism and Education: A Marxist Critique of New Labour's Five-Year Strategy for Education   103
   Mike Cole

Part III   Discourse, Postmodernism, and Poststructuralism

6. Indecision, Social Justice, and Social Change: A Dialogue on Marxism, Postmodernism, and Education   119
   Elizabeth Atkinson and Mike Cole

7 Textual Strategies of Representation and Legitimation in New Labour Policy Discourse 135
Jane Mulderrig

8 Marx, Education, and the Possibilities of a Fairer World: Reviving Radical Political Economy through Foucault 151
Mark Olssen and Michael A. Peters

Part IV   Politics, Divisions, and Struggle in Teaching, Learning, and Education Policy

9 The Feminist Standpoint and the Trouble with "Informal Learning": A Way Forward for Marxist-Feminist Educational Research 183
Rachel Gorman

10 Myths of Mentoring: Developing a Marxist-Feminist Critique 201
Helen Colley

11 Popular Press, Visible Value: How Debates on Exams and Student Debt Have Unmasked the Commodity Relations of the "Learning Age" 215
Paul Warmington

Part V   Labor and Commodification in Education: Theory, Practice, and Critique

12 Academic Labor: Producing Value and Producing Struggles 231
David Harvie

13 Marxist Political Praxis: Class Notes on Academic Activism in the Corporate University 249
Gregory Martin

14 The Making of Humanity: The Pivotal Role of Dialectical Thinking in Humanization and the Concomitant Struggle for Self and Social Transformation 267
Paula Allman

Contributors 279

Index 283

Acknowledgments

Thanks are due to several people, particularly for their support and encouragement during the latter stages of the production of this edition, most notably to Ruth Klassen and Ruth Rikowski, Tony and Glenn's partners. Elizabeth Molinari did brilliantly efficient and effective work for the final manuscript at a very critical stage. Also to our Palgrave editor, Amanda Johnson Moon, for both her continuing support and for her patience.

Finally, we would like to express thanks to the participants in the Marxism and Education: Renewing Dialogues (MERD) seminars for whom this is a *product* of our praxis.

Part I

Introduction and Overview: Marxism and Education—Renewing the Dialogues

Introduction and Overview: Marxism and Education—Renewing the Dialogues

Anthony Green and Glenn Rikowski

This book is the first in the Palgrave Macmillan Marxism and Education Series. Both the book and the series arose from the Marxism and Education: Renewing Dialogues (MERD) seminars initiated by us at the Institute of Education, University of London in 2002. The seminar series emerged at a point of our recognition of a somewhat bleak time for Marxist theory, research, and discourses in regard to education, despite some indications of small-scale revival. Our concern has been to address this context in the spirit of *taking Marxism seriously* and encouraging and supporting this in critical educational analysis, debate, and progressive practices. The MERD seminars have played a part in the ongoing revival.

The chapters included in this volume represent positive signs of this revival. They constitute a wide variety of themes and indeed styles of thinking and presentation for ideology-critique and substantive analysis, including complex philosophical, dialogical, critical discursive, empirical and concrete analysis and commentary, at times highly polemical, all associated with the preoccupations of diverse currents in Marxist analysis of education. There are inevitable tensions and differences of analytical emphasis which is as it should be for an *open* form of Marxism attempting to address a multiplicity of contexts around the continuing struggles for socialism in a world in which the value form of labor and commodification are central to neoliberal globalization of capital in all its educational-dimensions.

Here we will introduce and summarize the constituent chapters of the volume, while outlining the rationale for its division into five parts. Part I sets a scene with chapter 1 entitled "Education, Marxism, and Dialogue" with Green focusing on "pedagogy of critique," central to the working practices of the MERD seminars. The chapter aims to constitute a prefatory and loosely programmatic "opening." He draws attention to broad contexts and ongoing issues for Marxist theory and methodology as dialogistic, open-ended, and nondogmatic commitment to historical materialism in relation to themes continuingly significant for realist critical educational theory and practices as *Marxism embodied*. In doing so it also picks up major issues such as commodification, dialectical and class analysis, and the politics of education as restlessly in processes of "renewal."

The remaining four parts comprise levels or dimensions and contexts for working through issues in Marxism and education. In principle these parts are interconnected, mutually present, and dialectically related contexts of struggle and understandings-in-struggles ranging across the global framings of neoliberalism and transitional epochs, as addressed in part II, to very specific struggles and practices discussed elsewhere. Each issue requires articulation in the contexts of continuous concerns for critical epistemology especially signalled with the turns to postmodernist and poststructuralist discourses as significant moments of challenge and development for Marxist theory of education, as addressed in part III. Part IV refocuses on the politics of social divisions in relation to education policy and practices, and part V concludes the volume by addressing issues concerning the political themes of theory, substantive practice, and critique *in* struggles and *for* humanization.

Thus, part II, *Neoliberal Globalization, Crises, and Education in Transitional Epochs* focuses on the international and macro context of educational change. The constituent chapters explore the context of educational development and crisis in terms of capitalist globalization and neoliberalism. In chapter 2 Geraldine Thorpe and Pat Brady examine the *Political Economy of Education in the Transitional Period*. Their argument is that recent changes in education can be understood in the general context of *transition*, where, despite appearances to the contrary, capitalism is in *decline* and competing with the burgeoning movements of the potential future society. The authors outline Marxist political economy of education in the *transitional period* drawing upon interpretations of Marxist methodology inspired by Ticktin and Kennedy's approaches in which description gives way to explanation in Marxist mode of critique to identify what lies below the surface of appearances. And, as they conclude, this serves to underline the issue that it takes more than governments and capital to make education worthwhile.

In chapter 3, *The Role of Education in Capital Crisis Resolution*, Helen Raduntz complements the Thorpe/Brady analysis with elements of correspondence between developments in education and the evolution of the capitalist economy, also drawing on Marx's critique of capital and shows how the current economic crisis is influencing the nature of the changes in education in complex and possibly paradoxical ways. These reflect capitalism's dynamism and articulations in terms of constant restructuring around technological change. And so, paradoxically, capitalism's survival in crisis is emergent from alternate states of stability and instability. In the process education becomes redesigned in relation to new forms of production. This is especially true for the role of education in crisis management of capitalism's internal contradictions evident in many educational contexts today.

In chapter 4, on *What Neoliberal Global and National Capital Are Doing to Education Workers and to Equality,* Dave Hill argues, polemically in places, for the continued salience of social class analysis and political practices in the era of neoliberal globalization. The chapter provides substantive indications of how neoliberalism impels educational marketization and commodification, which in turn leads to widening educational inequalities, in particular class inequality. Hill advances principles that run counter to those of neoliberalism currently powering education policy across the globe. He argues that such principles appeal to anticapitalist forms of social justice within the context of socialist transformation in which Marxist class analysis requires clearer expression and differentiation from liberal progressive critical sociological analysis.

Mike Cole develops the themes of *Neoliberalism and Education: A Marxist Critique of New Labor's Five-Year Strategy for Education,* in chapter 5. He surveys educational marketization and its links to the ideology and practice in education today in the context of global neoliberalism. Specifically, Cole focuses on some of the neoliberal strands within New Labour's Five-Year Strategy for Education that was published in the summer of 2004. He provides a Marxist analysis and critique of these projected policy strands, working through issues of the capitalist agenda *in* and *for* education. Here he examines the debate from *within* Marxist critical analysis between Richard Hatcher and Glenn Rikowski on ways to understand recent private sector involvement in education policy as possibly a *Trojan Horse* for future development and leaves it to history to decide the outcome.

In part III, *Discourse, Postmodernism, and Poststructuralism* the focus is on the *dialogue* between and articulations *with* Marxism and modes of thought that also claim to "unsettle" conventional thinking, such as critical discourse analysis, poststructuralism, and postmodernism. Thus, in chapter 6, *Indecision, Social Justice, and Social Change: A Dialogue on Marxism, Postmodernism, and Education* Elizabeth Atkinson and Mike Cole take opposing positions on the critical efficacy of postmodernist and Marxist analyses. This lightly edited dialogue took place in "real time" at the first MERD seminar on 22 October 2002. The chapter draws out key themes from the original dialogue, using direct quotation and retrospective reflection to highlight some of the differences and a few similarities and points of connection between the two forms of educational theory and practice.

In chapter 7 Jane Mulderrig elaborates on *Textual Strategies of Representation and Legitimation in New Labour Policy Discourse.* She assesses the contribution of critical discourse analysis to critical education policy research. To this end, the political economy of New Labour education policy is examined using a combination of Marxist state theory (via Jessop), Marxist education policy analysis (through Hill), and close textual analysis. The focus is the discursive representation in the policy documents of key educational actors, and the discourse strategies by which that policy is legitimated. It is postulated that an instrumental rationality underlies educational actors, which is theorized as an indicator of the general shift toward the *commodification* of education that stems in part from a subordination of social to economic policy. Educational practices, argues Mulderrig, play an increasingly important role in the ongoing construction of the globalizing, knowledge-based economy, and the *learning society*. Given this context, it is argued that critical analysis of education policy discourse constitutes an important tool for research into the complex relationship between education and the economy, uncovering its inherent tensions and contradictions, and demonstrating the mechanisms by which contemporary educational agendas are legitimated.

In chapter 8 Mark Olssen and Michael Peters discuss *Marx, Education, and the Possibilities of a Fairer World: Reviving Radical Political Economy through Foucault*. The Atkinson/Cole dialogue has brought postmodernism and poststructuralism together in tension and contradiction with Marxism through opposing modes of analysis. Olssen and Peters offer an attempt to meld Marxism and poststructuralism, partly through Foucault and under an umbrella of *historical materialism* to suggest a synthesis for analysis of the educational implications of *knowledge capitalism* and democratic movement in the globalized world. Unlike Marxism, it is claimed, Foucault sees no one set of factors as necessarily directing human destiny and, echoing well-known critiques of *base superstructure modelling*, focuses on forms of articulation and determination potentially differing in relation to the relative importance of different nondiscursive (material) factors in terms of both place and time. However, not unlike much of Marxist analyses of cultural forms under capitalism a Foucauldian critique of political economy, they argue, shows how in the global economic era neoliberalism constitutes an authoritarian discourse of state management and control. Olssen and Peters thus argue that Foucault's approach supports a model of *global democracy*, with a "bottom-up" theory of the democratization of world order and the role of grassroots critical social movements, not least in relation to knowledge and education as a challenge to structuralist Marxism.

Part IV collects together themes around *Politics, Divisions, and Struggle in Teaching, Learning, and Education Policy*. This part focuses on Marxist and Marxist-feminist analyses of education or education policies (e.g., mentoring, informal learning, examinations). The *internal dialogues* within the chapters in this section move between feminism/Marxism, appearance/essence, policy/practice, and teaching/learning.

In chapter 9 Rachel Gorman examines *The Feminist Standpoint and the Trouble with "Informal Learning": A Way Forward for Marxist-Feminist Educational Research*. There are three main aims. First, it critiques the way in which current notions of informal learning both individualize and depoliticize the real lives and activities of learners. Second, it suggests that an educational research method that is both dialectical materialist and feminist is possible and appropriate. Finally, it is argued that radical educators must integrate an understanding of *oppression, exploitation*, and *political consciousness* into their definitions of learning and education. The arguments of the chapter are underpinned by two case studies of informal learning in the Greater Toronto Area (Canada). The first involves Kurdish women in diaspora. The second case study involves disability rights activists. The form of analysis is committed to specificity and complexity in relation to context. However, it argues that feminist and social justice oriented researchers, must be grounded in the concrete understanding of how power relations unfold over time, while in relation to global political economy, offering the possibility of avoiding the dangers of empiricist particularity and social relativity.

Helen Colley develops an analysis of *Myths of Mentoring: Developing a Marxist-Feminist Critique* in chapter 10. It is recognized that mentoring is highly popular today; much vaunted as the dominant solution to wide-ranging policy needs, from professional development to reengaging disadvantaged youth. However, it is weakly conceptualized. Both policy and academic literature promote approaches often defined by referring to mentoring in Homer's *Odyssey*. For Colley, these portray the mentor as a kindly sage and trusted advisor prepared to go beyond the call of duty for their mentee. This chapter argues that such appropriations are based on misreadings of Homer's myth. The Odyssey is in fact a very brutal story of a powerful prince mentored by an omnipotent deity. Colley uses *feminist theory* to present an epic interpretation of this original story as a parable of the establishment of male, military, political, and sexual domination. Her analysis raises a number of policy issues. Among these are concerns about the overly individualized responses to disadvantage, obscuring structural inequalities, and running the risk of social control over mentees; and about the additional, often unpaid burdens on mentors, as a post-Fordist, gendered form of exploitation. A *dialectical materialist* understanding (drawing from Marx and Marxism) of the interplay of relative essences and absolute appearances helps map the emergence of a discourse of mentoring as a regime of truth (drawing on Foucault) in this chapter.

Paul Warmington, in chapter 11, *Popular Press, Visible Value: How Debates on Exams and Student Debt Have Unmasked the Commodity Relations of the "Learning Age"* provides a detailed analysis of the media coverage of A-level and GCSE (General Certificate of Secondary Education) results. He draws on material from a project—the *Media Depiction Project*—undertaken by a team of researchers from the Schools of Education and Sociology and Social Policy at the University of Nottingham during August 2002 and March 2003. The chapter explores the scores announced in August 2002 through the conceptual frameworks and analyses of news templates, headlines, icons, and "gold standards". However, for Warmington, the key point is that these analyses start to tell us something about how qualifications function as a form of *exchange-value*. He argues that qualifications are *commodities* that express value in a *double form*: as exchange-value (e.g., in relation to graduate earnings post–A-Level) and as use-value (e.g., representing skills and knowledge that can be used in the labor process).

Part V picks up the broad political themes of *Labor and Commodification in Education: Theory, Practice, and Critique*. It is the final section and incorporates chapters that focus in various ways on expanding and evaluating aspects of issues pertinent to Marxist educational theory (in relation to social transformation) and educational Left politics within the academy.

Thus, David Harvie in *Academic Labor: Producing Value and Producing Struggles,* chapter 12, argues that those working within the education sector—schoolteachers, university and college lecturers, students, and others—are *productive* laborers, and offers renewal of analysis of teachers' labor process in relation to *value* and notably surplus-value for capital. The chapter discusses such labor in its relation to the production of new labor power, explores the extent to which it is *alienated* and thus explains how it produces surplus-value. Insofar as educators' labor is productive, they exist within capital. Educational institutions are also sites of struggle, however, and both students and educators resist the imposition of work. The chapter explores some of the ways in which teachers and students may refuse work and create space in order to pursue alternative projects that meet their own needs more effectively. To this extent, teachers and students are not productive laborers, as such. Instead they produce struggles; they exist against-and-beyond capital and, along with Rikowski, Harvie sees labor power as capital's necessary but "weakest link."

Chapter 13 is on *Marxist Political Praxis: Class Notes on Academic Activism in the Corporate University* in which Gregory Martin argues that a major task of academics who engage in Left politics today is to bridge the gap between scholarly research and activism, most importantly to link political causes inside and outside of the university. Regrettably, few radical pedagogues who unite their work with the issues of injustice and inequality operating in the world today have declared how Marxist educational theory is translated through embodied practice, especially through close contact with workers. Thus, crucial opportunities for making connections between theory and practice, teaching and learning in radical pedagogy are missed. As a departure from the normative strategies of political propagandistic-based practices produced within the university, this narrative provides a concrete account of academic/activism as the author sought to resist the reifying impulse of the corporate university through forms of collective engagement based in material interests and class conflict shaping political forces in Los Angeles. Here the focus of his one-year involvement at the Bus Riders Union (BRU) was his personal commitment to critical/revolutionary praxis organized around the need for individual and social transformation, which is the aim of revolutionary critical pedagogy. Written in the first person narrative, this chapter is an experiment in figuring out how academics interested in reactivating a radical pedagogy on the side of the oppressed can expand their involvement in some of the struggles and enterprises of community, and how they articulate with the division of labor.

Finally, in chapter 14, *The Making of Humanity: The Pivotal Role of Dialectical Thinking in Humanization and the Concomitant Struggle for Self and Social Transformation,* Paula Allman explores the meaning of *humanization* through dialectical analysis in the writing of Antonio Gramsci, Paulo Freire, and most especially Karl Marx. Thus, embedded in Marxist analysis, she argues for the significance of internal relations in dialectical critical analysis for democratic, critical *humanization*. She indicates how in capitalist society *dehumanizing* forms of humanity and *human nature* are developed, often through mainstream educational institutions, ideologies, and policies at the heart of these processes. Capitalism militates against the human project and humanity. To the extent that education aims to prepare people for collective action—to engage in revolutionary social transformation—there is hope against neoliberal individualisms. This requires transformed epistemologies to enable dialectical conceptualization of the real and of our *potential for humanization*. Herein, education contains the possibility and hope of developing the skills and powers to engage in revolutionary social transformation. This is a suitable note on which to close this volume, in the spirit of the ongoing struggle *in*, *through*, and *for* education as critical practices for collective emancipation.

Chapter 1

Marxism, Education, and Dialogue

Anthony Green

We must force the frozen circumstances to dance by singing to them their own melody.
—Karl Marx (1844/1992)

The materialist doctrine that men are products of circumstances and upbringing and that, therefore changed men are products of other circumstances and changed upbringing, forgets that it is men that change circumstances and that it is essential to educate the educator himself.
—Karl Marx (1845/1974)

Introduction

Let us note that while many have long been intrigued and inspired by these epigrammatic quotes from Marx and, of course, other quotable quotes from Marx's pen, others have become bored with what they perceive to be clichéd encrustations of an anachronistic social and political imagination. To my mind these are glorious statements of what we might take to be early intimations of the continuing power and resonance of Marx's work and Marxism for human possibilities. The first deploys a play of metaphors by which Marx released so lightly and succinctly the contemporary problematic for addressing emergent forms of the human condition and its strategic aspirations for progress. It is a philosophical statement, but the central motif here is euphonious, melodic (perhaps even lyrical) and insistently optimistic, therefore, on the theme of the search to compose the music of human time, not as contemplation and passivity or even as active in consumption, but in productive work oriented to collective *action*, itself renewable in ongoing reflexive analysis, further action, and debate, the central themes invoked by the second quotation on the essentially educative nature of these processes. Just as significant now as then, Marx draws us into critical spaces where art and science merge, where action, purposes, and being are tangled in possibilities and restraints, ever busy, restlessly constituting circumstances of openings and partial closures. The goal is collectively and mutually realizing humanity in its innumerable possibilities for identities and practices of human/humane *being*, as well as its, equally or perhaps more significantly addressing all forms of manifest ignorance, false starts, oppressions, arrogances, and barbarisms of *nonbeing* that stifle progressive ingenuity. These are circumstances to be transcended. Herewith, Marx propels us into dialectics and dialogues and into class dialogic about the unavoidability of these circumstances as both inspiration and as caution in productivity of the collective human essence. Thus, with Gramsci we must observe here through two mutually focusing lenses. One, of the optimism of the will, always, however, adjusted to its *other*, of rationalist pessimism and relentless questioning. Marx recognizes energies inquisitive for progressive possibilities in struggles against repressions emergent at all levels of circumstances in social reality. And these circumstances are not going simply to entertain us with their tantalizing potentials, nor *dance* without the forces of intellect, imagination, and collective will becoming engaged with the intricacies of social and political free-form choreography. Such circumstances are the humanizable terrains of fate, not inevitable ramifications of fatalisms, postmodern (Smith, 2001) or otherwise, but, as education and thus as enlightenment, and are to be appreciated, appropriated, and represented in relentless dialogue because there really is no alternative.

Such is class struggle in dialogue, and the second quote in the epigraph provides pause for more and other kinds of thought as Marx's centers and reinforces the notion that the revolutionary project is inevitably *educational*. Because class struggle is unavoidably located in the circumstances of time/space materiality it requires constant updating, renewal, and developing through forms of dialectical/dialogical practices as pedagogy of emergent mutuality immanent in critical diagnosis and prognosis. Herein, critical praxis emerges too, as with its products, out of and in relation to material reality, to make what is possible because immanent therein, not in an ethereal realm of ungrounded aspirations but in practicality, realistically utopian and allied to hope and demands for justice. Such considerations apply to all our practices, including those of the epigraph's author, Marx himself, and what we now consider to be his legacy.

This chapter is essentially in the spirit of these selective interpretive fragments of Marx observations and methodology by reflecting, pointing out, and musing upon some of the issues brought to my mind concerning mutualist pedagogy of critique, its objects and practices the MERD (the ironic euphony of the acronym is clearly recognized) project aims to work for and to articulate. The focus of MERD has been to provide a forum for Marxist analysis and discussion in and around education. It is a place to meet and consider materialist analytical representation for identifying, thinking, and connecting the dynamics of education as real social structural forms, and to try to examine the anatomy of their working processes in contingency, necessity, and no little ambiguity, while working for a socialist future. My offering is inevitably somewhat programmatic, open, and loose in many respects, aiming to capture something of the spirit of the MERD project without either necessarily pinning it down into sharp-edged forms and guarantees, or explicating in detail on specifics beyond broad perspectives and themes. Thus, reflecting MERD's dialogical format, this will be as much about identifying questions and continuing to open issues as settling upon explicit normative formats, unavoidable though this may be.

Now to the originating contexts of MERD and their source of inspiration briefly. The project began when Glenn Rikowski and I met in 2001 against the background of our recognition that, despite some small positive signs to the contrary (Rikowski, 1996; Willmott, 2001), Marxism had become largely marginalized in educational discourses during the 1980s and 1990s, so that by the early 2000s it no longer seemed relevant to many progressives involved in education and educational studies. We decided to call a one-day meeting under the banner Marxism and Education: Renewing Dialogues for October 2002. The response was remarkably positive and these meetings have continued twice yearly ever since with 60 or so coming to participate in each of these discussions from across the UK, Europe, and further including North America.

The dialogic format and rules of engagement, established from its inception, have aimed to avoid any kind of sectarian agenda setting and encourage open reconsideration and discussion of Marx and Marxism across the widest expanse of educational possibilities. The only prerequisite has been *taking Marxism seriously*. We intended to encourage participation not only for self-identified Marxists, but also for sympathetic Left critics of Marxism, and even radically anti-Marxist critics, so long as participation was compatible with this "serious" spirit. Thus, the form of the project is, perhaps in most immediate terms, apolitical and promiscuous, compared, for instance, to Marx's own work and writings, in that MERD is not directly active or activist, as such, at the "movement" level, either as formation, or strategies and tactics. Nevertheless, such issues are understandably significant items and topics for dialogue and renewal of Marxist appraisal, Marxist-Left socialist debates and rebuttal of neoliberal and other positions, as chapters in this volume, generally inspired by their author's presentations in MERD meetings, indicate. To the extent that MERD shows no signs of dispersing or of atrophy, it is alive and well, indeed, spawning at least one new MERD in the north of England as I write, it has so far succeeded in constituting some of its own conditions of production and indeed, reproduction! Much of this is thanks to the critical contributions, energy, and enthusiasm of Glenn Rikowski.

The central themes of *dialogue* and mutualist pedagogy of critique are historically somewhat vexed issues in the context of ongoing historical struggles about relations between theory and leadership, for instance, in debates and recriminations about the legacy of Leninism, in relation to Rosa Luxemburg, for instance, often depicted as embodiments of a pedagogy of centralist top-down transmission and control to further the needs of capital accumulation. Arguably, this is what had emerged as the material history of the Soviet Union's state capitalism, rather in contrast to its more socially radical democratic socialist/communist aspirations. While there are many historical achievements of the Soviet experience, their significance is often and all too shallowly seen as fuddled with tragedy and not a little with farce, tangible and consequential still into the historical present.

MERD has been concerned with radical analysis in and for education where the issues reemerge, for instance, around the legacies of Gramsci and Feire, as debatably offering and practicing bottom-up pedagogy in the face of tricky issues of organization, leadership, and expertise (Foley, 1999, 2005; Mayo, 1999). It has been important in our meetings to focus on a variety of material contexts and conditions of production of teachers and those with whom they teach, and to learn about what they aim to share and transform within themselves, their relations, and contexts of production and the wider horizons of opportunity and constraints for progressive practices, themselves to be collectively addressed, all significant in the chapters in this volume. It is vital to reflect, however, upon the experience of the MERD seminars as multilayered in the terms of formative analysis and intent, and as evolving through constructive critical work in the light of the complex nature of the relations between the *present*, the *prior*, and the emergent *new*. Thus, we have attempted to find ways forward. Each participant carries and, hopefully, discovers aspects of these complex layerings while we endeavor to work together toward renewal of Marxism in and through dialogue.

If this chapter has the beginnings of an argument, it is that these ways of approaching our concerns bring to light relational terrains of tensions and potential contradiction as well as coherences and the gelling of workable forms for dialogue. Little, however, is clear-cut and self evident, partly because we are constituents of what we are attempting to transform. Notable in this context, for instance, is *commodification,* vital to the guiding thread of Marx's work, ever-present and multiform as contexts, topic(s), and problematic(s) for us in and around education within capitalism rampant. Such phenomena and problematics are to be found in the very fabric of both modernist/postmodernist globalized educational relations of production and in the reproduction of these educational relations themselves, as capitalism powers increasingly into its stride through modes of performativity and educationally distractive managerialisms in the flows of *academic capitalism* (Slaughter and Rhodes, 2004).

Programatically, the keystones of *renewing dialogues* that characterize MERD concerns addressing issues: (1) within and between both Marxism and education, and (2) between Marxism and its constructive critics in and for education and progressive social change, not least, Marxism in debate with post-Marxisms, its Left democratic critics and/or forms of anti-Marxism. Below, I will raise some of the issues around the following themes: *Dialogue as Pedagogy of Critique*; *MERD, Marx, and Education: Renewal in order?*; *Class and Class Struggles: For Renewal in and through Education?*; and round off with *There Can Be No Conclusions to Taking Marx and Marxism Seriously: But Let Us Consider...*

Dialogue as Pedagogy of Critique

The loose programmatic framings indicated above echo the *forms* and *processes* of MERD, as *pedagogy of critique*. From its initiation, our meetings have been about synthesis and articulation through an open approach, free of restrictions, save *taking Marxism seriously*. This should generate items of discussion that are especially resourceful in ideology/critique in the face of manifold empiricisms constitutive of common sense moments of repressive hegemony. Thus, this work is also aiming to generate and deploy resources in and for our own critical "commonsenses." Of course these are part of the terrain of Gramsci's and Freire's arenas of struggles, but here to be facilitated through a form of communicative ethics, with mutual respects and openness not unlike that proposed by Habermas (1984). These substantive critical ramifications constitute the topics and resources for MERD worked upon and through pedagogy of mutuality in debate and emergent understanding, to explicate agreements, disagreements, and recognition of things yet to be resolved in a nondogmatic and nonsectarian spirit of openness. In some ways the meetings are out of time, precious, and unreal, of course. However, the tacit sense of suspension/separation and objectivity in relation to the "real" world of our many contexts of professional, social cultural, and political practices we invoke, describe, and analyze enables dialogues at all levels, from the analytical abstract to the practical and immediate. Nevertheless, this context of abstraction is simultaneously a source (a circumstance) of frustration in the face of the always pressing issues of *what is to be done* that arise all too frequently. The dialectics of *will/intellect-pessimism/optimism* are often expressed as ambiguities and indeed humor in recognition of the complexities glimpsed. As practices of renewal in and around Marxism and education, these are processes without guarantees. The products are as much in sharing information and mutual support, as developing clear breakthroughs in understanding and analytical forms of expression, let alone strategic and tactical aspects of struggles. The dialogue between Mike Cole and Elizabeth Atkinson (this volume) illustrates the process well.

Thus a key dimension of MERD is to inform each other and discuss these forms of action, exchange experiences, and theorize the articulation through dialectical/dialogical processes, while resisting temptations to simplistic reductions to, for instance, formulaic *race trumps class* in the mode of contemporary Critical Race Theory (Allen, 2004), or that *class overdetermines* in generally obvious ways any one of the many other social relations of differentiation and thus avoiding the needs to address and articulate complexity in every moment of potentially essentializing specificity in differentiation. Here, the format is always intended to be historical materialist, always already in the restless dynamics of the social, and *othering* has to be done with care and respect, realizing comradely/convivial forms of criticism rather than by personalization of debate, so commonly experienced on the "left" in forms of ambush for the satisfaction of the temporary embarrassment, derision, not infrequently for boundary and barrier contruction through "holier than thouist" or, indeed, arrogant claims to authority, which generally serve to foreshorten serious dialogue. The point is to articulate connections, to identifying spaces for mutuality and agreements to differ, in the face of polarizing, boundary-forming pressures and practices, of othering. MERD has generally been successful in this.

Nevertheless, we are sensitive to the emergent concerns that these contexts are merely providing spaces for self-identified non-Marxism to realize antisocialist possibilities, most notably the TINA (there is no alternative) of neoliberalism as global political economy immune, and its articulation with antiprogressive, neoconservative reactions and threats to its hegemony, all in the name of freedom and democracy, of course. In this context, for instance, Marxists may embrace the othering of Marxism as an honorable identity, one of the vital gathering points for collectivity and, in its turn, to be taken seriously in our critical practices. MERD certainly embodies this. Thus there are many points of tension and dilemma that revolve around the potential of robust dialogue actually reinforcing and strengthening positions on either side of more clearly drawn boundaries, perhaps through immunizing one or the other side against its opposition's sharpest criticisms, an uncomfortable process for Marxists when the forces of turbo capitalism are so evidently in the ascendant in almost everything.

Thus, some of the most delicate issues for MERD emerge around where and how to draw practical distinctions between *radical* analysis and *reformist* articulations, proposals, and practices. These connect to identifying important areas of alliance, mutuality, and coconcerns despite differences, thus working through articulations of *both/and* dialogic spaces rather than *either/or* forms of othering, and so avoiding negativist discourse giving rise to what we might dub intransigent discursive muscularity. It may be possible to glimpse intimations of this, perhaps, when McLaren seems to suggest that the North American progressive educational tradition and critical pedagogy are weak compared to a more robust and fundamental anticapitalism. Thus, while these forms may

> challenge[s] students to become politically literate so they might better understand and transform how power and privilege works on a daily basis...As a project of social transformation, critical pedagogy is touted as an important protagonist in the struggle for social and economic justice, yet it has rarely ever challenged the fundamental basis of capitalist social relations...[and] Marxist analysis has been virtually absent...unless class analysis and class struggle play a central role in critical pedagogy, it is fated to go the way of most liberal reform movements of the past, melding into calls for fairer resource distribution and allocation, and support for racial diversity, without fundamentally challenging the social universe of capital in which such calls are made. (McLaren, 2003: 65)

This kind of issue, focusing on the requirement of fundamental discursive radicalism as articulating a class theory with a social justice perspective, is thrown even more sharply, and problematically into relief when we consider the possibility of Marx's own *incompleteness* in relation to the vital question of class and class analysis in the social universe of capital.

This requires facing up to yet to be resolved emergent issues if we consider Marx's oeuvre not necessarily being a finished methodological or substantive analysis. Thus, authentic to Marx's own terms and practices we must recognize that, whatever Marx thought and wrote about education, say, both "was" and "is." In material terms, it is both of its own time and in our presently breaking time always in the making, our own analytical and practical creations. Herein it is important to consider that economic, social, and cultural changes have meant that both because of and regardless of the inevitable lacunae in his work, renewal of Marx on education is in order, indeed is vital (literally). Thus, development and redrawing is necessary to reinvigorate Marxism and education for the critique of contemporary conditions of production, circulation, and exchange, not least in what is conventionally regarded as education itself. Education may also, I am here adding more tentatively (though offered in the spirit of the second quote in the chapter-opening page), constitute a productive metaphor for representing a critical approach to the topic of the reproduction of social relations of capitalist production themselves, as interconnected processes of "curriculum," "pedagogy," and "assessments" in the production and reproduction of capitalist social relations, in labor power production. This may capture a dimension, therefore, of vital significance for these social relations' conditions of existence as life-worlds to be subjected to ideology-critique, much of which will be manifest in, for instance, social justice and/or "human rights" discourses, a crucial boundary area to be struggled over and through for Marxism in dialogues with its constructive critics around, for instance, issues that expose the complex antimonies of credentialized "merit."

In this context, while it is *sufficient* for Marxists to assume their work has contemporary resonance and relevance, there is a *necessity* to also demonstrate contemporary articulations. In this, it is not self-evident that Marx's own work, or Marxism more broadly, is readable as "truth" in a direct fashion from the texts, nor that it is a kind of religion of the word, a sectarian logocentric resource without ambiguities and tensions. It is to be dialogued with and through. The history of hybridity in Marxisms attests to the empirical, hermeneutic nonstarter here to those friends and foes alike when they seem to see the end of the history of Marxism, marking it as an irredeemably useless tool of the past. Such tools are there to be designed anew.

Thus the discussion and assessment of the legacy of Marx, most notably of its achievements for articulating the anatomy of capitalism in concrete abstract form continues, and there are many uncertainties and lacunae to be worked upon requiring developing and articulating themes, including prominently dimensions, particularly specific to the following:

- education in relation to ideology/hegemony for social and cultural reproduction in relation to commodification and ideology-critique;
- articulating the relations between class struggles and social class as *being in itself* and as life-world(s) being for *our/themselves;*

- for substantive historical/institutional analysis for tactical and strategic work on transformation and emancipation;
- in relation to tactical and strategic work on pedagogy and practices of resistance and emancipation.

MERD, Marx, and Education: Renewal in Order?

In his study, *Marx and Education*, Small makes the general point about Marx's approach to ideology and education that schooling is part of the way in which the working class mistakenly naturalize the social system, personalize success and failure, and, relating it to their schooling experiences, internalize the "demands of the methods of production as self-evident laws of nature" (2005: 86). Ideology, as the powers of *circumstances* of realizing class as relentless being *in itself*, is thus materialized through cultural forms and institutional practices. However, Small asserts that Marx does not explain this fully, leaving the terrain open to our further development and detailing.

In this context, MERD continues the processes of elaborating these issues in ongoing debate and development with (1) the Left functionalist and critical instrumentalist traditions of Marxist educational studies as well, via Bowles and Gintis (1976); Sharp (1980); Apple (1980, 1995); (2) critical pedagogy traditions of analysis of Giroux (1988, 1994); Giroux and McLaren (1994); (3) social and cultural reproduction traditions of Bernstein (2000; Moore et al., 2006) and Bourdieu (1990); (4) as well as postmodernisms (Cole, 2005; Edwards and Usher, 1994), by entering into, reporting upon, examining, specifying, and deconstructing these concerns for Marxist substantive analysis and ideology-critique. These kinds of ideology-critique continue to examine the manifold realities of lived illusions, for instance, of the *justice* of meritocracy as equality of opportunity in action and to explore specifications of the relations between the reproduction of the system of education and schooling. Much of this is achieved through deploying empirical analysis supported by a variety of critical conceptual frameworks with some explicit, and some more nuanced but traceable connections to Marx and the Marxist traditions. The target is ideology *effects* and their roles in realizing class structuration to reveal educational processes and practices as delusive in the process of the struggles for hegemony.

In line with many who have worked this area of critical analysis of education, Small makes the important point that Marx method, dialectical, totalistic, does not propose a *direct* link between school and economic production. Schooling does work on provision of the labor power required in production; however, the complexity and the possible relative institutional autonomy between the relations of schooling and relations of economic production must be acknowledged and examined. It is this *relativity* itself that must be at issue. Thus, even while, the time-honored polytechnic principle, for instance, necessarily combines education with productive work, as itself productive of both material use-value and of human *being*, we need to recognize the systemic features of institutionalized education "in its own right," dialectically generating both positive (progressive/enabling) and negative (regressive/alienating) emergents with respect to labor power. Thus, a central interest of MERD must be to continue to take up the challenges of understanding the articulation of schooling, education, and the economic productive in their ideological dimensions, their commodified forms of credentialing of identities but also in their productive capacities as progressive, human/humane use-values formation, as well. Thus, the generation of labor power is complex, dynamic, and expresses dialectical forms. Where it is suggested that it is inappropriate to directly transfer the method Marx uses to understand economic production to understanding the social relations of educational production, and we might add of understanding the *return of these into* relations of economic production, we need to recognize that Marx leaves the terrain for successors to develop. Critical Enquiry is required and scholars are edging these themes through in a spirit of renewal (Dinerstein and Neary, 2002; Postone, 1993; Rikowski, 2002).

Partly, so far as renewing Marx and Marxism on education is concerned, this is a case of time and space specificities to be investigated anew in historical analysis and thus to indicate the real transformations, despite many continuities between the mid- to later nineteenth century, and articulating them with our contemporary concerns, practices, and understandings. Partly, it is a matter of building on myriad contexts of Marxist and other progressively inspired educational practices, and partly, it is arguably reopening issues about the very approach itself, and identifying possible underdevelopments on analysis of the value form, labor processes, class formation, and dynamics of cultural and identity complexities in and through education as these relate to class and class struggles. Cans of worms may be revealed with far-reaching ramifications where Marxisms and serious anti-Marxisms/post-Marxisms have things to debate in and for educational contexts.

Nevertheless, this process is unlikely ever to displace either (1) the emphasis Marx himself placed upon the material intimacy of theory and practice, or (2) to recognize that any temptation to philosophize about the general features of the human, our very nature, must always be developed for the specific historical conditions prevailing in that time/space. Marx, in his political writings and the relatively few points he discussed education, directly, tended to focus on the education of the *working class*, in and for production for both economic and self-development. His attention to educating/pedagogy was addressed through the framework of developing disciplined attention to task-focused knowledge and skills in which undirected play, for instance, literally undisciplined activity, is of relatively little consequence, save as respite, or simply "childish," perhaps.

Thus, regarding pedagogy where play is more about activity, which in some real sense is self-directed by the "needs" of the child, Marx seems to differentiate this from, and endorse what appears to be a kind of, socialization and moral training that could also be seen as instrumental and narrowly functional, an educational means-ends model that might be very well suited to both a centralist policy of relations of educational production and state, and indeed, to contemporary globalized "liberal" capitalism and its labor power *requirements*. However, as Small argues, this issue needs to be seen in the context of Marx's epistemology of education. Here, Marx seems to draw a sharp distinction between play and work, and between play and *free labor*. Free labor must always involve subordinating will/mind to the disciplines of a task; say, as in the free labor of musical composition as improvization, and so on. Central to this process is recognition of objectivity in the form of the task; such disciplines set necessary limits to transformative activity and the development of knowledge and skills to articulate competent and creative practices, as such. As Small puts it, "work can become free rather than forced, but it cannot become play" (2005: 133). Social class formation in educational forms may well operate through distorted specificities of this kind of principle in mass schooling systems in the selective distribution of access to disciplinarity, thereby marking the educationally merited through graded credentials, along with the gentling symbolic violence (Bourdieu and Passeron, 1990) of social control, therapeutic or otherwise (Furedi, 2004; Sharp and Green, 1975) of the relatively unsuccessful. Tough love can be complex in these kinds of processes, too.

Thus, Marx seems to hold to a traditional progressive humanist model of all-round development via productive labor activity tied to the contexts of contemporary disciplines in production. Nevertheless, labor and education are to be combined in work/school as tasks/activities for economic, social, and moral purposes of all-round development, and fulfilling potential, rather than in direct preparation for relations of *commodity production in capitalist mode*, as such. On another note, this, however, is certainly not the decentered/fragmented subject of postmodernism, though close, perhaps to the multiplicity of the social subject, the articulation of the differential *I* and *me* in practices, not so far removed from the pragmatist activist tradition reminiscent of Mead (Goff, 1980) and Dewey (1922). In educational/pedagogic terms, Marx, in these respects, is understandably perhaps, of his time and seems to place relatively little regard for dialogue in pedagogy. Thus, the routes to the social self in materialist terms (Bakhurst and Sypnowich, 1995) and to contemporary Activity Theory, through Vygotsky via historical materialism involves long strings of intellectual articulation and controversy, with openings for further development for renewal (Daniels, 1996, 2001; Engeström, Miettinen, and Punamäki, 1999; Leontyev, 1979; Wersch, 1985) in which some originating inspiration of Marx may be found and continuted into considering working-class knowledges and labor process (Livingstone and Sawchuck, 2004; Sawchuck, 2003). It may be timely for us to renew Marxism and education on these kinds of themes.

Class and Class Struggles: For Renewal in and through Education?

Completing Marxism is, arguably, the task of realizing visions of democratic socialism in which the full range of humane potentialities, known and yet to come, are universally manifested, in their infinite variety by and for individuals and in splendid collective cultural forms. Thus, also included within this aspiration for continually expanding species being is the important work of addressing Marxism's analytical boundaries, its spaces and lacunae, it's yet-to-be-fully-articulated so far as Marx's own contributions were concerned. Indeed, this may need to be taken up in the context of class and struggle itself, as already mentioned. Here, the aspiration to *taking Marxism seriously* requires addressing points in the framework where it is arguably yet to be completed, while at the same time recognizing that Left critiques of capitalism may have significant points to debate with Marxism on these issues.

Lebowitz' analysis (2003), for instance, is interesting and challenging in these regards. Briefly, this perspective indicates there is a need to develop and "complete" Marx's work, specifically on theorizing the working class in class analysis. Lebowitz seems to suggest that we have a Marx who is analytically effective on critique of capital in economic production dimensions (the guiding thread) but is wanting on articulation of "class" and on real working-class people and their life-worlds. Their social relations have yet to be theorized and articulated with the analysis of capital, as such. Such a view undermines any simple economic and technicist (even possibly expertist) models of Marxist "dialogues" that take *Capital*, and read social and political themes and critical issues through what are taken to be self-evident in capital/logic.

Here, while MERD attempts to work dialectically on the internal and immanent, that is, applying Marx's method, there are many disputes, issues, and alternative tendencies about this with respect to class. Not least of these, implicit in citing McLaren above is how do we draw and then dialogue across Marxist/non-Marxist boundaries? Does this inevitably make our analysis become too loosely eclectic, undisciplined, and incoherent or stunted by incommensurability? Or, can we stick to Marx's totalistic, organic approach, that makes its claims to being essentially revolutionary because it is radical, going to fundamentals in its critique whereby everything has to be changed, and thus having the political implication of tolerating no truck with any kind of piecemeal reform. The optimism of the will may become a little daunted and look for respite for a historicist way out here, with assurances that TINA is inevitably moving in our direction, as Marx and Engels famously seem to have announced in the *Communist Manifesto* (1968). This raises issues, structural/materialistically and hermeneutic/materialistically, around the natures of class *in* itself/(themselves) and for itself/themselves. And, of course, such divisions are vital to the politics of the social relations of production, whether capitalist or through reformist progressive alternatives.

Marx's methodology in *Capital* produces a deliberately one-sided "economic production" account; this is because, most coherently, that methodologically was its point and guiding thread as a *capital* standpoint analysis. By the same token that analysis remains underdeveloped on other and complementary aspects of "production" and indeed, as Lebowitz argues, of the worker/working class, itself. Moreover, recognizing the culturally contexted nature of real class struggles, the "circumstances" in which history is made, opens consideration of opportunities, for instance, to address the lacuna in Marx of the terrain of educational experiences and pedagogy. Thus, what is the working class in educational terms and can modern schooling realize potentials to being effectively humanist? Or, is it tethered to performativist and commodificatory modes, unredeemingly implicated in oppression and reproducing capital in exploitation, as and through human capital formation, and/or as habituses of compliance and resignation, even, in and through "resistance," as Willis' famous study (1977) proposes?

Here, should Marxist perspectives address much of liberal sociology and progressive empirical educational analysis, as partially helpful but flawed, nevertheless coming into their own in several respects, each of which is important to be in dialogue with? The tricky issue then may emerge that class struggle is always historically/culturally realized in contexts of circumstances, with deeply complex specificities in terms of moving horizons of meaning, of possible wants and needs generated by capital and its adjunct processes, themselves. The present is portrayed as a moving wave to work with and against. This raises a specter that might seem to leave Marxism methodologically unravellable, and becoming transformable into an adjunct role in Left critique but with nothing distinctive of value to contribute or develop because of the relentless return to what might be termed the "labyrinthine path strewn with historical accidents" (Pamuk, 2005: 231).

Another negativity may arise, as indicated above where this is articulated in lived practices by cultural and other divisions amongst productive workers (including critical/intellectual workers) with respect to both identifying creation of "false needs" and, on the "radical" wings, around protoessentialist logics already mentioned, of "trumping" in Critical Race Theory (CRT), where race trumps class. Here unproductive boundary building between critical positions emerge, for instance, around Marxism in relations with any of the other significant standpoint positions: gender/patriarchy; race/ethnicity/xenophobia; religious intolerances, and so on. Emancipation from each of these oppressions requires the demise of capitalism, no doubt. However, capitalism can and does still operate very effectively with these as reformably in commodifiable play, precisely as issues to be addressed as cultural "needs" and thereby constituting oppressions as social problems whose role is to be ever-in-the-process-of-being-reformed through professionalisable practices. Indeed, these contexts are routinely capitalized upon as emergent "products" through being the focus of attention by the local state; as commodities in reform packages for voluntary sector practices; as objects for public/private partnerships; as think tanks topics; or as commodified in expressive leisure as consumption in all manner of "art," music, and much more besides; through multilayering of identities as commodities, images, self-expressions, constantly addressed as issues all to indirectly reproductive effects by performing the illusion of demonstrating the human face of "caring" capitalism trickling down its creative beneficence and liberalism.

Each of these are possible moments in capitalisable creativity and chaos operating potentially to divide, to deflect from forms of solidarity and thus inverted, while appearing as progressive contributions to dispersing oppressions they leave pernicious class formation and social injustice untouched. They serve as illusions of hierarchies of "achievement," legitimating reproduction of the dominant relations of production managed in defusion through central governmental "target setting" in education, for instance. However, in taking this line do we here then begin to move too deeply into the terrain of, say the Left Weberian tradition of liberal progressive critical sociology on the hermeneutics of power/domination and oppression, where articulations with capital formation and social and cultural reproduction can be demonstrated as existential fate consequent on the workings of "power" but thereby not seriously challenged? Thus, for instance, in education is the tremendous reputation and the continuing legacy of Bourdieu for providing critical analytical tools and vital sensitizing concepts, epithets of symbolic violence, and so on in this respect, part of the problem rather than the solution? Thus, do these methodologies of radical analysis, inevitably draw critique onto the terrain of Left liberal sociology and thereby returning to the cosy middle grounds of safely, irresolvable Third Wayisms?

In turn arises an issue for educational moments of the critique of political economy: for Marxists, can it ever be possible to overdo "production" because Marx was, after all, very enthusiastic about the historical mission of modernity/capitalism as the high point of the history of class struggle to his day? And, is this the same as myopic attention to the "economic"? Marx was engaged in a critique of political economy, certainly, but are we still left with analytical and practical work to do on vital issues such as when and where is the articulation of class in *class struggles*? It may well be that, despite attempts to lay down a normative framework for dealing with class, and a particularly engaging attempt is made, for instance, by Hill most recently reformulated in a collection devoted to "Zeitgeist Analysis" (Moisio and Suoranta, 2006), and in this volume, class in contemporary contexts, as ever, is realized always and only in any and every moment of the relations of the capitalist production, their reproduction, *and most vitally in the practices of critique* of powers in these relations and their reproductive consequences. Any analysis will inevitably involve framing and developing ideology-critique by addressing, either explicitly, or by implication, the currently dominant form of answers to the ever-present questions of equity and distributive justice. Namely, how do our present educational frameworks of practices offend against the principles *from each according to his/her abilities and to each according to his/her needs*? This is, of course, immensely challenging and complex in the face of both evident "progress" in the creation of the possibilities of identifying and realizing "needs" satisfaction in relation to many areas of knowledge, want, and health. Seen also in the contexts of polarization in many distributional respects, both within the "developed," and between that and the developing world, and the knowledge complexities of imagination effects of such "flows" (Fahey and Kenway,

2006) such themes become ever more pressing. Wherein does class formation, solidarity, and class struggle lay in these globalizing processes? In turn, there is the urgency of all this as the *beast* is churning onward in a variety of capitalist modes with TINA tags attached, in order to deflect and defuse radical inquiry into the ways in which the current solutions may be reproducing rather than resolving, or indeed dissolving by transcending, the problems.

Capital's TINA is that it must *both* divide and separate workers to render labor power effective, *and*, simultaneously assemble them both in technical relational terms, and as a productive commodity, itself a condition for its own development/existence/expansion. However, within this process there are both radical opportunities and constraints. Where might the contemporary capacity for (1) managerial redesign and its combination with (2) modern/postmodernist restless new identity formation, self-individualization in circuits of both production and consumption, continue to serve the development of humans for fulfilment through ever-heterogeneous complexity? And (3) where, in this process is it simply, overdeterminedly, one of homogenisation, an originating point of inspiration for Frankfurt School Critical Theory tradition (Adorno and Horkheimer, 1947)? For class struggles, however, if the anticapitalist class is to have a common denominator or, indeed an essence, what, where, and when are these points of identification for class struggles? Tricky issues of essentialism in analysis emerge here that are best addressed by recognizing that the question is always a matter of identifying forms that are relational, open, potentially dynamic. For instance, in terms of identity formation around heterogenization/homogenization, class struggles are about making links in moment of recognition and discovering the ideological/political common denominators to be gathered into organic, complex class solidarity. Such processes involve creation in sharing ideology-critique that are constitutive of the class structuration through mutualist *education/pedagogy of critique in action,* and "educating the educators" locally through to the emergent global and back again.

Thus, how can and should we do more than "simply" focus upon capitalism's interest in promoting the "collective worker" *in itself* as labor power in the dominant relations of production, while deflecting and dispersing *for* itself in transcendence? For capital, these social relations must both assemble and divide human beings to counter threats to the reproduction of these relations in and through their own collective actions and forms. Thus, the Janus faces of trade unionism, for instance, and varieties of modes of corporatism as contexts of struggles. Never, for capital, must the essence coalesce into a critical mass of collectivity. Of course, both the long history of capitalist state formations and their articulations with globalization across many dimensions exist to continue to do sterling work for capitalist renewal here (Harvey, 2005), while, nevertheless, continuing to provide possibilities of alternative assembly points and processes, not least through the virtual materialities of ICT (intermediate communications technology), including such tender and apparently personalizing domains, as MySpace, YouTube, FaceBook, and more. And, of course, in relation to complex ecological concerns, amongst others, spelling potential for total disaster globally, unless the cycles of production/consumption are not redesigned, barbarism beckons.

In this perspective the key to going beyond class *in itself*, for the working class to express class *for itself*, is thus in articulating class struggles in which self-transformation is the only possible way forward, self-education, class autopedagogy of deconstructing divisions, building connections and moving toward creating expansive relations of production for the play of innovative forces of production. Nevertheless, however salutary it may be in the face of what some recognize as reductive, stultifying economism, to repeat over and over again the phrase *class struggle*, the issue seems to be that, class articulated solely *in economic production* continues to be not enough. We need to work on the theorization, the conceptualization, and articulations between the economic and the noneconomic to examine and expose their relational dynamics and examine how and when they create divisions that weaken the progressive class formation: via sexism, racism, xenophobia, and so on to overcome divisions, forge material unities but not to reduce diversities in life-worlds, or reinforce sectarianism through theoretical separations for emergent articulations and flexible interconnections.

These concerns raise a multitude of issues for renewal and development of Marxism in relation to education around many points of potential focus for MERD, not the least

- the possibly essential incompleteness of Marx's account of the working class and its relations to its identity formations in struggles;
- the implications of a working theory of the state;
- the problems of a methodological perspective that urges us to view the critical problems in a totalistic form, as social and cultural production and reproduction, articulated in a complex whole itself a methodological antidote to empiricism, where the everyday, commonsenses continuously reproduce illusion, and critically disabling ideological effects;
- the ever-expansive problem of the centered limits of the imagination, the almost certainly extremely challenging, possibly even essentially mindboggling implications that tend, perhaps to incline deploying simplistic formulations in debate, and thus as tokens in dialogues in which concrete abstractions as models tend to be set up against each other as competing paradigms with sharp boundaries, precipitating othering procedures and refusal of common ground;
- the productive role of leisure and networked communications in class struggles as "education" through cultural forms of mutualist pedagogy of critique as well as critical attention to fun as both commodity and its critical opposite in humorous and unsanctimonious ideology critique in theater, music, song, and poetry, even while experienced as respite for renewal;
- thus, coming full circle in this listing, to what extent can and do we recognize resources in each of these in class formation and class struggles

themselves? How do these enable us to address the issues of the arguably still open questions: what is class in Marxist terms, and when is class struggle? Marx's legacy is for us to renew and realize progressive class forms in struggle. This is unavoidable if it is not to be abandoned as much Left liberalism, indeed post-Marxism (Laclau and Mouffe, 1985), would seem to counsel.

There Can Be No Conclusions to Taking Marx and Marxism Seriously: But Let Us Consider...

Straws in the popularising breeze?

- *Why Read Marx Today?* by Jonathan Wolff published in 2002, described in a blurb from the *Economist* as an "engaging read";
- AERA established a Marxist special interest group in 2004;
- Marx voted Britain's favorite philosopher in BBC listeners' poll in 2005;
- Marx's *Das Kapital: A Biography* by Francis Wheen published in 2006, publisher's blurb courtesy of Karl Marx himself—"Capital is dead labour which, vampire-like, lives by sucking living labour."

Question: Why, if Marx appears to be so popular, there also appear to be so few Marxists around?

While Marx leaves an incredibly inspiring legacy of analysis, brilliant insights and connections through exuberant writing and commentary on specific critical themes and historical developments, nevertheless it is a corpus of work that at the general theoretical level is about "tendencies," modeling of concrete abstractions, to be taken subsequently as inspirational sensitizing concepts within an overall intellectual and political project that is totalistic in aspiration to render the dialectical whole fully in its relational forms, a total critique of political economy. The anatomy of capitalism is laid out in *Capital* through accounts of the value form, commodity form, and so on, and the necessity to struggle against capitalisms tendencies is clearly announced, not in the sure knowledge of inevitable success, though parts of the *Communist Manifesto* and *Capital* inspires this hope, but in real historical circumstances that are more than likely to be dancing largely to other, and often quite catchy tunes.

There are thus several questions, ever-present issues, and contexts for MERD to work upon:

- class in formation and struggles;
- surplus-value exploitation;
- commodity forms and commodification of everything;
- social, economic, and cultural polarizations, both in capitalist centers and the rippling layers of its peripheries becoming ever more closely articulated by the flows of economic globalization;

- conditions and consequences of poverty and underdevelopment despite abundance, that is, social injustices of distribution and consumption;
- globalization of communications: opportunities, constraints, and struggles in relation to the uneven development of the structuration of the global in class struggles;
- where gender, religion, race/ethnicity, are key terrains of antagonism, and the local/global nexus is ever-present, the issues are about the theory and practices of and their articulations with and through critique of capital forms and thereby to resist polarization and othering. Thus the requirement to treat them as contexts for *both/and*, rather than competitive or combative *either/oring* as we develop analysis in ideology-critique of the relations of ideas to repressive powers and to empowering possibilities articulated across all dimensions of capital in the educational dimensions of the social;
- the processes, and this is where a dialogical model such as MERD comes into its own, are around identifying nodes and apexes of current antagonism and contradictions around education: globally to locally in all aspects of policy, institutional forms, working practices in the production and reproduction of knowledges, policies, and practices for realizing and inhibiting species being. This includes ongoingly recognizing and resisting commodification. It is, of course, as Marx put it, "men and women making history, not in conditions of their own making" that counts (Marx, 1845) and it is the open conditionality of such material practices and understanding that must form the focus of our endeavors as educational projects collecting toward a humane democratic socialism.

Whimsically, then, would that the words were magic and the teleology would be self-realizing, an unfolding as one moment in an essentially autonomously connected reversal of *capital logic,* perhaps, merely in the act of speaking but then, that would not be historical, or materialist, nor could we be addressing circumstances to be understood, acted within and upon, and thereby changed and readdressed in their consequences, known, intended, and yet to be discovered, as "new" and unanticipated.

Let us return to both our opening quotes and the straws in the wind just mentioned and close by reconsidering commodification. Few can doubt that Marx's main enemy, the value form, is in full flow, perhaps only now coming fully into its own. For those taking Marx and Marxism seriously this is not the time to slink back into any shadows of neoliberal history, or even acknowledge the end of history itself, but to continue to reappraise and develop in the sure knowledge that Marx's legacy includes a continuous reexamination of the internally contradictory and life-negating mechanisms of capitalism. While Marx brilliantly recognized the historical magnificence of capitalism, its achievements cannot include transcending the need to choose between socialism and barbarism, both globally and personally, between life-affirming emancipation and the multiple corrosions and corruptions of the human spirit and its potentials for the good. Putting flesh and bones in action to the specter that continues to haunt now ever more fully globalized capitalism is the Marxist project. MERD aims, in an inevitably somewhat modest fashion, to be part of that process by encouraging and facilitating renewal in several specific ways.

Here, MERD gives pause to consider forms of commodification in the academy for articulating, academic capitalism. This involves amongst many other processes, the lived world of the competitive struggles to acquire reputation and distinctions in identities for being the most eminent, for instance, the leading Marxist *scholars* in ever-diversified and fragmenting fields of inquiry. Attention to these structures and processes may reveal some other dimensions of the way capitalism works behind our backs through creating fronts of *institutionalized expertise*, in the form of marketable professors, articulating with commercial publishing and the corporate nexus connecting this and the performative busyness of higher education as business, not least in the writing and reading of *these* words! Reputations become commodities in interinstitutional competition for central funding according to relative positioning on league tables, in systems weighted and enacted largely, of course, according to the Mathew principle, "for whomsoever hath, to him shall be given" (Matthew 13:12). While the labor theory of value in its pristine form may be somewhat hard put to fully explicate such phenomena, nevertheless social relations of exploitation and social class formation are unlikely to be far away working in and through our own practices of commodification. The point about the educator requiring education is that no one is outside the whale, even its most radical critics.

The commodified ghost of the author of *Capital* may well roar with ironic applause at the most recent exuberant product of the Marx industry *Das Kapital: A Biography* (Wheen, 2006: 120–121) where, expressing deep ambiguity quoting a 1997 *New Yorker* article to the effect that Marx's

> "books will be worth reading as long as capitalism endures." Far from being buried under the rubble of the Berlin Wall, Marx may only now be emerging in his true significance. He could yet become the most influential thinker of the twenty-first century.

The irony would not be lost on Marx, partly because he had so much difficulty with those in his own day who designated themselves *Marxist*, whose de/reconstructions of *Capital* he found more than wanting, to the extent that, perhaps not wholly unlike Groucho many years later who quipped that he would not join a club that would have him as a member, if that was to be the form and fate of Marxism, then he, Karl is reported by Engels to have declared, "Ce qu'il y a de certain c'est que moi, je ne suis pas Marxiste" [If anything is certain, it is that I myself am not a Marxist] (Engels, 1888).

Where thinking, or perhaps substitutes for thinking, have become a commodifiable *serious* leisure activity in the affluent developed and developing world, it makes sense that Wheen's book comes as a neatly packaged item, itself keenly priced for its market, just so at £9.99. But is it rather more than a congealed alienating commodity? The human spirit ensures that it has more than this in its potential in relations of critical production. Thus, while "all that is solid melts into air" (Marx and Engels, 1968) applies as much to the material and social relations and products of precapitalist forms of production and reproduction of their relations, it must, with positive irony apply to Marxism, *as a commodity*, too. The commodification of Marxism as a capitalizable good constitutes both a circumstance of constraint and of opportunity for any project of Marxist renewal, not least in, through and *as* education. Such is one moment in Marxism and Education: Renewing Dialogues.

References

Adorno, T. and M. Horkheimer. (1947/1973) *Dialectic of Enlightenment* (London: Allen Lane).
Allen, R. (2004) Whiteness and Critical Pedagogy. *Educational Philosophy and Theory*, 36 (2): 121–135.
Apple, M. (1980) *Ideology and Curriculum* (London: Routledge and Kegan Paul).
———. (1995) *Education and Power*, 2nd ed. (London: Routledge).
Bakhurst, D. and C. Sypnowich. (eds.). (1995) *The Social Self* (London: Sage).
Bernstein, B. (2000) *Pedagogy, Symbolic Control and Identity: Theory, Research Critique*, rev. ed. (London: Routledge).
Bourdieu, P. and J.-C. Passeron. (1990) *Reproduction in Education, Society and Culture*, 2nd ed. (London: Sage).
Bowles, S. and H. Gintis. (1976) *Schooling in Capitalist America: Educational Reform and the Contradictions of Economic Life* (New York: Basic Books).
Cole, M. (2005) *Marxism, Postmodernism and Education* (London: Routledge).
Daniels, H. (1996) *An Introduction to Vygotsky* (London: Routledge).
———. (2001) *Vygotsky and Pedagogy* (London: Routledge Falmer).
Dewey, J. (1922) *Human Nature and Conduct* (New York: Henry Holt and Company).
Dinerstein, A. and M. Neary. (eds.). (2002) *The Labour Debate: An Investigation into the Theory and Reality of Capitalist Work* (Aldershot: Ashgate).
Edwards, R. and R. Usher. (1994) *Postmodernism and Education: Different Voices, Different Worlds* (London: Routledge).
Engels, F. (1888/1924) From Letters of Fredrick Engels, Engels to Eduard Bernstein. In *Marx Engels Archives* (Moscow). Available at http://www.marxists.org/archive/marx/works/1882/letters/82_11_02.htm.
Engeström, V., R. Miettinen, and R.-L. Punamäki. (eds.). (1999) *Perspectives on Activity Theory* (Cambridge: Cambridge University Press).
Fahey, J. and J. Kenway. (2006) The Power of Imagining and Imagining Power. *Globalisation, Societies and Education*, 4 (2) (July): 161–166.
Foley, G. (1999) *Learning in Social Action: A Contribution to Understanding Informal Education* (London: Zed Books).
———. (2005) Educational Institutions: Supporting Working-Class Learning. *New Directions for Adult and Continuing Education*, 106: 37–44.
Furedi, F. (2004) *Therapy Culture: Cultivating Vulnerability in an Anxious Age* (London: Routledge).

Giroux, H. (1988) *Schooling for Democracy: Critical Pedagogy in the Modern Age* (London: Routledge).
Giroux H. and P. Mclaren. (1994) *Between Borders: Pedagogy and the Politics of Cultural Studies* (London: Routledge).
Goff, T. (1980) *Marx and Mead* (London: Routledge and Kegan Paul).
Habermas, G. (1984) *The Theory of Communicative Action*, Vols. 1 and 2 (Boston: Beacon Press).
Harvey, D. (2005) *A Brief History of Neoliberalism* (Oxford: Oxford University Press).
Laclau, E. and C. Mouffe. (1985) *Hegemony and Socialist Strategy: Towards a Radical Democratic Politics* (London: Verso).
Lebowitz, M. (2003) *Beyond 'Capital'. Marx's Political Economy of the Working Class*, Second Edition (Basingstoke: Palgrave).
Leontyev, A. N. (1979) The Problem of Activity in Psychology. In J. V. Wertsch (ed.) *The Concept of Activity in Soviet Psychology* (Armonk, NY: M. E. Sharpe).
Livingstone, D. and P. Sawchuck. (2004) *Hidden Knowledge: Organised Labour in the Information Age* (Aurora, Ontario: Garamond Press).
Marx, K. (1844/1992) *A Contribution to the Critique of Hegel's Philosophy of Right*. In *Early Writings* (Harmondsworth: Penguin).
———. (1845/1974) *Theses on Feuerbach in Marx and Engels's*. The German Ideology (London: Lawrence and Wishart).
Marx, K and F. Engels. (1848/1968) *The Communist Manifesto* in *Marx and Engels: Selected Works* (London: Lawrence and Wishart).
Mayo, P. (1999) *Gramsci, Freire and Adult Education: Possibilities for Transformative Action* (London: Zed Books).
McLaren, P. (1995) *Critical Pedagogy and Predatory Culture: Oppositional Politics in a Postmodern Era* (London: Routledge).
———. (2003) Critical Pedagogy and Class Struggle in the Age of Neoliberal Globalization: Notes from History's Underside. *Democracy and Nature*, 9 (1) (March): 65–90.
Moisio, O.-P. and J. Suoranto. (eds.). (2006) *Education and the Spirit of Time: Historical, Global and Critical Reflections* (Rotterdam: Sense Publishers).
Moore, R., M. Arnot, J. Beck, and H. Daniels. (eds.). (2006) *Knowledge, Power and Educational Reform: Applying the Sociology of Basil Bernstein* (London: Routledge).
Pamuk, O. (2005) *Istanbul: Memories and the City* (London: Faber and Faber).
Postone, M. (1993) *Time, Labour and Social Domination: A Reinterpretation of Marx's Critical Theory* (New York: Cambridge University Press).
Rikowski, G. (1996) Left Alone: Endtime for Marxist Educational Theory? *British Journal of Sociology of Education*, 17 (4): 415–541.
———. (2002) Education, Capital and the Transhuman. In D. Hill, P. McLaren, M. Cole, and G. Rikowski (eds.) *Marxism against Postmodernism in Educational Theory* (Lanham: Lexington Books).
Sawchuck, P. (2003) *Adult Learning and Technology in Working Class Life* (Cambridge: Cambridge University Press).
Sharp, R. (1980) *Knowledge, Ideology and the Politics of Schooling: Towards a Marxist Analysis of Education* (London: Routledge).
Sharp, R. and A. Green. (1975) *Education and Social Control* (London: Routledge and Kegan Paul).
Slaughter, S. and G. Rhodes. (2004) *Academic Capitalism in the New Economy* (Baltimore, MD: Johns Hopkins University Press).
Small, R. (2005) *Marx and Education* (Aldershot: Ashgate).

Smith, M. (2001) *Reading Simulaca Fatal Theories for Postmodernity* (Albany, NY: State University of New York Press).

Wersch, J. (1985) *Vygotsky and the Social Foundations of Mind* (Cambridge, MA: Harvard University Press).

Wheen, F. (2006) *Das Kapital: A Biography* (London: Atlantic Books).

Willis, P. (1977) *Learning to Labour: How Working Class Kids Get Working Class Jobs* (Farnborough: Saxon House).

Willmott, R. (2001) The "Mini-renaissance" in Marxist Sociology of Education: A Critique. *British Journal of Sociology of Education*, 22 (2): 203–215.

Wolff, J. (2002) *Why Read Marx Today* (Oxford: Oxford University Press).

Part II

Neoliberalism, Globalization, Crises, and Education in the Transitional Epochs

Chapter 2

Toward a Political Economy of Education in the Transitional Period

Geraldine Thorpe and Pat Brady

Introduction

The argument in this chapter is that recent changes in education can be understood in the context of transition, where capitalism is in decline and competing with the burgeoning movements of the potential future society. We cannot explain the context by *describing* how it appears to us. We have to find out what lies beneath the surface, what underlying movements and forces are shaping and changing our world. It is a matter of *explanation*.

Marx aimed to show how the laws of motion of political economy operated in mature capitalism. Using his materialist dialectical method he provided a profound critique of the classical political economists' treatment of the economic categories of the capitalist mode of production and therefore of modern society. In so doing the laws of motion were revealed. This method allows us to analyze the mode of production today. He and Engels saw political economy as the science of civil society (1869). One of Marx's greatest contributions according to Engels was his discovery of the law of development of civil society (1883/1993).

Capitalism, like all social formations has a beginning, maturity, and an end. In its mature phase it operates according to its own laws of motion but in its decay class struggle sharpens and the system increasingly needs mediation in order to survive. The state as the instrument of the ruling class is brought in to regulate the economy in the interest of that ruling group or class.

A Marxist political economy of today should explain change. To this end, we draw on theories of decline and transition, particularly those of Hillel Ticktin and Peter Kennedy. The law of value is the defining law of capitalism and is crucial to the law of accumulation. In decline they need mediation. The forms this takes, however, can only temporarily arrest decline. Social forms that herald the beginnings of a new kind of society but are not of the new society also appear; their form is transitional, an example being laborism. Kennedy's theory of this is discussed below. The fundamental law of socialism is democratic planning based on common property and the direct association of producers. Today there are many forms of planning through monopolies, government, and bureaucracy but, as Hillel Ticktin has pointed out, they are transitional forms, neither capitalist nor socialist ones (1983: 23–42, 1998: 21–42).

While phenomena such as so-called globalization, privatization, and commodification may suggest that capitalism is developing and invading more and more areas of society we argue that this is more appearance than reality. In order to distinguish appearance from reality, a method is needed. If Marxist, it has to be the materialist dialectical method of Marx. Marx saw his treatment of political economy as his dialectical method (1976: 102). As the title implies, this chapter is not presented as a complete theory. Rather, in the constraints of limited space, it points to areas of Marxist political economy that educationists may seek to develop in order to explain change in education.[1]

Marx's Method of Political Economy

Marx's dialectic can be seen as a means to explain the effects of socially organized labor interacting with nature. The task is to discover how these are mediated and why and how they change. The aim is to reveal the condition of the laws of motion of society and of the historical relations that give rise to them. By way of abstraction and analysis of the economic categories he showed these relations were contained in them in the form of contradictions (see below). The dialectic exposes the movement of a mode of production through analysis of material historical facts. Marx's method is thus the science[2] of political economy.

Although for Marx the aim was to reveal the laws of motion of bourgeois society, he could not approach these directly. But as they characterized a mode of production such as capitalism or feudalism they must arise from the way labor is organized in production, how the surplus product is extracted, and how it is distributed. As this is always carried out in a particular historical context, these laws depend on the social relations of the time for their movement. Historical relations would thus put their stamp on all that they were involved in. They would affect the social character of the economic categories such as labor, the product, the value form, capital, rent, and so on. Of utmost importance to Marx therefore was the analysis of these categories. Through them he could then reveal the historical character of the relations that were contained in them in the form of contradictions, how and why they changed, and ultimately the movement of the laws of motion.

The social substance common to all economic categories is labor. It is the basis of value relations in any mode of production because these are the relation between different kinds of labor but of the same duration of effort, that is to say value. The relation between quality and quantity of labor is the basis of the relation between its concrete and abstract forms. The social form that the value relation takes differs from one mode of production to another and again in the transitional periods between them (see below).

The economic categories also have physical properties comprising the raw materials that human labor works on and these too are characterized by quality and quantity. However, the "raw materials" that are of concern to education are the *social* forms of "language" and "consciousness"; these are historical results of change in the mode of production and extraction of the surplus product and thus in historical relations. Although they reside in people and their products (Marx, 1976: 44) they are also shaped by the needs of the ruling class. For example, science is the form of consciousness that developed in an organized way from the early seventeenth century in Europe, and earlier in the East. It is crucial to the mastery of the natural, physical, and social world and was increasingly applied to industry from the nineteenth, century. As such, language and consciousness could not be seen as economic categories due to their development in the wider society. Nonetheless, part of this is through the education of potential labor power,[3] an economic category, and involves the cost of wages of teachers.[4] All economic categories contain such aspects as quality and quantity, concrete and abstract, natural and social, in the form of contradictions. The ways these interact with one another not only differentiate one category from another, but also reveal change in social relations. The category labor, for example, would not have the same social form in socialism as it did in capitalism, which in turn differs from that which had been found in feudalism. The prevailing social form of the product of labor in precapitalist societies was the use-value. In capitalism, the commodity prevails. Today, both forms of the product are present. These are accompanied by transitional or quasi-forms of one or the other, particularly in privatized areas of the public sector. Hugh Willmott sees the quasi-commodity and the quasi-market in higher education (1995: 995).

Where property is held in common, contradictions require little mediation. Where private property is involved, social forms such as exchange-value, money, markets, and the state develop to mediate the value relation. Where the economic forms of mediation are under pressure, either because they are too immature as in the long transition to capitalism or because they are in decline, as we shall argue is the case today, the state is brought in to assist. This places bureaucracy backed by armed force in the leading mediatory role.

Marx's treatment of the categories exposed their internal contradictory relations and revealed how they were mediated. Human labor is always for a particular purpose and is thus of a particular kind. Duration of the effort of labor is measured by socially necessary labor-time. Marx was the first to emphasize that labor has both a concrete and an abstract form. Thus the mode of extraction and distribution of the surplus product is always in the context of particular social relations. It is the social form that the economic categories take that characterizes a mode of production.

In the production of a use-value, concrete and abstract labor interact to make, for example, a chair or a coat; they have been given a common purpose. In conditions of property held in common, they do not separate into an antagonistic relation. The development of the opposite in the form of private property was the result of an increased surplus product. It led to change in relations of exchange and distribution, changing the form of the market from one on the margins of social life regulated by social decree such as the "just price" to one regulated by "competition." Through such change, concrete labour and abstract labor in the product are brought into an antagonistic relation. This is expressed in the contradiction between use-value and exchange-value that turns the use-value into a commodity. It cannot be used until social relations of exchange have been entered and a transaction involving its exchange-value has been successfully completed. Thus the economic categories are bearers not only of different kinds of labor in the form of a contradiction but also of the relations in which this was performed.

In a letter to J. B. Schweitzer, 24 January 1865, Marx wrote,

> "The secret of scientific dialectics" depends upon comprehending "economic categories as the theoretical expression of historical relations of production, corresponding to a particular stage of development of material production." (Bhaskar: 1991: 146–147)[5]

The secret lies in Marx's focus on the objective expression of relations in the economic categories. These are in fact the sum of living subjective activity in a particular phase of history. It is this insistence on such fact that makes his dialectic scientific. He did not neglect empirical data but his dialectic is not confined to this. Kennedy correctly argues that Marx placed greater emphasis on the ontological aspect of the categories than on their history (2003: 168). In other words his concern with these was to reveal the relations hidden in them. As the so-called commodification of education is of current concern, we focus on the commodity. How did Marx approach his subject?

First, Marx used a process of abstraction in his research (1973: 100–108) to find the essential, material category that was the basic cell of the system. By abstraction we mean the separation of the object into its constituent parts. He noted at the end of *Grundrisse* that the section on value should be brought forward (1973: 881). In the Foreword, Martin Nicolaus notes that *Grundrisse* shows Marx's method of working while *Capital* (Volume 1) shows his method of presentation (1973: 7–63). Marx saw that the secret of the value form—money—could only be revealed by an analysis of the material cell of the capitalist system, the commodity.[6] It was the cell because living labor itself, once the organizing unity of the labor process as for the craftsperson or only a condition of production as for the slave and the serf, had been transformed into a commodity, labor power (Marx, 1976: 954).[7]

Second, it is necessary to show how the primary category relates to others. Social relations and the accompanying forces in production decide how the economic categories relate to one another, not the order of their historical appearance. For example, in feudalism, labor and land, not money, were the forms that value took. Thus the internal relations of a system are found in how the contradictions interact and how the categories relate to one another.

In his presentation Marx analyzed the commodity further to find other categories in order of their connection to it. From the commodity he abstracted use-value in which he found concrete labor and abstract labor. The former is qualitative and useful; use-values are different kinds of labor. The latter is quantitative, a magnitude of effort of a particular duration; it is therefore a value measured by socially necessary labor-time (1976: 137, 954). In a use-value these aspects of labor interact with one another relatively freely where exchange relations are not present; an example is in systems of free education at the point of consumption. State funding and the accompanying bureaucracy are both dependent on revenue from the surplus product. Control of teachers and researchers is bureaucratic: it is not by the law of value and the valorization process.

However, in exchange relations based on private property, use-values have to take the commodity form of use-value and exchange-value in order to accommodate change in the historical form of value. The free interaction of the dual aspect of labor has ceased and the two have separated into a relation of both antagonism and inversion. Abstract labor has separated from and subordinated concrete labor to it. Concrete labor has to wait for abstract labor to be realized before it can be. The relation in the commodity is therefore use-value and *value:* concrete labor and abstract labor; it appears in exchange relations as use-value and *exchange-value*. This relation was long known before Marx's time. His contribution was to analyze the category exchange-value for the first time (1976: 139). His aim in this was to explain the origin of the money form (ibid.). The mass of commodities that he referred to in the opening chapter was of the mature commodity, the immediate result of the capitalist production process; it was not the simple commodity that the isolated producer took to local markets or annual fairs. In capitalism money is the value form of the commodity. Its embryonic form is exchange-value. This was formed in relations of barter when one commodity, was exchanged for another of a different kind, for example, a barrel of wine for 10 pelts of fur. They were from different kinds of labor but were of the same magnitude.

Marx analyzed exchange-value and found the relation between the relative form of value and the equivalent form of value. He explains how this relation developed as a result of human activity in production and trade from one that lacked fixity because commodities could act either as the relative or the equivalent forms of value as is the case in barter, to one of fixity in which the equivalent form was money, measured by the weight of the precious metal gold.[8] The relation was now antagonistic because the equivalent form of value (money) excluded all the relative forms of it from acting as money. It was inverted because the equivalent form of value appeared to command the exchange and consumption of use-values. Marx explained this in detail in Chapter 1 of *Capital (Volume 1)* (1976: 125–177). This chapter needs to be studied in order to see how Marx revealed in these categories the material and social foundation of the law of value and its relation to the laws of development and accumulation. It may be difficult but it allows insight into how Marx used his materialist dialectic, as well as contributing to the means to test critically what is said about commodities and markets today. It might be added that revolution was prerequisite to the establishment of capitalism just as it is to establishing socialism. In short, it requires new social forms of the economic categories and emancipation from the historical forms of antagonism and inversion brought in by private property.

Marx explained that the relation in the commodity between these two forms of value would necessarily break down. In this process commodity fetishism would change from an ideology premised on real markets to an ideology that was premised on a consciously regulated market and decaying law of value. In other words commodity fetishism loses its material foundation with the separation of the relative and the equivalent forms of value. He continued his analysis of the mature commodity in the Appendix (1976: 943–1084). Its production and exchange are not measured by the single, simple commodity but are treated as aggregates of which the individual commodity is an aliquot part (1976: 954–957). That is to say, the value of any one commodity is a fraction (without remainder) derived from the total value divided by the total number of individual parts. The immediate result of commodity production is surplus-value in the form of capital.[9] The capitalist production process is now the labor process that is the bearer of and subordinate to the valorization process.

Marx could not have pursued the logic of the separation of the commodity's internal relations were it not the case that the foundations for its decline and new social form in socialism were already appearing, for example, in credit (1981: 566–573). As he pointed out, socialized labor necessarily led to the socialization of capital. It forced capital into joint stock companies, credit, and monopolies (1981: 567). The last was the death knell of competition and the free market. The relative and equivalent forms of value had separated and were held together only by the mediations of monopoly control, which involved industry, the state, and the banks.

The bourgeois state, according to Istvan Meszaros, is "mediation *par excellence*" (1995: 491). It is required when the systemic movements cannot overcome barriers and are thus in trouble. It means that the poles of the contradictions in the categories need mediation if they are to interact. The alternative, as Ticktin indicates throughout his work, is that the poles of the categories separate. In the former Soviet Union, even use-values could not be produced without defects (1992: 12) In the mature phase of capitalism, Ticktin points out, the movements were simple because they revolved around a single law, the law of value. At this time, there was very little state intervention in education. Once universities were approved by the state, they were left to themselves. Most schooling in Britain was private, that is, provided by capital or by independent tutors consuming unproductively the revenue of individuals, groups, families, and so on, not by the state. As Marx said,

> The highest development of capital exists when the general conditions of the process of social production are not paid out of deductions from the social revenue, the state's taxes... but rather out of capital as capital. (1973: 532)[10]

State intervention signals that relative and equivalent forms of value are separating and can no longer interpenetrate without mediation.

Was the separation of the relative and equivalent forms of value inevitable? According to Nicolaus,

> Certainly. The entire work [*Capital*] is addressed precisely to the historic, economic, political conditions on which this initial identity [of opposites] depends; more: the main purpose of the work is to demonstrate that the contradictions within this identity necessarily lead to the suspension of these same conditions and hence to the break-up of commodity production, and to the rise of a system of production founded on use values. For Marx, the identity of opposites is conditional; but their non-identity, their struggle, antagonism and break-up are inevitabilities. Just the opposite in Hegel. It is the difference between a conciliatory, harmonising "dialectic"... and a revolutionary subversive method. (1973: 41)

So Marx had to discover the relations of all categories to one another as well as the relations that operated within each category. For example, if labor is the social substance of all products, then it is the measure of value. Value of the total product would of course virtually disappear in an automated production process and its real price would be close to zero.

Seeing it this way, the new society appears in its embryonic form in its predecessor. The same was as true for capitalism as it is for socialism. Exchange-value, the earliest expression of the value form, appeared when two commodities encountered one another in historical relations of barter. It is a historical category that developed into the money form. Another example is the commutation of peasant service into money in the feudal system. This was an early sign of the wage labor and capital relation but it remained in feudal social form. Marx's analysis of the relation between the relative form of value and the equivalent form of value in Chapter 1 of *Capital* (Volume 1) yielded the secrets of money and commodity fetishism and the source of their decline.

Third, Marx's dialectic is a theory of change. It explains why the value relation changes over time. In other words, the law of value is subject to the law of development and always has a historical form. Engels believed that the discovery of the law of development was Marx's greatest contribution to political economy because it allowed him to explain change in the relations and forces of production (1883). In his words and deeds, Marx was "a revolutionist" (ibid.). The logic of the law of development is demonstrated by the long process of the division of labor. This was accomplished by a process of separation of the constituent aspects of not only the tasks of the labor process but of the nature of the human species as well. The separation of natural attributes such as mental and physical labor, the sexes, the old and young, and of politicogeographical categories such as nationality and race into antagonistic relations acts as barriers to the full socialisation of labor. This historic process of separation was completed in the wage labor and capital relation (Marx, 1973: 489) that ushered into being the category labor power as a commodity. Labor power was bought and sold in earlier societies, in, for example, the Roman army, although unlike the labor of the peasant it was unproductive. But the wage labor and capital relation did not prevail in production until the industrial revolution. From the discovery of the economic category, labor power, Marx could develop his theory of surplus value.

Views of Political Economy

Marx drew from the work of earlier bourgeois political economists, for example, William Petty, Adam Smith, and David Ricardo. In their day, they were revolutionary in developing political economy as a science. They had to counter the landed aristocracy's view that land was the only source of value and therefore of wealth. They did this by showing that manufacturing was more productive because nature imposed fewer restrictions on it. Manufacturing was not subject to the quality of the soil or weather. But, they did not pursue it as a science as far as Marx did. Once the bourgeoisie became the dominant class, they began to claim that the economic categories and laws of motion of the capitalist system were natural and eternal. Harry Braverman, for instance, exposed the one-sided character of bourgeois "scientific management" of the labor process (1974: 86). It could be added that the work in human resource departments is to serve the employer and not the employee[11] (Kellaway, 2006).

Many writers on education have criticized observed changes in it but without analyzing them in the context of political economy. Although they use political-economic concepts drawn from this, many tend to use them in a way that suggests that they have been frozen in the time that Marx used them. In other words, there is no hint that relations within capitalism as a system go through phases and that accordingly the relations in its categories undergo change. If this is not taken into account, it is easy to "prove" that Marx was wrong in much of what he said, as it does not apply to today.

Political economy, for Marxists, is the science of civil society as noted above. However, this for many is highly contentious. Marx said that it roused fury and passion in the bourgeois economists (1976: 92). Hugh Willmott noted reluctance on the part of academics to dwell on it (1995: 994). John Smyth noted problems in explaining change in academic work (1995: 1).

For Paul Samuelson, one of America's most widely read economists, "political economy" and 'economics' were synonymous. According to Fredy

Perlman, in his introduction to Rubin's *Essays on Marx's Theory of Value*, this amounts to a "great evasion." Perlman says,

> If economics is indeed merely a new name for political economy, and if the subject matter, which was once covered under the heading of political economy is now covered by economics, then economics has replaced political economy. However, if the subject matter of political economy is not the same as that of economics, then the "replacement" of political economy is actually an omission of a field of knowledge. If economics answers different question from those raised by political economy, and if the omitted questions refer to the form and quality of human life within the dominant social-economic system, then this omission can be called a "great evasion." (1973: ix)

For Marx as noted above, political economy involved a critique of the economic categories in order to show their internal relations at a given time and how they changed. The aim was to show the effects on them by the laws of motion of society and class struggle. It required a scientific approach to economic facts. Through his method, he discovered that there were categories, contradictions, and laws of motion that corresponded not only to a specific historical mode of production but also to phases within that mode.

The reviewer of *Capital* (Volume 1), N. Sieber, professor of political economy at the University of Kiev, wrote (with Marx's approval) in the *European Messenger* of St. Petersburg:

> The one thing that is important for Marx is to find the law of the phenomena with whose investigation he is concerned; and it is not only the law which governs these phenomena...that is important to him. Of still greater importance to him is the law of their variation, of their development, i.e. of their transition from one form into another, from one series of connections into a different one. (Marx, 1976: 100)

Among the laws that governed the phenomena that Marx was investigating were those of development, value, and accumulation.

The law of value is premised on socially necessary labor-time as the measure of value. It involves the contradictions between use-value and exchange-value (the relation of the relative and equivalent forms of value) premised on concrete labor and abstract labor. Marx showed that the law of value varied in that it had its birth, maturity, and decline. It was presupposed in the simple commodity through the law of equivalence—a quantity of one kind of labor exchanged for the same amount of a different kind of labor. It matured in capitalism when labor power prevailed as a commodity enabling capital to become self-expanding value. In mature capitalism, the objective laws of motion appeared to govern society. As Marx said, class struggle appeared to be latent or sporadic in a mature phase (1976: 96). In other words, the objective laws appeared to dominate the subjective action of individuals. In decline, the relation, for example, of abstract and concrete labor, polarizes and would separate were it not for mediation. This meant the state had to intervene.

The 1848 revolutions in Europe were an early indication that the law of value needed mediation. As Marx and Engels wrote in the *Communist Manifesto*, the relations of production were at odds with the forces of production and on the brink of separation (1977: 41). To look at this more historically, the development of the forces of production socialized labor in the factory system. The struggles that resulted from this forced the small private entrepreneur to find cash in order to develop the means of production. The Limited Liability Act 1857 allowed the savings of society as a whole to be utilized in the development of industry. Hilferding drew attention to this aspect of the process of socialization of capital in Germany (1910, 1981: 218). Joint stock companies were able to develop as a result and thus began the rapid socialization of capital. Expansion of credit led to the development of monopolies. Banks played a key part in this, especially in Germany and the United States (Heilbroner and Milberg, 1998: 86–92).

Marx on Education

What is the relation between a particular phase of development of the value form and education? Marx rejected the notion that precapitalist conditions preceding the industrial revolution could serve as the foundation for a socialist education system because production developed only spontaneously. Factory legislation was the "first conscious and methodological reaction of society against" such spontaneity (1976: 610). In mature British capitalism, education was run by the church, the state, and capitalist entrepreneurs. Martin West argues for a return to entrepreneurial education in a free market. He notes that children learned what their parents wanted them to and attendance was not compulsory (2000). However, his argument begs the question as to why the state had to intervene. Marx, pointing to a report by a royal commissioner, noted that the tendency in such schools was for illiterate teachers, overcrowded classrooms, and incomprehensible babble from the children (1976: 523–524). In other words, in order for profit to be made in education, wages and costs had to be at the minimum and fee-paying pupil numbers at the maximum. As Marx pointed out, surplus-value could be made in private education only by raising the school's income (mainly fees) and lowering wages of the teachers (1976: 1044). Similarly, education should not take place in the home or the factory because parents and employers wanted the children to work (1869). Marx's view was that school attendance should be compulsory (1976: 613). The workers had to ensure that the state enforced this as well as funding education but in every other respect the state should be kept out of (Marx, 1866: 6). The workers had to convert state force into their own power. Only in this way, Marx argued, can children and juvenile workers be freed from the "crushing effects" of the system and be guaranteed an education that develops their potential as all-round social individuals (ibid.). The report on the conclusions of the International Workingmen's

Association based on a speech by Marx[12] noted that the solution for workers lay in "converting *social reason* into *social force*" because

> under given circumstances, there exists no other method of doing so, than through *general laws*, enforced by the power of the state. In enforcing such laws, the working class do not fortify governmental power. On the contrary, they transform that power, now used against them, into their own agency. They effect by a general act what they would vainly attempt by a multitude of isolated individual efforts. (1866; IWMA: 5–6, Original italics)

One example of the need to do this today is in the enforcement of the clause on academic freedom (Education Reform Act, 1988, Section 202: [2] [a])[13] to counter the effects of increasing intensity of bureaucratic management on teaching and research.

Marx disagreed with Robert Owen that a socialist education in capitalism was possible. However, he used Owen to show that the factory system provided the necessary "germ" of the education system of the future and "fully developed human beings" (1976: 614). Why was this so? In our view, in addition to the points above, one, labor was socialized and cooperative; it was the embryonic form of the directly associated producers. Two, organized workers thus had the political strength to compel the state to fund education but otherwise to keep a distance from it. Three, the factory system required science for its development; its regulation was an illustration of heightened social consciousness. It thus had the potential means to raise the language and consciousness of workers to a high level. Marx advocated that the children studied subjects with only one conclusion, such as grammar and science. In that way, the politics and religion of the teachers would not interfere with the education of working-class children. He rejected subjects like religion and bourgeois political economy on the grounds that they admitted more than one conclusion, and children would get plenty of those debates at home (IWMA: 1869). His view was that education was to do with the all-round development of the individual. Thus, physical education, training in production techniques, and actual work all had their place alongside intellectual development. Indeed, he argued that children would come fresh to their studies in the afternoon after exercise and work in the morning. He argued that through such an education the working class would prepare itself for emancipation and raise itself far above the other classes (1976: 613–614).

The history of education shows that this still has to be accomplished. The universities in Britain in the mature phase of capitalism had little to do with the development of the value form and industry unlike those in Germany. The ancient universities in England, Scotland, and Dublin were not freed from the Church of England's Thirty-Nine Articles until Gladstone abolished the Universities' Tests in the 1870s. The 16 universities enjoyed virtual laissez-faire until World War II. It was not until the postwar years that they formed a "university system" (Simon, 1946–1947: 79; Shattock, 1996: 31).

A nine-point agreement was struck between the Universities' Grants Committee and the Treasury on how the universities and the state would relate to one another (Shattock, 1996: 3). The seventh point that "public money should not be used to compete against itself" is a notable casualty of today (ibid.).

The Declining Value Form

Finance Capital

Marx saw "financial capital" as part of the unity of productive and circulatory capital. Circulation begins in production. In conditions of private property, it is a cost on production and on society. It is necessary for capital to complete its moment of turnover. Marx gives the example of the hunter and the fisherman (1973: 632–633). When they stand on private property, exchange involves a cost. They must cease production in order to execute the exchange of their commodities in circulation. If they employ someone to act as merchant, they must pay for it. Financial services, banks, insurance companies, law firms, marketing, advertising, and others are part of these costs of production. Marx then gives the example of the hunter and fisherman standing on communal property. Where property is owned in common, the costs of exchange vanish. Production is for need and association is directly between the producers. There is no need for money, advertising, and all the services that grow out of circulation. There is therefore no need to create a surplus product to cover these costs. Finance capital is a description of all the above services, which help the total capital to circulate and reproduce.

Marx saw the beginnings of the developing role of financial capital in *Capital* (Volume 3). It took Hilferding, who coined the term finance capital (1910, 1981) and saw that it was an abstract form of capital, Lenin who saw it as "parasitic," and Ticktin who critiqued his predecessors and argues that it is the organizing, conscious force of the capitalist system in decline (1985: 2), to develop Marx's observations.

Ticktin (1986) distinguishes between the forms finance capital took in Germany, Britain, and the United States. The form it took around 1870 in Germany was a close working together of banks, industry, universities, and the state (1986: 9–15). This was a natural consequence of coming late into the industrial revolution and having to protect its infant industries. Protectionism was the rule. The United States did the same and developed under protection. Britain as the leading industrial power preached the doctrine of free trade, just as its successor, the United States, is today. Britain as a result of class struggle took the easy option and exported capital in the forms of loans and equipment to its rivals, including the United States and Germany, allowing its domestic industry to run down. This is the beginning of decline of British industry and the context for the mass unemployment of the interwar years (Hobsbawm, 1969).

What were the implications of the polarizing of finance capital from industrial capital?

One. Finance capital developed imperialism and the world market. Although profits were maintained, the rate of growth of new capital, especially in Britain, fell into decline. British industry was neglected as finance capital resorted to earlier forms of extraction of surplus labour in the colonies.

Two. The difficulty in maintaining the relation between the relative form of value and the equivalent form of value based on the gold standard increased. The general strike of 1926 in Britain was triggered by the insistence of finance capital returning to the gold standard at the prewar rate at $4.80–£1.00. This led to the coal miners having to take a cut in pay in order for British coal to be sold in the world market. The unrealistic exchange rate[14] eventually led to abandonment of the gold standard by the leading powers in the early 1930s and the return to national money and protectionism. But it was too late to contain the crisis. The result was depression, fascism, and World War II. The law of value could no longer regulate the system without mediation.

Three. The postwar Bretton Woods Conference 1946, which set up the International Monetary Fund (IMF), was an attempt to restore international money and exchange rates based on the U.S. dollar. This was abandoned in 1971. These illustrations serve to emphasize that failure to establish international money is a sign of the decline of the relation in the value form and thus of the system itself.

Decline

How do we know a mode of production is in decline? We have pointed to Marx's theory of the value form and to the decline of money from its international form to its national one. The effect of decline on the laws of motion of the system is that they function less well. This means that the system consistently "underperforms in relation to its potentialities" (Ticktin, 1994: 69). Ticktin says, "It may produce more goods per capita but the systemic stimuli—laws, contradictions—operate less well. They have to be supplemented with new forms." Most important, he says, "To the extent which these supplementary forms are successful, it appears that the old mode of production has overcome its old age and acquired a new lease of life. There is therefore a problem with the recognition of the real decline itself" (ibid.: 69–70). What Ticktin means here is that in a transitional epoch supplementary forms in addition to the declining old ones and the immature rising new ones have to be consciously created in order to mediate the complex movements. This might explain why most, including many Marxists, assume development of capitalism today.

Let us explain this. The effect of Keynes was to give the appearance that capitalism was restored to health in the postwar years. Keynesianism, however, represented mediation on the part of the state. It could not therefore be seen as a *development* of capitalism. In fact, the resulting full employment is not a feature of mature capitalism, which depends on unemployment of part of the workforce (the reserve army of labor). Full employment led to increased working-class militancy.

The 1930s depression and the abandonment of gold as the universal equivalent form seemed to suggest that the system was finished. The law of value could not prevent mass unemployment. Noncapitalist forms under fascism and national governments (as in Britain in 1931 and in both world wars) had to come in operating in a similar way to Stalinism. In other words the mode of extraction and distribution of the surplus product had to be managed. The laws of motion of capitalism and the relations that gave rise to them had to be suspended. This explains why earlier forms of labor such as forced labor, for example, in German factories and in German and Russian concentration camps, were used. Finance capital was the dominant form of capital but its solutions were different from those of today. Mass unemployment and fascism are no longer options today. The postwar landslide victory of the Labour Party was evidence that people were not prepared to accept mass unemployment. This was realized by the ruling class during the war, hence the model for the welfare state was set out in the Beveridge Report, 1942. The Beveridge aim of full employment was achieved by the economic policies of John Maynard Keynes.

The Keynesian era seemed to show that capitalism could always solve its problems. The state had, however, to intervene to an unprecedented extent. It had to restore the collaboration of labor and did so through labourism (Kennedy, 2004). According to Kennedy, this involved the revival of social democracy. Drawing on Hilferding, he shows how the general principles of social democracy are, one, the socialization of capital and the subsequent political control of finance capital; and two, the imperialist expansion abroad underwritten by agreements reached between Western powers over the partition of Africa at the Berlin Conference, 1884 (2004: 66). Kennedy says,

> For Hilferding, social democracy was finance capital's strategic policy option to contain the growing threat that a more collective and powerful labour movement posed for the continuation of capital accumulation. (ibid.)

This involved a coalition of the state, industrial and finance capital and the trade unions. The extraction of the imperial tribute from the old colonies was greatly reduced with the liquidation of overseas assets to pay for world wars, national liberation movements, and the end of Empire. Class struggle during World War II forced a retreat from finance capital. Industrial capital was restored along with progressive taxation, full employment, and the welfare state. For education, this period saw the 1944 Education Act implemented and the expansion of the higher education sector through Robbins. A return to industry led to a high demand for scientists and engineers. The ruling class had to curb finance capital and turn to industrial capital and expanded domestic consumption in order to extract surplus-value. In our view, Keynes represented a class compromise, which by its very nature was bound to collapse. It explains why the ruling class had to concede progressive taxation, full employment, and the welfare state including free university education and student grants. All these measures reflected the motion of capital and the contradictions thrown up that required political responses.

In a political context, Keynes was a response to Stalinism and the cold war. Stalinism represented a controlling mechanism based on the ideology of socialism in one country and top-down control by party and state. Capitalism had to be seen as better than so-called communism. As Ticktin says, it did not take much to convince the working class that it was better than the labor camps and empty shops of the Soviet Union. The underlying motion was the continuing decline of the law of value and the need on the part of the ruling class to suspend this decline. Finance capital had to invest in production as the source of its value. The postwar restrictions placed on its activities abroad compelled it to invest in its home industry. The cold war justified arms production at the expense of the taxpayer. The postwar consensus served to keep the workers down but full employment and the increased socialization of labor had their own contradictions. The working class raised their demands and were not prepared to accept the tightening of its belt implied by the Labour government's acceptance of the conditions imposed by the IMF in the mid-1970s hence the Winter of Discontent in 1979.[15]

All this contained contradictions for the ruling class. Mass production accelerated the socialisation of labor and led to increased working-class demands. As Kennedy says, "the theorists of labourism tend to omit its class basis" (2004: 72). He outlines "three principle class foundations of labourism as: the management of money; the partial decommodification of labour power; and management of the labour process" (ibid.). Sheila Cohen notes that the labor process theorists neglected theory of the "valorisation process" and "exploitation" (1987: 34–50).

The planning involved and the growing demands that came out of full employment were moving the system toward the creation of use-values at the expense of exchange-values with nationalization and expansion of public services. That is to say it was moving in the direction of socialism. Decaying laborism was inadequate for disciplining the workers (Kennedy, 2004: 72). As Kennedy argues, the contradiction between concrete labor and abstract labour was polarizing and the former superseding the latter. Productive labor continued to yield abstract labor and thus relative surplus-value but in return, capital had to expand the provision of use-values to meet growing needs, for example, public services (1994, 2004). This is the theoretical context of the struggle over the public services today. New Labour rhetoric acknowledges the demand for better services but because it is trying to serve the need of finance capital, it is caught in a contradiction. It cannot possibly deliver what it promises the electorate. Kennedy's argument, based on his doctoral thesis (1996), shows that we can talk of a tendency to supersession of abstract value-producing labor by concrete, useful labor. The contradiction in labor, most of which in Britain is unproductive because it does not create new value, is now more to do with that between surplus labor-time and disposable (free) time. Workers in all sectors are forced into long, stressful hours to save capital.

The present phase of the rule of finance capital began in the Labour government of Callaghan and Healey in the mid-1970s. So-called Thatcherism was a logical extension of this. Thatcherism instead of rolling back the state represented greater intervention than ever before outside wartime. The Blair governments went even further. The increasing contradictions forced them to do so (Kennedy, 2004: 65–95). One example is in the field of taxation. Monetarism, a tool of finance capital, involved a reversion to regressive taxation. How has New Labour responded to this?

Prem Sikka (2003) showed how corporations cost society approximately £25 billion a year in tax avoidance. He also shows how the tax burden over four years (1999–2003) has been shifted dramatically from business onto workers. The tax burden for companies for 2003 was £6 billion lighter than in 1999, falling from £34.3 billion to £28.8 billion, while for individuals it increased from £93.05 billion in 1999 to £105.1 billion in 2003. In percentages of the total, the contribution from companies fell from 36.9 percent to 27.4 percent and from individuals, it rose from 63.1 percent to 72.6 percent. Sikka wrote,

> In the land of "reverse socialism" people on the minimum wage, bus drivers, nurses, pensioners and debt-ridden graduates pay a higher proportion of their income in tax than millionaires. (ibid.)

Martin Wolf writes on a similar form of polarisation in the United States (*Financial Times*: 15). On top of this, there are all the hidden regressive taxes, which are passed off as "consumer choice" in the market. Examples are tuition fees, congestion charges in cities, parking fees, and road tolls. These raise revenue for the state and are nothing to do with purchasing goods or services like higher education. The gambling of pension funds on the stock exchange to maintain finance capital might force local authorities to raise council taxes to cover the losses. Again, the burden falls regressively on the taxpayer. If capitalism can only be kept going by transferring money from taxpayers to capital, it cannot be said to be developing, because these processes are to do with the mode of production whose fundamental law is the law of value. If that is in decline then by definition the mode of production is in decline. As already indicated, the present era is to do with competing social forms from the movements of decline, development, and transition in social relations.

We see daily reports on fraud and corruption in the world of finance but as Sikka points out, governments have no stomach for tackling tax avoidance by corporations, the reason being that political parties are dependent on business for donations (2003), which is a corruption of the whole political system. One effect is loss of confidence in parliamentary democracy.

Recently there have been questions raised about the sale of honors for cash. This is not new; Lloyd George faced similar questions at the beginning of the twentieth century. What is new is that the police are looking into the present scandal. Indeed, one of the biggest beneficiaries of government contracts in education has been a company named Capita. The chief executive of Capita resigned because he was one of those who provided loans to the Labour Party.

Finance capital represents a redistribution of the surplus product in favor of capital. It requires cuts in the public services and privatization. Privatization does not mean that the public services become private enterprise. It is the state opening up its coffers to private capital through outsourcing, tax breaks, Private Public Partnership (PPP), Private Finance Initiative (PFI), and so on. The use of the engineering firm, Jarvis (which donates to the Labour Party) for education consultancy is another example of the corruption of the whole project. The contradiction of record spending in the health service and nurses in revolt is explained by private companies feeding off the health budget.

The more the private sector intervenes the worse the level of service as is so evident in health, education, and transport. Ticktin pointed out in 1983 that finance capital was feeding off pensions and that this is an example of a declining form of a declining form, that is to say, of total capital. In decline, capital increasingly has difficulty in continuing its movements. It meets more barriers to realization of value from production—it cannot sell all that is produced and automation lowers the value produced. In circulation, it meets more barriers to movement and to replenishing and increasing its value. There is a massive surplus of idle capital today. It turns to construction, services, credit cards, and others to survive. Governments prop up failed banks, and stock markets are consciously organized when they falter. Privatization was a mediation to stimulate capital to work. It provided the ideology of private property in the public services in order to perpetuate the ideology of exchange-value and the market. The only basis for the economic category of exchange-value, as we explained above, is private property. Markets cannot exist without it. It is increasingly transparent that the state provides "profit" and that privatization is not the solution to decline. The government persists with the ideology of the Third Way, the PFI, and PPP because it cannot find an alternative. These are transitional forms, not capitalist or socialist ones. Transitional forms are necessarily short-lived because they lack the social relations required to establish a systemic form. For example, usurious capital and merchant capital were transitional forms with different modes of extracting surplus labor from that of capital proper. It required the industrial revolution for industrial capital as the basis of the capitalist system to prevail. It transformed the formal subordination of labor that existed under the earlier forms of the capital relation to the real subordination of labor in the wage labor and capital relation.

Today, there is a surplus of capital with no home. This means that capital is lying idle in banks. Capital in money form cannot accumulate; it can only waste. It has to interact with labor in order to accumulate because labor is the source of surplus-value. Capital's natural habitat is in production using productive labor. That is the only environment in which it can form. The postwar expansion of industry in Britain was abandoned because large-scale industrial production while providing full employment, capital accumulation, and taxes for the treasury, also created the contradiction of increased socialisation of production and increased worker demands for use values such as education, health, pensions, and so on. This is the background to the Winter of Discontent. Capital is therefore exported to countries like China where most workers have recently arrived from the countryside. In terms of exploitation, it is like the early years of the industrial revolution but it is the central state, not the law of value that plays the crucial mediatory role. The United States and UK capital form a significant percentage of investment in China whose exports are putting U.S. and European workers out of work.

Idle capital is loss of value and of labor. It is waste. It is unable even to realize the potential that is present today, for example, automation, increased leisure, and freedom from want. This is the context for change in education. We have tried to show the dialectical relation between production and circulation, industrial capital and finance capital, and the relation between the objective laws of motion and subjective movement and class struggle as expressed in the economic categories. Marx focused on the development of industrial capital and showed the importance of capital in its circulatory form for the realization of value. The two were complementary in mature capitalism. Today it is different. Finance capital is the dominant partner. At the same time it is a declining, parasitic, and abstract form that needs eventually to return to production. It tries to avoid this by going abroad. The traditional mediations like depression, mass unemployment, and drawn-out wars are now no longer options for capital in the long term. In Britain, finance capital dominates the whole of social life and is leading to increased unrest. The huge surplus of capital is not being spent on public services where it is needed. Instead, private companies are plundering them.

Education Today

Education involves mediation between the individual and society. It has to take on board all the contradictions. This is why it seems to be facing both to the market and to the state. The Education Reform Act, 1988, allowed for this. The legacy is a bloated bureaucracy on the one hand; instability, corruption, and waste on the other. The local state has virtually been abolished in education and the state has centralized its power in it. The effect on the labor process of teachers is the tendency to destroy use-value. The squeeze to increase surplus labor-time to make savings impacts negatively on teaching and research. The gap between the potential use-value and the actual use-value of their activity widens. Many academics are now demoralized, stressed, and silent in their "factories of credentialisation." Academic freedom appears to have been abandoned in all but name; many academics seem too fearful for their careers to make a stand. However, the negative aspect of decay always has its positive aspect. A. H. Halsey's research on the condition of the Oxbridge professoriate points to the widening of access to higher education and to the spread of literacy (1992). However, in terms of standards, this development is very uneven.

Finance capital requires a flexible workforce, not an educated workforce in the old sense. It needs a small number of elite universities but it will not pay for a general, worthwhile education for the majority of individuals. Most attend modular programmes and short courses that are focused increasingly narrowly on a particular area of service work. University departments concerned with production such as science and engineering are being closed down. The humanities, history, and philosophy are meeting a similar fate. There are signs that students will vote with their feet. Universities that promote themselves too far as "modern" and "entrepreneurial," ready to do service for capital and the government, will find themselves with plummeting recruitment and soaring dropout rates. Education has to be worthwhile for the individual if he or she is to pursue it but it will take more than the government and capital to make it so.

Conclusion

The changes in education have a context. We have argued that Marx's dialectical method of treating political economy provides an explanation that allows us to see that capitalism is in decline and with it all its institutions. We have outlined an example of his dialectic in the commodity. Another example is found in the historical process of the division of labor, which he saw as process of separation of the unity of humanity and nature that was "posited" in its opposite in the wage labor and capital relation (1973: 489). Political economy involves the dialectical movement between objective laws of motion arising from the contradictions in the economic categories and the subjective actions of individuals and classes.

Contemporary society is in a transitional period from capitalism to socialism. We have argued that capitalism has a beginning, middle, and end. It is today in the process of decline and competing with the burgeoning movements of the future society. The mediations holding the system together are working less well in the present period as seen in the need to combine the contradictory forms of market and state. One reason for this is the collapse of mediations such as Stalinism, the cold war, and with them, social democracy, all of which held back the working class. The system needs unemployment and war but cannot not return to the 1930s or any earlier form.

Finance capital is the abstract, parasitic form that commands capitalism in decline. It subordinates industrial capital and in so doing destroys its host. Its sphere is circulation through which it develops global markets. These are in crisis, forcing finance capital to feed off pensions, construction, and the public sector. The dominance of finance capital explains government seemingly contradictory policy and educational change today.

The internal dynamics of capitalism and class struggle necessarily bring about the demise of its laws of motion. Its governing law of value is brought into decline by competition, concentration, and automation. In the same process, it creates the possibility for humanity to liberate itself by eliminating living labor from the production process. In other words, it creates its own graveyard and its own diggers—the modern proletariat. Marx summed up the struggle as

> Thus at the level of material production...we find the *same* situation that we find in *religion* at the ideological level, namely the inversion of subject into object and *vice versa*. Viewed historically this inversion is the indispensable transition without which wealth as such, i.e. the relentless productive forces of social labour, which alone can form the material basis of a free human society, could not possibly be created by force at the expense of the majority. This antagonistic stage cannot be avoided....What we are confronted by here is the alienation...of man from his own labour. To that extent the worker stands on a higher plane than the capitalist from the outset, since the latter has his roots in the process of alienation and finds absolute satisfaction in it whereas right from the start the worker is a victim who confronts it as a rebel and experiences it as a process of enslavement. (Marx, 1976: 990, Italics in original)

The task for the working class and the rest of humanity is to raise subject over object in the struggle for emancipation. Education plays a crucial role in this by which the educator becomes educated.

Notes

We are indebted to the work of Hillel Ticktin (editor) and Peter Kennedy (managing editor) of *Critique Journal on Marxist Political Economy of Today*. Any errors of interpretation are ours.

1. For a useful political economy of education see Rachel Sharp's "Introduction" to *Capitalist Crisis and Schooling* (1986: ix–xxxiii). Also Madan Sarup's *Marxism and Education* (1978).
2. See Edgley (1991: 419–424).
3. See Rikowski (2000) for a view of labor power as the product of education.
4. The implications of this for "praxis" in education are in Marx's theory of education outlined below. Marx noted the "difficulty" in relating the "practical social relations of education" to historical relations because of "uneven development" (1973: 109). Bureaucratic relations fostered by the presence of the state further complicate this.
5. It is also in a serial part of the letter in *Der Social-Demokrat*, No. 17, 3 February 1865.
6. Those who argue that education is now involved in commodity production might revisit Marx (1976: 125–177) and the Appendix "The Results of the Immediate Process of Production" (943–1085).
7. Marx appears to have first named this category in *Grundrisse* (Nicolaus, 1973: 21).
8. Silver became a secondary equivalent form, allowing for coin of less value than gold coin.
9. This is after wages, costs of reproduction, taxes, and so on have been deducted.

10. These deductions from social revenue—state taxes—create the situation "where revenue and not capital appears as the labour fund, and where the worker, although he is a free wage worker like any other, nevertheless stands economically in a different relation" (Marx, 1973: 532). See also Mandel (1975: 270).
11. Lucy Kellaway frequently draws attention to this in her columns in the *Financial Times*.
12. The Geneva Congress, 3rd to 8th September 1866 reported by Wilhelm Eichoff.
13. With regard to the function of the University Commissioners,

> In exercising those functions, the commissioners shall have regard to the need—(a) to ensure that academic staff have freedom within the law to question and test received wisdom, and to put forward new ideas and controversial or unpopular opinions, without placing themselves in jeopardy of losing their jobs or privileges they may have at their institutions. (Education Reform Act, 1988, Section 202: [2] [a])

14. Britain was forced off the gold standard by the unrealistic rate of $4.80: £1.00. Others also left because they did not want economic policy dictated by something outside their control.
15. The chancellor of the exchequer, Dennis Healey, accepted IMF terms in return for a loan. In effect, these were to pave the way for reduction in public spending, unemployment, monetarism, the end of laborism, and the restoration of finance capital's dominant role in the management of the system.

References

Bhaskar, R. (1991) Dialectics. In T. Bottomore (ed.) *A Dictionary of Marxist Thought*, 2nd ed., 143–150 (Oxford: Blackwell).

Braverman, H. (1974) *Labor and Monopoly Capital* (New York: Monthly Review Press).

Cohen, S. (1987) A Labour Process to Nowhere? *New Left Review*, 165 (September/October): 34–50.

Edgley, R. (1991) Philosophy. In T. Bottomore (ed.) *A Dictionary of Marxist Thought*, 2nd ed., 419–424 (Oxford: Blackwell).

Education Reform Act. (1988) (London: Her Majesty's Stationery Office).

Engels, F. (1883/1993) Speech at the Grave of Karl Marx, transcribed by Mike Lepore. Available at http://www.marxists.org/archive/marx/works/1883/death/burial.htm (accessed 26 September 2005).

———. (1869) *Biography of Karl Marx*, trans. J. and T. Walmsley. Available at http://www.marxists.org/archive/marx/bio/marx/eng-1869.htm (1 of 4) (accessed 26 September 2005).

Halsey, A. H. (1992) *The Decline of Donnish Dominion* (Oxford: Clarendon Press).

Heilbroner, R. and W. Milberg. (1998) *The Making of Economic Society*, 10th ed. (New Jersey: Prentice Hall).

Hobsbawm, E. J. (1969) *Industry and Empire* (Harmondsworth: Penguin).

Kellaway, L, (2006) Business Life: Lucy Kellaway: The Answer. *Financial Times*, 26 April, 10.

Kennedy, P. (1996) *The Decline of Capitalism and the Rise of Labourism in Britain: A Theoretical Exposition*. Doctoral Thesis, University of Glasgow.

———. (2003) Market Socialism as Market Fetishism. *Critique*, 34 (May) (Glasgow): 168.

Kennedy, P. (2004) Labourism and Social Democracy Post-1945. *Critique*, 35 (Glasgow): 65–95.
Marx, K. (1866) Speech to the International Working Men's Association. Available at http://www.marxists.org/history/international/iwma/documents/1866 (accessed 7 August 2005).
———. (1869/1985) Speech to the IWMA. *MECW*, Vol. 21, 398 (London: Lawrence & Wishart).
———. (1973) *Grundrisse* (Harmondsworth: Penguin).
———. (1976) *Capital*, Vol. 1 (Harmondsworth: Penguin).
———. (1981) *Capital*, Vol. 3 (Harmondsworth: Penguin).
Marx, K and F. Engels. (1976) *The German Ideology*. *MECW*, Vol. 5 (London: Lawrence & Wishart).
———. (1977) *The Communist Manifesto* (Moscow: Progress Publishers).
Meszaros, I. (1995) *Beyond Capital* (New York: Monthly Review Press; London: Merlin).
Perlman, F. (1973). Introduction. In I. I. Rubin (ed.) *Essays on Marx's Theory of Value*, ix–xxxviii (Montreal: Black Rose Books).
Rikowski, G. (2000) That Other Great Class of Commodities: Repositioning Marxist Educational Theory. Paper presented at the British Educational Research Annual Conference, Cardiff University, 9 September 2000.
Sarup, M. (1978) *Marxism and Education* (London: Routledge and Kegan Paul).
Sharp, R. (ed.). (1986) *Capitalist Crisis and Schooling: Comparative Studies in the Politics of Education* (Melbourne: Macmillan).
Shattock, M. (ed.). (1996) *The Creation of a University System* (Oxford: Blackwell).
Sikka, P. (2003). Socialism in Reverse. *The Guardian*, 15 April.
Simon, E. (1946) The Universities and the Government. *Universities Quarterly*, 1: 79–95. Also published in M. Shattock (ed.) (1996) *The Creation of a University System* (Oxford: Blackwell).
Smyth, J. (ed.). (1995) *Academic Work* (Buckingham: SRHE & OU).
Ticktin, H. (1983) The Transitional Epoch, Finance Capital and Britain. *Critique*, 16 (Glasgow): 23–42.
———. (1986) Towards a Theory of Finance Capital Part 2: The Origins and Nature of Finance Capital. *Critique*, 17 (Glasgow): 1–16.
———. (1992) *Origins of the Crisis in the USSR* (New York: Sharpe).
———. (1994) The Nature of an Epoch of Declining Capitalism. *Critique*, 26 (Glasgow): 69–94.
———. (1998) Where Are We Going Today? The Nature of the Contemporary Crisis. *Critique*, 30–31 (Glasgow): 21–48.
West, M. (2000) State Intervention in English Education, 1833–1891: A Public Goods and Agency Approach. Discussion Papers in Economic and Social History, No. 37, October, University of Oxford.
Willmott, H. (1995) Managing the Academics: Commodification and Control in the Development of University Education in the U.K. *Human Relations*, 48 (9): 993–1027.
Wolf, M. (2006) A New Gilded Age. *Financial Times*, 25 April, 15.

Chapter 3

The Role of Education in Capital Crisis Resolution

Helen Raduntz

Introduction

The simple yet complex reality that we have to face is that capitalism actually survives and thrives on economic crises; that crises act as circuit breakers in capitalism's otherwise headlong rush to self-destruction; and that expansion is the chief means of resolving them. Paradoxically, therefore, it can be said that whatever its intensity and wherever it occurs at a local or global level, a crisis functions as a stimulus driving capital accumulation and expansion. The cumulative effect is that today there are few avenues of human activity and social life, including education and associated information services, that are beyond subservience to the absurd and irrational whirlpool of capitalism's dynamics.

This is the key issue confronting those who seek to develop a practical theory of education for social change, who see that there are possibilities for a more socially oriented existence beyond the limits of the horizon set by capitalism, and who are determined to work toward it.

The following analysis presents a critical account of the dynamics within capitalism that are driving education toward marketization and its exploitation as a crisis management strategy for capital. In this task I draw on Marx's critique of capital and on the analyses of those Marxists (Harvey, 1982; Mandel, 1968; Rubin, 1972) who have sought to extend his analysis within the historical materialist frame of reference, and to complement the critical discourses on education, for instance, government business policies for education (Allen et al., 1999), the commodification of public education (Kenway and Fitzclarence, 1998), education markets (Marginson, 1997a), equity (Hill and Cole, 2001), social justice (Fitzclarence and Kenway, 1993), pedagogy critique (McLaren and Farahmandpur, 1999), ideology-critique (Hill et al.,

1999), resistance critique (G. Rikowski, 2001), and the impact of trade policies on information services (R. Rikowski, 2005).

Toward Education for Capital Crisis Resolution

If education is to become a crisis resolution strategy for capital it must be subjected to marketization. Marketization under the historical conditions specific to the capitalist economy or mode of production is a process in which things are severed from their social connections and transformed into reified, atomized units of measure as commodities. In the process they acquire a market or exchange-value based on the amount of labor expended on their production, the only form in which their value can be converted into money capital—which is the whole point of the exercise.

The separation, or alienation, of things as discrete entities from their social settings, which explain their nature and their existence, is a precondition for their commodification and privatization within the capitalist economic system. It is therefore an important part of understanding the current directions of educational change to refer to education's historical development involving its alienation and commodification. The development can be seen as a series of transitional phases that if current trends are maintained will see education's integration into the mainstream of the capitalist economy as a capitalist mode of education (Raduntz, 2001: 317–347ff).

Education's historical record shows that from its beginnings it has been inextricably bound up with class divisions serving the interests of ruling elites, and that though in the modern era a trend can be detected toward education's social democratization it still remains linked to the interests of the capitalist ruling class. At the beginning of the modern era with the disintegration of feudalism and the influx of new learning into Europe pressures began to mount that were to challenge the church's monopoly on education. Thus began education's separation as a function within the institutional setting of the medieval church and its structural reconstitution into schools, universities, and systems of education solely dedicated to education.

At the same time there began the segmentation of the body of knowledge into discrete disciplines that Mészáros (1970: 109–111ff) links to the alienating influence of the capitalist economy. In these forms each discipline could be ascribed a marketable value rated according to its economic and political usefulness upon which the stipends of professors could be determined. For example, in the emerging capitalist market economy professors of highly valued medicine and rhetoric could command the highest salaries, philosophers the lowest (Ganss, 1956: 39).

Corresponding to these trends and as towns began to set up schools the knowledge to be taught became segmented and graded according to the degree of difficulty. A teacher's stipend therefore could be calculated not only on the degree of difficulty but also on students' results. It so happened that in the town of Treviso its schoolmasters' stipend was fixed not only according to pupils' results but also according to four categories of difficulty.

The "elementary stuff" attracted half a ducat while the fourth category attracted two ducats (Ariès, 1973: 173).

At this stage the provision of education took on local and regional characteristics and was spread across a range of informal domestic and formal independent institutional settings. It was from these existing provisions of education that the emerging capitalist economy was able to draw for artisans' knowledge and skills and on the public reservoir of scientific and technological knowledge, which was to become the basis of its economic growth.

In the industrial revolution period, between 1750 and 1850, such was capitalism's success in applying the newly created scientific knowledge and technology in the sphere of production that the technical efficiencies thus engendered began to be systematized in education. Modelled on organizational arrangements similar to those existing in commerce and industry, national education systems were established by state authorities mainly because in capitalism's formative years the scale of education required to support the phenomenal growth of the economy demanded funding beyond the resources of private investment capital. There was, however, an additional factor. The existing private provision of education could not supply the numbers of skilled workers, scientists, technicians, and bureaucrats increasingly demanded by a nation's burgeoning industries and state instrumentalities.

In this regard, the Prussian education initiative in the early nineteenth century provided a model that gave Germany an economic advantage and which other European nations sought to emulate (Musgrave, 1967). Prussia's initiative also included the setting up of the state-sponsored University of Berlin with a mandate to conduct inquiry free from political interference, and to make provision for teacher training. However, with regard to the latter initiative, by mid-century reactionary, Prussian governments were to blame its highly educated teaching force for instigating the social unrest that came to a climax in the revolutions of 1848. In the eyes of conservative and reactionary governments too much teacher education, it seemed, could undermine their regimes.

As a result of these developments education generally became highly systematized, centralized, and bureaucratically administered by the state. From the beginning therefore the public through the offices of the state has borne the costs of modern systems of education that feeding into the economy as part of its infrastructure have underpinned capitalism's expansion.

From a function monopolized by the medieval church, education has since become reconstituted as a state monopoly. However, I submit that in capitalism's historical trajectory punctuated by economic crises of varying magnitude state monopolization of education systems can be regarded as a transitionary stage in education's evolution toward its integration as an adjunct of the economy proper. In the event capitalism's expansionary tendencies will demand that education systems become segmented so that the potentially lucrative sectors can be privatized and all that this means in education's subjection to the fiscal, structural, and efficiency regimes that have been spectacularly successful in raising labor's productivity and promoting economic growth at least in the economically advanced nations of the world. The gradual introduction of a utilitarian and instrumental approach to learning and incursions of market forces into education's governance are indicative of this trend, and the subsequent carving up of education systems is likely to leave governments to administer a residual and poorly resourced public sector.

However, such a utilitarian makeover of education has its countervailing undertow that is beginning to cohere around the concept of citizenship and civic values (see, e.g., Marginson, 1997b). Motivated by humanist ideals proponents see education as a means of developing human potentialities and advancing the creation of a society that would, toward this end, promote and reflect the principles of what it is to be human (see Bowen, 1981: 440–443ff). A project of this kind requires universal education for social change; a focus on the shared creation and acquisition of knowledge; and the exercise derived from liberal values of critical reason—aspirations that appear to be the very antithesis of current trends toward education's capitalization.

The expansionary economic conditions of the post–World War II era under the regime of Keynesian economic management policies conspired to bring the economic and social imperatives of modern education together. Education was seen as a means not only of personal development fulfilling the aspirations of the majority of people for a better future, but also of national and global economic development. Under these circumstances government funding of education occupied a large slice of the gross national product reflecting education's growing prominence in guaranteeing an ever-expanding economy (ibid.: 526–529ff).

During the 1960s in response to a decline in economic growth education became increasingly utilitarian to the detriment of the ideals of human development, a trend that led to the tertiary student uprisings particularly in France in the late 1960s. Following the economic crisis of the 1970s, however, the utilitarian trend became more marked as reflected increasingly in the proletarianization and deteriorating conditions of teachers' work (Ozga, 1988). It gathered momentum as state policies under the influence of neoliberal ideology particularly in English-speaking nations began to restructure their education systems along corporate lines to become more flexible and responsive to the needs of their economies in the face of global competition and free trade regimes.

The corporate model borrowed from the world of business is one in which education systems, while administratively centrally controlled and regulated by fiscal expediency, are decentralized and fragmented as competitive, semi-autonomous units of productive educative activity. On this basis it is not inconceivable that a collection of schools could exist as franchisees within a transnational corporation's stable of franchised product lines alongside, for instance, a health clinic, a mining venture, and a military hardware supplier.

This brief historical excursion is designed to facilitate an understanding of the rationale underlying education's reconstitution as a capitalist enterprise. That rationale is to be found in contemporary capitalism's attempts to overcome its current, yet inherent, economic crises through expanding into untapped sources of economic growth, of which education is one of the most lucrative in an economy increasingly dependent on knowledge generation and information.

The Inner Dynamics of the Capitalist Economy

The historical overview of education reveals elements of correspondence between developments in education and the evolution of the capitalist economy. What has now to be explored is the nature of the current economic crisis and how this is influencing the nature of the changes in education. For the exercise the following analysis draws on Marx's critique of capital in which he employs the value relation as his governing principle and the relation between the forces and social relations of production as his organizing framework.

In the opening analysis of capital Marx establishes that the source and substance of value that permits the value exchange and equalization of commodities of different kinds is the amount of labor power expended in their production carried out under specific social conditions in which "private individuals or groups of individuals...carry on their work independently of each other" (Marx, 1954: 77). Marx also establishes that because commodities of equal value only can be exchanged on the market, the source of profit or capital accumulation must be found in the sphere of production. In this sphere there exists social conditions that lend themselves to the exploitation of workers. That exploitation has its genesis in historical circumstances that witnessed the separation of workers from their means of making a living. On that account workers are forced to sell their labor power on a competitive labor market for what they can get in the form of wages to owners of capital whose interest in the transaction is to employ workers to produce commodities for sale primarily for profit.

If capitalist employers can recoup their labor costs in the sphere of production in less than the contracted time of, say, eight hours, then in the remainder of the time a worker produces a surplus of value that, when the commodities produced are exchanged, constitutes the margin of profit (Marx, 1954: 188). By employing a number of technological strategies and instituting labor processes, structures, and management techniques designed to maximize efficiency employers can substantially raise workers' productivity and lower their wage costs, thereby increasing their profit margins in relation to their original capital investment.

In a market based on the capitalist form of private property it is assumed that commodities produced by workers after their wages have been covered legitimately belong to their employers. As an illustration of this kind of arrangement, in a contract between de Jelly, master-weaver, and one Nicholas Cornélis in 1634, there is stated bluntly that the latter will be paid half of what he makes, the other half being the master's profit (Mandel, 1968: 132), which de Jelly could legitimately claim according to the rules of market exchange. There is therefore a symbiotic or dialectical relationship between the production of commodities and their exchange in which commodities and the value embodied in them circulate between each "pole." Their value is sourced in production according to the amount of labor power expended on them and then realized in their exchange as expanded capital or profit.

The circulation process, however, is fraught with problems. Any disruption, whether through labor unrest and technological breakdowns, transport and delivery delays, or slow sales, constitutes a crisis in capital accumulation and expansion which is the primary motivating force historically specific to the capitalist economic system. In Marx's account (1954: 529 and 1956: 25–26) capital is a circulating process of which production and exchange are merely moments. Through the medium of money, in the sphere of exchange money capital is invested to purchase labor power and means of production. In the form of production capital it then traverses through the labor process in which means of production are converted into commodities as materialized value containing the value of the capital originally invested plus a surplus value. Finally, the commodities are brought to the market, sold, and their value realized as expanded money capital. And so the process begins in a continuous cycle of reproduction and accumulation. At every stage there can be problems that become apparent if we follow the capitalization process from its beginning in the market place.

In the sphere of exchange a capitalist enterprise faces not only competition from other enterprises, but also uncertainty with regard to prices. It must therefore grab as large a market share as it can. This can be achieved in the sphere of production by producing commodities more efficiently and productively and therefore at a cheaper price in order to undercut the prices of competitors. The competitive advantage, however, is shortlived because in order to avoid bankruptcy competitors must themselves follow the lead and introduce cost saving strategies and technology similar to those employed by our enterprise.

However, this strategy rather than grant above average profits to all enterprises merely averages out prices across competing enterprises that serves to reduce profit margins to practically zero. It is an impossible situation of crisis proportions that forces each enterprise in order to escape bankruptcy to renew efforts to restore its profitability mainly through introducing technological change, extending existing markets, or expanding into new ones.

The result is a pattern of enforced and continuous leapfrogging among enterprises as new technologies are adopted independently of the will of any particular capitalist enterprise for the reason that in the final analysis the expansion of capital depends on above average profit making (Harvey, 1982: 120–121). The response of individual enterprises to this chaotic state of affairs in the marketplace over which they have little influence is to increase their control in the sphere of production in order to raise productivity in any way they can.

Since its introduction in the industrial era of capitalism mechanization together with scientific management techniques and segmented detail work regimes have been refined to afford maximum control over the labor process and increased productivity (Braverman, 1974). Such efficiencies included the systematic separation of mental and manual labor and the subdivision of work processes into constituent specialized operations. The mobilization and concentration of science and technology as powerful forces of production in the hands of management is an outcome of the division of labor that provides an enterprise with the organizing ability and capacity to revolutionize production almost at will.

The operative word is "almost" because with continual rounds of restructuring in their workplaces and corporate organizations to conform to new production processes, enterprises are confronted with the perpetual antagonism of workers who face the threat of redundancy, falling wages, and the progressive deterioration of their working conditions, for example, the erosion of autonomy over their work practices, the intensification of their work, and the reduction or obsolescence of their skills as they become mere appendages in a mechanized labor process. Marx (1954: 372–451ff) and Braverman (1974) have recorded the deleterious effects of these conditions on workers.

These circumstances present a paradox or contradiction that is consistent with capitalism's highly dynamic mode of production. On the one hand, the reduction of labor's input and consequently its value in the drive for efficiencies and greater productivity minimizes the very factor, labor power, that creates the surplus-value on which profitability depends. On the other hand, competition in the marketplace as well as worker antagonism can constitute motivational factors for an enterprise to engage in continuous technological change and organizational restructuring in order to maintain viability.

Together these two forces can operate to raise productive capacity beyond the ability of the market, as it currently exists, to absorb the commodities that are produced. One reason is the absence of the majority of potential consumers with purchasing power because of the constant pressure to reduce their wage levels. The eventuality of a market collapse constitutes a crisis in capital realization of major proportions that immediately prompts action on the part of enterprises to resolve it. These conditions are reflected in the daily price fluctuations that are a feature of the capitalist market. If the circulation of commodities through production and exchange is taken as a whole we can see why, paradoxically, capitalism actually survives and thrives on alternate states of stability and instability and equilibrium and disequilibrium.

As we are all aware prices fluctuate daily, but over time it can be noted that they oscillate around an average price that in idealized circumstances would be proportional to the value of the labor-time materialized in the commodities presented for exchange. In this ideal eventuality a state of equilibrium would exist throughout the economy reflecting the equal distribution of labor among enterprises according to demand. In the capitalist economy, however, the average price is proportional to the costs of production for a given product plus the average profit on the capital invested (Rubin, 1972: 63–75). Furthermore, because the capitalist economy is supply rather than demand driven, and because its division of labor consists of producers working independently and separately from each other, there is no one controlling the distribution of labor and individual producers have no way of gauging how much to produce or how much their competitors are producing. There is therefore the constant tendency to overproduce or underproduce.

In this event, if demand does not rise to meet supply, then overproduction and a corresponding fall in prices will occur. Consequently, in response as production contracts the capital and labor surplus to requirements will gravitate toward those enterprises enjoying a period of expansion. The reverse is true for underproduction. The seesaw distribution of capital and labor allows expanding enterprises to produce above average surplus-value and thereby to realize above average profits at the expense of enterprises whose production is in decline. The effect is to create conditions of disequilibrium, instability, and uncertainty on the market that paradoxically at the same time are the very conditions necessary for capital accumulation and expansion.

Taken as a whole we can see how the imperative for individual enterprises constantly to revolutionize their forces and social relations of production, while stimulating economic growth and profit making, can at the same time jeopardize the continuing reproduction of capital accumulation. The market is the only mechanism available to correct this threat in the absence of any regulation of production (a circumstance that would compromise the freedom of enterprises in the conduct of their business). In planning their production targets, enterprises are influenced only by the market where their products are equalized, and where it becomes clear as indicated by the rise or fall in prices that supply has outstripped demand. In this case, enterprises are induced to expand or cut back on their production to avert a crisis in an attempt to the equilibrium demanded by the market.

The foregoing account has sought to show how instability and disequilibrium paradoxically stimulate as well as threaten economic growth, creating a roller coaster pattern in which uncertainty and instability follow certainty and stability while at the same time sustaining a society based on a form of private property and wealth creation. It is a pattern punctuated by economic crises that constitute the mechanism impelling the capitalist economy to expand on an ever-increasing scale in order to maintain its economic viability.

The Expansion of Capitalism: The Ultimate in Crisis Management Resolution

Capitalism's development is characterized by periodic crises derived from the unlimited development of the forces of production through constant technological change. These forces continually press against the barrier of capitalism's social structures and relations. In these circumstances the cycle of accumulation threatens to come to a halt. The ensuing economic stagnation is expressed in the forms of overproduction, speculative investment, crises, unemployment, and the overaccumulation of noninvested capital (Harvey, 1982: 190).

The causes of economic crises internal to the working of capitalism are many and varied and include factors within the spheres of production (machinery breakdown, strikes by workers, labor shortages, delays in deliveries of raw materials) or exchange (poor consumer demand, inflationary price fluctuations, problems associated with credit and insolvency), and in the circulation between and within the spheres of production and exchange (slow turnover times due to inefficiencies in transport, trade barriers, and the slow pace of or inadequate structural reform) in relation to developments in the forces of production.

The crises cause the devaluation of labor power and skills together with the devaluation, depreciation, even destruction, of existing capital value invested in means of production, machinery, and fixed assets. The economic wastage is enormous, and the repercussions are socially devastating in terms of political tensions and social conflict.

The wastage in labor and the devaluation of surplus capital paradoxically, however, serve to provide the basis for a recovery of the accumulation cycle by clearing away obsolete technologies, practices, and structures and instituting structural adjustments, rationalizations and reforms, new monetary systems, new policy initiatives and new organizational forms (Harvey, 1982: 431).

Although many social institutions and organizational structures—what Harvey calls social infrastructures—particularly those under state control, such as education—are not directly productive in terms of capital accumulation, nevertheless capital value in the form of revenue circulates through and is modulated by them. They become part of capitalism's continuous exploitation and the modification of organizational arrangements that can alleviate and contain the tensions arising from capitalism's inner dynamics. The circulation of value through social infrastructures on this account can be regarded as momentary circuits in the total accumulation process. Such flows for instance, have supported scientific and research and development (see Harvey, 1982: 398–415).

Economic crises thus possess a dual function not only of devaluing and destroying the "old" but also of preparing the ground for economic recovery and the "new" in a continuous round of economic instability and stability. In this way, the crises mediate a "space" between production and exchange in which the forces and the social relations of production can adjust to and resolve the tensions arising in the sphere of exchange. For unlimited development of the forces of production in response to competition in the market place creates stresses that can be resolved only by organizational restructuring and social adjustment. However, if the basic class relation remains unaltered, the contradictions between the forces and the social relations of production are not resolved but merely displaced and recreated on a different plane (Harvey, 1982: 326).

Crisis Management

In capitalism's history, economic crises have given rise to what may be described as crisis management strategies that function not only to resolve the consequent economic instability, but also to stimulate economic recovery, that again prepares the way for another crisis. In order to clarify the process, credit and finance capital can play a positive role in providing loans to further economic expansion. In the form of debt, however, credit can cause bankruptcy as well as stimulate speculative fever to the point of economic collapse. In the sphere of production, for instance, enterprises can draw on a reserve army of unemployed and therefore cheap labor to aid economic recovery and to counter the inflationary pressures that scarce expensive labor can help to create. But too many unemployed workers causes not only social instability but also a drain on public revenue that could more profitably be employed to support the private sector of the economy.

In other examples, a necessary degree of economic stability can be achieved by means of organizational restructuring, raising interest rates, changing government policy to provide "a level playing field" through greater regulation, and ensuring competition to counteract monopolistic tendencies. Eventually, though, these kinds of measures will end in economic stagnation because economic growth and capital accumulation and expansion is dependent on economic inequity, labor exploitation, the freedom to create labor redundancies, competition in the market place, constant organizational and societal restructuring, and the entrepreneurial freedom to expand by the privatization of any resources whatsoever which have capitalization potential. These are now classic crisis resolution and management tactics.

Crises or threats of stagnation send shockwaves throughout the capitalist economy and the time-honored means of resolving them has been expanding either by revamping existing markets or venturing into hitherto "undeveloped" or "underdeveloped" economies or potentially lucrative spheres of social life. The contradiction between the two imperatives, economic stability on the one hand, and instability on the other, accounts for capitalism's extraordinary dynamism and expansionary tendencies.

In our contemporary situation in the wake of the 1970s economic crisis the contradiction, as Mandel (1975: 562) points out, and as foreshadowed by Marx (1959: 247–250ff), is reflected, on the one hand, in the enormous productive capacity and economic growth that the capitalist market driven economy has generated. On the other hand, it is reflected in the barrier to ongoing capital accumulation and expansion represented in the class relation between capital and labor based on largely exclusive private property ownership. Rather than finding a mechanism that would equitably distribute the resulting huge capital surpluses throughout society, the capital-labor social class relation, which Marx identified as the fundamental social characteristic of the capitalist economy, causes the surpluses to gravitate toward and become concentrated in the hands of a relatively few capitalists and corporate enterprises. As a capitalist class and to maintain their profit-making class interests these few must find a way to reinvest their capital returns mainly through expanding into new markets.

As a consequence, since equity in distribution without a return in the capitalist economic system is unthinkable, and because the competitive pressures to reduce wage costs precludes the majority of the working population from effectively becoming consumers above and beyond satisfying their basic subsistence needs, expansion is the only option in managing and resolving capitalism's tensions that are manifested in the rapid succession of economic crises, the growing disparity between wealth and poverty, and an escalation of social conflict, instability, and strife.

Today, predictably, the capitalist system is responding through expansion, driven primarily with increasing speed by technological innovation and by the restructuring most facets of the economy. In the process barriers to free trade, competition, and the exploitation of labor are gradually eroded and few sectors of social life including education have escaped the predatory overtures of corporate appropriation and privatization facilitated by compliant political leaders and governments.

The Direction of Educational Change: For Social change or for Capital

It is within the matrix of capitalism's contradictions and the role played by economic crises that the current direction of educational change can be understood. For education has, in the current period of capitalism's evolution toward an information-based economy, taken center stage as a crisis management strategy in its roles as a productive force, as a consumer of surplus capital, as a means of warehousing the army of the potentially employed (see Shor, 1980), and of rotating surplus labor through cycles of employment and unemployment in a lifelong educative process. Furthermore, education has become significant in terms of its role in research, in staff development and training in order to ameliorate the excesses and social stresses of constant organizational restructuring, and in assisting employees to adapt to the needs of their employers under these circumstances.

These roles can be effective only if education is closely tied and responsive to the economy's swiftly changing needs, and this means education's marketization and integration into the market economic regime itself. In the event education becomes subject to all the contradictions, instability, and uncertainty that are inherent in the capitalist economy and that have plagued its evolution. It will also mean the introduction into education of mainstream capitalist working conditions and its subjection to the rapid succession of structural changes that are a feature of capitalism's contemporary development.

The recognition of education as a productive force has since the 1940s emerged out of capitalism's growing dependence on technological innovation not only in the development of electronically controlled automation (Mandel, 1975: 207), but also in what is considered to be the new basis of economic growth, information and communication technology. The growing dependence on these technologies has seen a shift away from manual to intellectual labor in the corporate workplace, and increasingly massive investments in science, technological innovation, and research. In turn these developments have created an enormous demand for highly skilled intellectual labor power that accounts for the expansion of the tertiary education sector in the 1950s.

As the cumulative growth of science and the greater acceleration of research and development gain momentum capitalist processes of increasing divisions of labor, rationalization, and specialization in the drive for private profit penetrate the spheres of intellectual labor and scientific education (Mandel, 1975: 249 and 261–263). There follows the proletarianization of intellectual labor and the instrumentalization of curricula where humanities barely rate in terms of market-value as tertiary qualifications are tailored to meet the needs of corporations and the labor process (see Noble, 2002).

These developments can be ascribed to the enormous buildup of private surplus capital that not only puts pressure on cash strapped government-funded education systems to instigate public-private partnerships, but also motivates business interests to invest in education as a means of absorbing their excess capital, in an attempt to revitalize a stagnating economy. In these circumstances education is doubly lucrative because it also provides opportunities for educational goods, consultancy and professional development services, online curricula packages, and information technology. Finally, there is enormous profit potential represented in the repositories of information and knowledge and library facilities (Rikowski, 2001/2002), if privately owned, and if intellectual property can be privatized and leased or rented out in perpetuity.

As education becomes remodeled along business lines features common to commodity production, mechanization and automation, standardization, overspecialization and the parcellization of labor will increasingly penetrate education as it has in other sectors of social life subjected to privatization (Mandel, 1975: 387). It is perhaps in the area of crisis management, in the overall containment of capitalism's internal contradictions, however, that education plays a significant role, a role that is hitherto somewhat underresearched.

Conclusion

The analysis has attempted to demonstrate not only the close reciprocal relationship between the capitalist economy and education, but also that the contemporary developments in education can be explained only with reference to the inner dynamics of capitalism and its drive to resolve its current economic crises.

Today, education and associated services represent potentially a multimillion dollar industry to a capitalist economy awash with capital desperately seeking investment opportunities. Its progressive integration into the capitalist economy proper is an indication that education and its social responsibilities is being sacrificed in order to keep the market-driven economy afloat for private capital accumulation. As its most lucrative services become integrated into the economic mainstream education, its workers will inevitably become subject to the same efficiency and productivity raising dynamics that have traditionally characterized the capitalist mode of production.

Moreover, what the analysis has shown is that in the process education becomes another tool in capitalism's crisis management armory involved in stimulating economic growth on the one hand, and simultaneously absorbing the surplus capital that accrues from that growth, on the other. In other words, education is being shaped to function for capital as a crisis resolution and management strategy.

The analysis also reveals the nature of capitalism's inner dynamics to be cold, ruthless, and inexorable, responding only to the law of the market, as Harvey (1982: 203) and Marx before him have observed. Yet, as a social relation capitalism is a product of human historical development but it has come to dominate, dehumanize, and delimit the very human creative attributes, open expression, and exchange of ideas on which the economic system depends for its continued existence and growth, and which education is supposed to develop and encourage.

The practical question then becomes how might educational research and effective action be employed positively, not to pose an alternative approach to education within the existing capitalist social relations, but to redirect the current trends in education and to overcome the capitalist education social relations being forged for it so that an education and a society might materialize that is not driven by the private profit motive. The project requires a determined struggle against the limits capital is imposing on education, and for an education the primary function of which is focussed on positive social change and development.

References

Allen, M., C. Benn, C. Chitty, M. Cole, R. Hatcher, N. Hirtt, and G. Rikowski. (1999) *Business, Business, Business: New Labour's Education Policy* (London: Tufnell Press).
Ariès, P. (1973) *Centuries of Childhood* (London: Methuen & Co. Ltd.).
Bowen, J. (1981) *A History of Western Education: The Modern West, Europe & the New World*, Vol. 3 (New York: St. Martin's Press).
Braverman, H. (1974) *Labor & Monopoly Capitalism: The Degradation of Work in the Twentieth Century* (New York: Monthly Review Press).
Fitzclarence, L. and J. Kenway. (1993) Education and Social Justice in the Postmodern Age. In B. Lingard, J. Knight, and P. Porter (eds.) *Schooling Reform in Hard Times*, 90–105 (London: Falmer Press).
Ganss, G. (1956) *Saint Ignatius' Idea of a Jesuit University* (Milwaukee, WI: Marquette University Press).
Harvey, D. (1982) *The Limits to Capital* (Oxford: Basil Blackwell).
Hill, D. and M. Cole. (eds.). (2001) *Schooling and Equality: Fact, Concept and Policy* (London: Kogan Page).
Hill, D., P. McLaren, M. Cole, and G. Rikowski (eds.). (1999) *Postmodernism in Educational Theory: Education and the Politics of Human Resistance* (London: Tufnell Press).

Kenway, J. and L. Fitzclarence. (1998) Consuming Children? Public Education as a Market Commodity. In A. Reid (ed.) *Going Public: Education Policy and Public Education in Australia*, 47–56 (Canberra: Australian Curriculum Studies Association).

Mandel, E. (1968) *Marxist Economic Theory*, Vol. 1 (London: Merlin Press).

———. (1975) *Late Capitalism* (London: New Left Books).

Marginson, S. (1997a) *Markets in Education* (St. Leonards, NSW: Allen & Unwin).

———. (1997b) *Educating Australia: Government, Economy and Citizen since 1960* (Melbourne and Cambridge: Cambridge University Press).

Marx, K. (1954) *Capital*, Vol. 1 (Moscow: Progress Publishers).

———. (1956) *Capital*, Vol. 2 (Moscow: Progress Publishers).

———. (1959) *Capital*, Vol. 3 (Moscow: Progress Publishers).

McLaren, P. and R. Farahmandpur. (1999) Critical Pedagogy, Postmodernism, and the Retreat from Class: Towards a Contraband Pedagogy. In D. Hill, P. McLaren, M. Cole, and G. Rikowski (eds.) *Postmodernism in Educational Theory: Education and the Politics of Human Resistance*, 167–202 (London: Tufnell Press).

Mészáros, I. (1970) *Marx's Theory of Alienation* (London: Merlin Press).

Musgrave, P. W. (1967) *Technical Change, the Labour Force and Education: A Study of the British and German Iron and Steel Industries 1860–1964* (Oxford: Pergamon Press).

Noble, D. F. (2002) *Digital Diploma Mills: The Automation of Higher Education* (New York: Monthly Review Press).

Ozga, J. (ed.). (1988) *Schoolwork: Approaches to the Labour Process of Teaching* (Milton Keynes, UK: Open University Press).

Raduntz, H. (2001) *A Contemporary Marxian Critique of Trends in Education and Teachers' Work in an Era of Major Structural Change*. Unpublished doctoral thesis, University of South Australia, Adelaide.

Rikowski, G. (2001) *The Battle in Seattle: Its Significance for Education* (London: Tufnell Press).

Rikowski, R. (ed.). (2001/2002) Introduction: Globalisation and Information. *Information for Social Change*, 14: 3–5.

———. (2005) *Globalisation, Information and Libraries: The Implications of the World Trade Organisation's GATS and TRIPS Agreement* (Oxford, UK: Chandos Publishing [Oxford] Limited).

Rubin, I. I. (1972) *Essays on Marx's Theory of Value* (Detroit, MI: Black & Red).

Shor, I. (1980) *Critical Teaching and Everyday Life* (Chicago: University of Chicago Press).

Chapter 4

What Neoliberal Global and National Capitals Are Doing to Education Workers and to Equality—Some Implications for Social Class Analysis

Dave Hill

Introduction: Education, Neoliberalism, and Class War from Above

Education is being neoliberalized globally. Privatization, deregulation, decentralization, reduction in public social, welfare and educational spending, the shift in the tax burden from rich to poor, from the middle and upper classes to the working classes, all disproportionately affect the working class.

Neoliberalization of education has five major effects. These are the

1. increases in inequalities both within and between states;
2. reduction in the quality and standards of schooling and higher education overall for working-class populations;
3. assault on democracy;
4. suppression and compression of critical space in education (and other state apparatuses (see Hill, 2001, 2005b, 2007b); and
5. decrease in education workers' pay, rights, and conditions (see Hill, 2005a, 2006a, 2007a).

Academic labor itself is becoming both more proletarianized and more straightjacketed into increased subservience to the interests of capital (see Hill, 2001, 2003, 2004a, 2004b, 2007a, 2007b).

In this chapter, I attempt to argue from a classical Marxist analysis, the continued salience of social class in terms of it being the *essential* and the major victim of neoliberalism and of intensified capitalist exploitation. I address three aspects of class denial in the media and schooling ideological state apparatuses, and in the work of conservative, postmodernist, and some Left academics ("revisionist Left"[1] or "Left liberal") such as Michael W. Apple. These analyses variously hide or deny the class nature of capitalism, or, in Apple's case, relegate class to just one form of structural exploitation, with equivalence to race and gender exploitation/oppression.

First, the working class is not dead in objective terms—the number of wage laborers globally is growing. Second it is the working class that is being damaged most by neoliberal policies by attacks on "the social wage" and by fiscal policy—the increasing tax on workers and decreasing tax on business and the rich in a policy of educational and social/economic triage. Third—and this is connected—workers' pay, rights, and conditions of employment—and the rights, powers, and the organizing/negotiating ability of trade unions are also under systematic attack from neoliberal policies. These policies comprise an intensified "class war from above" (Dumenil and Levy, 2004; Harvey, 2005).

In this discussion of class, I recognize the internally differentiated nature of social class; argue the necessity of developing class consciousness—a necessary state of mind for engaging in a successful socialist transformation of capitalist economy, society and politics; and assert the necessity of the leading role of the working class and its organizations in the global anticapitalist struggle. I conclude that such a Marxist analysis is necessary both theoretically and organizationally if capital is to be radically transformed.

Part One: Social Class Analysis

The Objective Salience of Social Class

Academic and media arguments that "class is dead," or, if not dead, is declining relative to preceding periods in capitalist development, are common. It is commonly suggested that this "disappearance" of class has occurred in objective terms; that "fast capitalism" (Agger, 1989) has transmogrified the working class into a service, hitech, increasingly self-employed and/or embourgeoisified workforce.

The Capitalist Class Knows It Is a Class

It is interesting, and rarely remarked upon, that similar arguments are *not* advanced regarding the disappearance of class signifiers and identifiers among the capitalist class. They appear to know very well who they are. Nobody is denying capitalist class consciousness.

The "class struggle from above" is relatively invisible to many outside the Marxist tradition. There is a general failure to recognize the class nature of antiunion legislation, changes to the taxation system, privatization of public and of welfare services. Combined with the decline of visible resistance/"class war from below" in the 1980s and 1990s in the UK, following the defeat of

The Great Miners' Strike in 1985, this led many commentators to announce, sometimes rather hopefully, the end of class society. Pakulski and Waters (1996) argue that (working class) consciousness and the motivation/desire for recognition of "class war" and class struggle have (largely) disappeared. Thus Pakulski argues,

> The key assumption of class analysis—that all important social conflicts have a class basis and class character because class represents the key social dimension of modern (capitalist) society—does not withstand critical scrutiny. (1995: 75)

I now want to argue against the three major forms of arguments that social class is either dead or diminishing in importance as the essential and salient objective structural cleavage in capitalist society.

First, I want to make some comments about postmodern analysis of class. Second, to refer to analyses that we now live in a hitech, service industry, deindustrialized world, where anticapitalist struggle is now focused on and by "social movements." Third, about the work of Left theorists such as Apple who assert, inter alia, the equivalence of "race," class, and gender as forms of oppression under capitalism, refuting the salience of class as the essential form of exploitation in capitalist society.[2]

Postmodern Analysis

Together with others (e.g., Cole and Hill, 1995, 2002; Cole, Hill, and Rikowski, 1997; Hill and Cole, 2001; Hill et al., 2002. See also Allman et al., 2005; Cole, 2007), I have argued against the postmodernist analysis that an appreciation of multiple subjectivities necessarily negates the salience of class as a form of structural analysis, and as the necessary center of anticapitalist resistance. Some critical or "resistance" postmodernists fear that "multiple voices" will be lost or oppressed in communities or in mass movements and have therefore suggested that the most that might be hoped for is a gathering of voices within increasingly small and homogenous groups. Beyer and Liston contest such analyses:

> personal and social conditions need to be continually created, recreated, and reinforced that will encourage, respect, and value expressions of difference. Yet if the valorization of otherness precludes the search for some common good that can engender solidarity even while it recognizes and respects that difference, we will be left with a cacophony of voices that disallow political and social action that is morally compelling. If a concern for otherness precludes community in any form, how can political action be undertaken, aimed at establishing a common good that disarms patriarchy, racism and social class oppression? What difference can difference then make in the public space? (1992: 380–381)

Social Movements

Similar, though not identical, arguments are advanced by some representatives of the "social movements" concerning the demise—or at least, the demise in importance (for the anticapitalist project)—of class and of class-based movements such as trade unions. For example, the debate at the (2004) World Social Forum between Chris Harman and Antonio Negri (Harman, 2002) stated both positions clearly. Harman's salient points against Hardt and Negri's (2001) conception of "multitude" as a category to replace "class" are set out below in the discussion on the size of the working class and an assessment of changes in its internal differentiation.

Left Liberals/Revisionist Leftists Such As Michael W. Apple

A third source of attack on class analysis comes from perspectives, such as that of Apple. Apple attacks, though sometimes specific personalities Marxists such as Peter McLaren, and writers within what some describe as "the British Marxists" for having an outmoded class analysis, rooted in the age of the white, male (indeed, masculinist) industrial worker—the proletariat. He depicts class analysts as people who have learned nothing since the reproductionist Marxists of the 1970s such as Bowles and Gintis and Louis Althusser. (See, e.g., Apple 2005a, 2005b, 2006a, 2006b.) For example, in his 2005 American Education Research Association (AERA) speech, Apple (2005b) attacked Marxists who have returned to a two-class model, citing "my good friend Eric Olin Wright...we should use Erik Olin Wright" instead of "outmoded models."[3]

While detailed sophisticated classificational analysis of social class into multiple categories can be useful, it has a number of drawbacks—which have a theoretical as well as a political dimension.

I want to make a fourfold criticism of Apple's position. These relate to

1. his analysis of *social class classification*;
2. his "tryptich" *race, class, gender model of equivalence*;
3. his *relative autonomist, antireproductionism* critique of some contemporary Marxist thought; and
4. the implications of Apple's analysis for *political organization and action*.

Social Class Classification

First, let us look into the complex (whether from the Left or the Right) models of class, where there are 6 class, or 16 class or 66 class categorizations are based on Weberian notions of class, which are based on patterns of consumption rather than relationship to the means of production.

Social class analysts who are Marxist, such as Erik Olin Wright (e.g., 1985, 2005; Wright et al., 2001), theorize *strata* in the class structure, in particular those in "contradictory class locations" (middle or junior managers, and workers with some autonomy over their working conditions such as doctors, university lecturers). Complex models of categorization, whether from a Weberian consumerist/lifestyle categorization, or whether from an attempt such as Wright's at Marxist analysis (of combinations of relationships to the means of productions, and the social relations of production) customarily perform a number of disabling functions for class analysis and subsequent political project and political organisation.

First, they hide the capitalist class as a class. Nowhere in the lists of classified strata or "classes" appears the capitalist class. Second, they hide the working class as a class, they segment the commonality and *objective* unity of the working class in itself. These distinctions of/between layers, or strata propagate and present a picture of a fragmented working class with divergent interests. Third, they thus work to fragment working-class solidarity, class consciousness, and its sense of solidarity, of social and political cohesion. They impede class consciousness. As such, they hide the class struggle.

They erase both the proletariat and the capitalist classes as antagonistic entities unified in the contradictory and exploitative social (property) relations of capitalist production (Hill, 1999; see also Hill and Cole, 2001; Kelsh, 2001; Rikowski, 2001; McLaren and Scatamburlo d'Annibale, 2004; McLaren and Jaramillo, 2006). Paula Allman, Glenn Rikowski, and Peter McLaren (2005) also critique the various types of complex and sophisticated (and in some ways useful)[4] "Box People" classifications—putting people in what could, in an orgy of sophistication, result in endless subclassification. The trouble is, you can get so sophisticated, and so focussed on the micro that you cannnot see the wood from the trees or a way though the orchard.

Raced and Gendered Social Class

Apple criticises class analysts for ignoring race, gender, and sexuality. He suggests that we need a much more nuanced and complex picture of class relations and class projects to understand what is happening in relation to "racial dynamics" as well as those involving gender (e.g., Apple, 2005a: 392, 2006a: 116). In this, Apple presupposes that the Marxist theory of class cannot address differences such as those of race and gender—essentially, that it can address only the "economic." He has, for a long time, asserted, inter alia, the equivalence of race, class, and gender as forms of oppression under capitalism, refuting the salience of class as the essential form of exploitation in capitalist society.

Apple accuses traditional Marxists of *class reductionism*, a view in which all the features of human social experience are reducible to the class division in society, where everything can be explained by its function for the reproduction of capital, ignoring the specificities of that oppression such as that based on gender or race. Apple's accusation is that classical Marxists "privilege" class and marginalize race, gender, and sexuality. *But the concept of class, the existence of class, the awareness of class are themselves sometimes buried beneath, hidden by, suffocated, displaced in the work of Michael W. Apple.*

Martha Gimenez (2001: 24) succinctly explains that "class is not simply another ideology legitimating oppression." Rather, class denotes "exploitative relations between people mediated by their relations to the means of production." While Apple's "parallellist," or equivalence model of exploitation (equivalence of exploitation based on race, class, and gender, his tryptarchic model of inequality produces valuable data and insights into aspects of gender oppression and race oppression in capitalist United States; however, such analyses serve, as Gimenez (and Kelsh and Hill, 2006) suggest, to occlude the class-capital relation, the class struggle, to obscure an essential and defining nature of capitalism, class conflict.

Marxists recognize that class is raced and gendered and characterized by other forms of personal, institutional, and structural discrimination and oppression. Many Marxists have sought to argue this since the white male-dominated Marxism of the Stalinist era and the Third International of the 1950s. In my own work with Rikowski, McLaren, and Cole (e.g., Cole et al., 2001; McLaren et al., 2001; Hill et al., 2002; see also Smith, 2007) we take great pains to recognize and combat both in writing and in political action, the current ferocity and historical nature of race and gender oppression and discrimination based on sexuality and disability.

With respect to one aspect of structural inequalities reproduced within the education system in England and Wales, in terms of educational achievement, Gillborn and Mirza (2000) show very clearly that it is the difference between social classes in attainment that is the fundamental and stark feature of the education system in England and Wales.

In their analysis of attainment inequalities by class, race, and gender for the years 1988–1997, 5 or more higher grade General Certificates of Education (GCSEs—the exam usually taken at age 16 in England and Wales relative to the national average), the gender difference between girls and boys is half that relating to race (comparing white students with African Caribbean). This in turn is less than half of the social class difference—the difference between children of managerial professional parentage on the one hand, and children from unskilled manual working-class homes (Gillborn and Mirza, 2000: 22).

There is a specific race factor involved. Dehal (2006) points out that the impact of economic disadvantage does differ significantly across BME (Black and Minority Ethnic) groups. He concludes that "economic disadvantage is the key driver of ethnic disparity." He notes that the ethnic groups of poor children—(again, as defined as those receiving free school meals)—performing least well are, respectively, starting with the lowest attainers of ethnic groups, the Gypsy Roma/Travellers, then the white British, then white and black Caribbean, and then the black Caribbean. In contrast, poor children from the Chinese, Indian, and Bangladeshi ethnic groups achieve the highest attainment results, with the poor Chinese children the highest achieving ethnic group of all (as denoted by 5 or more Grade C to Grade A* at the GCSE exam stage, at around age 16) of all poor children, and Gypsy Roma/Traveller, and white British poor children the lowest attainers of all children in poverty.

Thus, Dehal's conclusion, based on empirical data collected for the period 2003–2005, confirms that of Gillborn and Mirza (2000). There is a specific race factor involved—some ethnic groups of 15–16-year-olds in receipt of free school meals—such as white and African-Caribbean and Roma children—do perform/attain more poorly than the average for all 15–16-year-old children in receipt of free school meals, and considerably more poorly than Chinese and Indian group of such children. However, the race effect, the effect of being part of a particular ethnic group, has less impact on achievement and underachievement than does social class. Class analysis is more reliable as a measure of achievement/underachievement than race analysis.

To return to the broader relationship between race, gender, and social class, and to turn to the United States, are there many who would deny that Condoleeza Rice and Colin Powell have more in common with the Bushes and the rest of the U.S. capitalist class, be it white, black, or Latina/o, than they do with the workers whose individual ownership of wealth and power is an infinitesimal fraction of those individual members of the ruling and capitalist class. *O povo, unido, jamais sera vencido*—the workers, united, will never be defeated. In a class war there are many victories and defeats. With a unity, recognizing and affirming difference, but also affirming class solidarity, such defeats are less likely. As is shown in the millions of men and women workers, white, black, Mahgrebian on the streets of Paris and Rome against neoliberalization of public services in recent years.

Objectively, whatever our race or gender or sexuality or ability, whatever the individual and group history and fear of oppression and attack, the fundamental form of oppression in capitalism is class oppression. While the capitalist class is predominantly white and male, capital in theory and in practice can be color and gender blind—even if that does not happen very often. African Marxist-Leninists such as Ngugi Ngugi Wa Thiong'o (e.g., Ngugi wa Thiong'o and Ngugi wa Mirii, 1985; Ngugi Ngugi Wa Thiong'o and Ngguggi, Moses Isegawa, 2005) know very well that when the white colonialist oppressors were ejected from direct rule over African states in the 1950s and 1960s, that the white bourgeoisie in some African states such as Kenya was replaced by a black bourgeoisie, acting in concert with transnational capital and/or capital(ists) of the former colonial power.

Apple writes that his analyses (of the multilayered and Gramscian lessons about hegemony learned by the Radical Right in the United States) show that the Left and progressive forces have much to learn from the success of the Radical Right in the United States. Perhaps what we, and he, should learn, is that the capitalist class, the male and female, black, brown and white bourgeoisie, actually do recognize that they survive in dominance as a class, that whatever their skin color, or dreams, or multifaceted subjectivities and histories of individual or group hurt and triumph, precisely because they do know they are a class. They have class consciousness. The capitalist class does not disable itself by negating its class consciousness over issues of race and gender. It gets on with exploiting the labor of the raced and gendered working class.

Relative Autonomist Antireproductionism

Apple does not only attack Marxist class analysts. He also expends energy in critiquing those who point out the social class reproductive nature of capitalist education. In this respect, his books advance a "relative autonomy" thesis. Thus, Farahmandpur's critique of Apple is that Apple's (1993, 1996, 1999) "recent work on class...views social class as a subjective phenomenon that is culturally determined...Michael Apple...relegates class as an objective force to a subjective phenomenon that is by and large culturally determined" (2004).

As Kelsh and Hill (2006) elaborate, Apple's analysis of "the social" "where neither the economic, the political, nor the cultural sphere is determinant" (1993: 25), and instead determination by any "sphere" is seen to be "historically contingent" (ibid.) advances a view of the social in which local analyses are the only ones possible."

Following Ebert (1996), Kelsh and Hill continue,

> For historical materialists, however, "cultural and ideological practices are not autonomous but are instead primary sites for reproducing the meanings and subjectivities supporting the unequal gender, sexual, and race divisions of labor, and thus a main arena for the struggle against economic exploitation as well as cultural oppression." (Ebert, 1996: 42–43, Italics added)

The work of the knowledge industry (those knowledge workers who update the ruling ideology and naturalize it), however, is to augment ideology rather than critique it and expose it as a site of class struggle.

The intensification and extension of capitalism and the unprecedented commodification of the human, bear out Marx and Engels' *The Communist Manifesto*. They analyzed the impact of global capital on the worker:

> it has resolved personal worth into exchange value, and in place of the numberless indefeasible chartered freedoms, has set up that single unconscionable freedom—Free Trade.

They pointed out that capitalism

> left remaining no other nexus between man and man than naked self-interest, than callous "cash payment" and "egotistical calculation." All would be reduced to "paid labourers."

This unprecedented commodification and capitalization of the human would seem to demonstrate that that the relative autonomy attaching to the political region of the state from the economic, the autonomy between various vertical and horizontal levels of the state apparatus, and within state apparatuses, is far less than that claimed by Apple in his attacks on those Marxists who use forms of reproduction theory.[5] While there is (growing) resistance, (e.g., see Hill, 2008a) state apparatuses at various horizontal levels, apparatuses such as schools, vocational colleges, universities, in nations are being rolled over by the juggernaut of neoliberal capital. (Some) teachers, schools, universities do resist—but not many, and within very, and increasingly constrained spaces for resistance, for counterhegemonic practice.

The increased (raced) social class hierarchicalization of education provision and results, which I describe nationally and globally below, are indeed exemplification of the broad thrust of the analyses made by Bourdieu, Bowles and Gintis, and by Althusser. Of course, each of these writers can be critiqued, for example, for internal inconsistency, for leading to a pessimism in the face of the iron fist of capital, and for downplaying resistance. But the capitalist class system *does* reproduce existing patterns of exploitation and domination. And the iron fist of capital *is* tightening. The bell of economic determination is now tolling (see Hill, 2001, 2004a, 2004b, 2005a, 2006c, 2007a, 2007b). As through the control and surveillance of schoolteachers and teacher educators and other university educator through the mechanisms of new public managerialism, through heavily prescriptive, assessed, and monitored curricula in schools and teacher education. And more widely, through the Patriot Act in the United States and decline of civil liberties in the UK.

Recognition that the iron fist of capital is tightening nationally and globally, and inward, into the human (Rikowski, 2002a, 2002b), and recognition that there is a class war from above, can inform resistance and anticapitalist action.

Indeed, Farahmandpur (2004) suggests that "[a]long with a number of other scholars on the Left, Apple has dismissed the centrality of class struggle in efforts at educational reform," together with "the vanguard role of the working class in the arena of social change."

In contrast to the Leninist strategy of "democratic centrism," in which the vanguard party operated as the "ideological and political compass" of the proletariat, indeed, in contrast to a workers' party per se (whatever its internal political organization and form of democracy) Apple firmly espouses the notion of a "decentered unity" that consists of an alliance among feminists, multiculturalists, lesbians, gays, antiracists, environmentalists, peace activists, progressives, and neo-Marxists. Apple asserts that the Radical Right "has recognized that to win in the state, you must win in civil society.... It created a decentered unity, one where each element sacrificed some of its particular agenda to push forward on those areas that bound them together. Can't we do the same?" (2001: 194–195). And so Apple calls for a more widely based, but *not class-based*, broad progressive movement.

While, in Simon's phrase, "it is ... necessary to recognize that a pluralism of social movements is a condition for a fully developed democracy" (1982: 104), there is the possibility that that class consciousness, socialist consciousness, the transition from a Gramscian "common sense" to a Gramscian "good sense" can be lost, indeed the belief not only in paths to (peaceful or violent defensive) socialist revolution can be displaced and disappeared—to the extent that many communist parties throughout the world actually dissolved themselves following the collapse of the Soviet-style state capitalist (or on an alternative view "deformed workers' state") systems.

That was a loss of confidence, of belief, of class action, of socialism, in the name of a more widely based, but not class-based, broad progressive movement. This particular "war of position" has, empirically, led to the (incorrectly termed) center-left of the Clinton and Blair Third Way transmogrifying—becoming deeply colonized by capital, becoming, indisputably, part of the historic bloc of the capitalist class. As Bellamy notes, the diminution of class analysis "denies immanent critique of any critical bite," effectively disarming a meaningful opposition to the capitalist thesis (1997: 25). And as Harvey notes,

> neoliberal rhetoric, with its foundational emphasis upon individual freedoms, has the power to split off libertarianism, identity politics, multiculturalism, and eventually narcissistic consumerism from the social forces ranged in pursuit of justice through the conquest of state power. It has long proved extremely difficult within the US Left, to forge the collective discipline required for political action to achieve social justice without offending the desire for individual freedom and for full recognition of particular identities. (2005: 41)

Class Analysis Has Political Implications

The classic and current task as Marxist activists is develop a subjective awareness of class—class consciousness—if we are to play a role in the transformation from the working class being a class *in itself* into a class, to a class (acting) *for itself*. In *The Poverty of Philosophy*, Marx states that the working class is "already a class against capital, but not yet for itself. In the struggle...this mass becomes united and constitutes itself as a class for itself" (1977: 214).

> In the capitalist countries, the only class which can accompany a real, deep social transformation is the working class. Marxism properly interpreted emphasizes the primacy of class in a number of senses. One of course is the primacy of the working class as a revolutionary agent...[T]he primacy of class means...that building a multiracial, multi-gendered international working-class organization or organizations should be the goal of any revolutionary movement so that the primacy of class puts the fight against racism and sexism at the center. The intelligibility of this position is rooted in the explanatory primacy of class analysis for understanding the structural determinants of race, gender, and class oppression.

Class analysis and class consciousness provide not only the understanding and the will—they enable class action, which in turn can enable the class appropriation, equitable utilization, and democratic governance of the orchard, of the economy, polity and society, and the ending and transforming of capitalism into a democratic socialist society. In the process, the nonclass analyses of a few rotten apples will be thrown aside.

The Growing Global Working Class

> The working class [exists] as never before as a class in itself...with a core of perhaps 2 billion people, around which there are another 2 billion or so people with lives which are subject in important ways to the same logic as this core. (Harman, 1999: 615)

If the total size of the working class includes not only those engaged in waged labor, but also those who are dependent on income that comes from the waged labor of relatives or savings and pensions resulting from past wage labor—that is, nonemployed spouses, children, and retired elderly people, then the worldwide total figure for the working class, those completely dependent on wage labor, comes to between 1.5 and 2 billion, notes Harman (2002).

In addition, there are, globally, millions of "semiproletarians" (which Harman describes as "semiworkers" or "peasant workers"). In China, for example, more than 100 million people from peasant households seek at least temporary wage labor in the cities each year. Together, suggests Harman, the working class and semiworkers constitute somewhere between 40 and 50 percent of the world's population.

Deindustrialization and the Service Sector in Advanced Capitalist Economies

Arguments that the working class has disappeared usually rests on superficial impressions about what is happening to the old industrial working class, at least in some advanced economies. To take one example, Polly Toynbee writes, "We have seen the most rapid change in social class in recorded history: the 1977 mass working class, with two thirds of people in manual jobs, shrunk to one third, while the rest migrated upwards into a 70 percent home-owning, white collar middle class" (2002; see also 2003).

Industrial employment has fallen noticeably since the 1970s in quite a number of countries, in Britain and Belgium by a third, and in France by more than a quarter. But, as Harman (2002) notes, these do not represent a deindustrialization of the whole of the advanced industrial world.

The number of industrial jobs in the advanced industrial countries as a whole was 112 million in 1998—25 million more than in 1951 and only 7.4 million less than in 1971. In the United States, for example, the number of industrial workers in industry in 1998 was nearly 20 percent higher than in 1971. "Old" industries have by no means disappeared, or moved abroad. Baldoz, Koeber, and Kraft note, "more Americans are now employed in making cars, buses and parts of them than at any time since the Vietnam War" (2001: 7, cited in Harman, 2002: 9). In Japan the industrial workforce more than doubled between 1950 and 1971 and was another 13 percent higher in 1998.

Furthermore, many jobs classified as "service" jobs are classification/data collection changes only (see also note 15). As Harman notes,

> some of the shift from "industry" to the "service sector" amounts to no more than a change in the name given to essentially similar jobs. Someone who works in a factory putting food into a tin so that people can warm it up to eat at home is a "manufacturing worker"; someone who toils in a fast food shop to provide near-identical food to people who do not have time to warm it up at home is a "service worker." (2002: 10)

In Britain the proportion of people in manual jobs is, in fact, much higher than a third. The Office for National Statistics' Living in Britain 2000 shows 51 percent of men and 38 percent of women as being in its various "manual" occupational categories in 1998 (2001, Table 3.14).

Similarly, in the United States (in 2001) the total service-related occupations of 103 million people included 18 million in routine "service occupations" with a decidedly manual cast to them (including nearly a million in "household services," 2.4 million in "protective services," 6 million in "food services," 3 million in "cleaning and building services," and 3 million in "personal services"). Then there were 18 million in routine clerical jobs and 6.75 million sales assistants.

Together these groups constitute well over half the "service sector." Add to them the 33 million workers in traditional manual industries, and you have some three quarters of the U.S. population made-up of workers (Harman, 2002: 15).

Proletarianisation of Teachers, College Lecturers, and Other Service Workers Such As Health Workers

In addition to these numbers, a number of major occupations, sometimes classed as "services" (e.g., Rowthorn, 2001) should be recognized, for the most part (not at the upper managerial echelons) as working class—in the sense that their labor is essential for the accumulation of capital, groups such as health workers and education workers.

These workers' labor has been proletarianized, in terms of intensification of work, loss of autonomy over the curriculum and pedagogy, being subject to the surveillance and rigors of "new public managerialism," payment by results/performance-related pay, increased concern with timekeeping, and tighter and more punitive discipline codes. Part of this proletarianization has been an increased level of identification by such service workers and "education professionals" with the working-class movement, workers' struggle, and industrial action. That is by increased working-class consciousness.

Social Class, Economic Class, and Class Consciousness

There are significant issues concerning intraclass differentiation and about class consciousness. It is important to recognize that *class*, for Marx, is neither simply monolithic nor static. Marx conceived of classes as internally differentiated entities. Under capitalist economic laws of motion, the working class in particular is constantly decomposed and reconstituted due to changes in the forces of production—forces of which members of the working class are themselves a part.[6] Furthermore, Marx had taken great pains to stress that social class as distinct from economic class necessarily includes a political dimension, which is in the broadest sense of the term "culturally" rather than "economically" determined.

And, class-consciousness does not follow automatically or inevitably from the fact of class position. The *Poverty of Philosophy* (1847) distinguishes between a "class-in-itself" (class position) and a "class-for itself" (class-consciousness); *The Communist Manifesto* explicitly identifies the "formation of the proletariat into a class" as the key political task facing the communists. In *The Eighteenth Brumaire of Louis Napoleon* Marx observes,

> In so far as millions of families live under economic conditions of existence that divide their mode of life, their interests and their cultural formation from those of the other classes and bring them into conflict with those classes, they form a class. In so far as these small peasant proprietors are merely connected on a local basis, and the identity of their interests fails to produce a feeling of community, national links, or a political organisation, they do not form a class. (1852/1999)

Thus social class exists in a contingent rather than a necessary relation to economic class. The process (and conceptual category) that links economic and social class is that of "class-consciousness." This is arguably the most contentious and problematic term in the debate over class.

To repeat, if there is one class that does not lack class-consciousness, the subjective appreciation of its common interest, and its relationship within the means of production to other social classes, it is the capitalist class. Currently, in this period of neoliberal capitalism, there is, in advanced capitalist countries, class war from above. Here, the "crucial protagonists are usually those who own or control the main means of domination in capitalist society" (Miliband, 1991: 56, 117).

In Britain the period between 1970 and 1985 was one of intense class struggle from below, as well as from above, marked by a series of major industrial actions: strikes by miners, dockers, and construction workers in the early 1970s, the Grunwick photo development workers' strike of the late 1970s, the so-called Winter of Discontent that preceded the defeat of the Callaghan Labor government in 1979, the Great Miners Strike of 1984–1985 and the Wapping printworkers' strike against the Murdoch Press, which marked the end of this sequence. Following the Thatcher government's victory over the miners in 1985, sympathetic strikes were banned, cooling-off periods and compulsory postal ballots prior to strike action were enforced and widescale union derecognition occurred.[7]

In the current period of neoliberal capitalism, major aspects of class war from above have been the gamut of neoliberal fiscal, welfare, and employment policies. I now describe such policies in the education sector.

Part Two: Neoliberal Education Policies

Neoliberal Policies: Educational Marketization, Commodification, and Privatization

The national faces of neoliberal policies vary in detail and extent. This is partly a function of the balance of class forces in any given polity, what ruling leaderships in government, acting on behalf of their national capitals, think they can and try to get away with, and how effectively they are resisted. In general, privatization of education, the private taking over of public institutions, and the growth of the private sector, tends, as yet, to characterize neoliberal policy in developing and less developed countries, such as Chile, Brazil, Haiti, Pakistan rather than developed countries (See Hill, 2005a, 2006a, *The Rich World*; Shugurensky and Davidson-Harden, 2003; Mukhtar, 2008; Hill et al., forthcoming; Hill and Kumar, forthcoming; Hill and Rosskam, forthcoming). However, as yet slow trends, for example, in the United States, Spain, England, and Wales indicate this will very likely become typical in developed countries, too.

Developed Countries

Privatization

In most developed countries, outright privatization of publicly funded schools has not taken place. The Academies program in England and Wales is close to but not the same as privatization. Rikowski (2005a, 2005b, 2005c, 2006b)[8] notes that policies that can be seen to be preprivatizing, in the sense of both "softening up" public opinion for private control, and establishing an exacerbated hierarchy of schools, are the voucher system and charter schools, as, for example, in the United States—and, indeed, the scheduled 200 academies in England and Wales (Beckett, 2007).

Educational Management Organizations and Charter Schools in the Unites States

In 1999, 13 education management organizations (EMOs) such as Edison Schools, Inc., were attempting to run schools "for-profit." They managed 135 for-profit schools in 15 states. Today, 51 companies manage 463 schools in 28 states and the District of Columbia, 81 percent of which are charter schools. The for-profit management of public schools generally takes two major forms: local school districts contracting with an EMO for the management of existing traditional K-12 public schools (termed "contract schools") or EMOs managing public charter schools either as the charter holder or under the terms of a contract with the charter holder. Charters have been expected to grow exponentially under the George W. Bush's federal education law, "No Child Left Behind," which holds out conversion to charter schools as one solution for chronically failing traditional schools (Schemo, 2004). Hursh (2002) is one of many who point out that most schools, under current U.S. measures of measuring "failure," are "born to fail." It is likely that the number of charter schools will grow dramatically.

These are means of softening up the education service to business control and various forms of profit making by capital. Rikowski goes further, and suggest that any degree of privatization and private involvement acts as a "profit virus," that once a public service such as education is infected ("virused") by private company involvement, then it will inevitably become liable to GATS regulations, and open up to free trade in services by national and by multinational and foreign capital (2003). Schools can enter into deals with private sector outfits. They can also sell educational services to other schools.

In Britain there is currently what may be seen as the "hidden" preprivatization of state schools in England by enabling schools to function as "little businesses" through increased autonomies and business-like managements and corporate aspects, and the ability, within the 2002 Education Act and the 2006 Education and Inspections Act 2006 (DfES, 2006) for schools to act as capitalist enterprises in terms of their ability to merge and engage in takeovers of other schools (Rikowski, 2005, 2006a).

Privatization takes many forms other than the outright control (whether "for profit" or "not for profit." In Britain, the Centre for Public Services' booklet of 2003, *Mortgaging Our Children's Future*, analyzes the various policies and initiatives underway in secondary schools in England and Wales (see also Rikowski, 2005a; Hill, 2006c). It discusses Making Markets, City Academies, and Specialist Schools, School Companies, the Excellence in Cities Program, Privatizing Local Education Authorities, the Private Finance Initiative (PFI), Outsourcing/Restructuring of School Meals, the Education Action Zones Policy.

Commercialism and Marketization

One aspect of profit taking in education is *commercialization* of education. *Direct* commercial penetration is evident in the increasing use of commercially sponsored materials in the classroom and around the school. Whitty (2000) notes that the growing influence of commercial organizations as consultants in public provision can itself contribute to a change in the ethos of the sector. He comments,

> while they might strictly be regarded as elements of marketization, they could also be considered a prelude to privatization in the fuller sense of the privately funded and privately provided education Many critics of devices like devolved budgeting, internal markets, cost-centres and self-governing state schools competing in the marketplace, have been seen as examples of "creeping privatization." (2000: 3)

Whitty suggests that some aspects of marketization contribute to privatization in an ideological if not a strictly economic sense, even where quasi-markets are confined to public sector providers. Aspects of ideological privatization include fostering the belief that the private sector approach is superior to that traditionally adopted in the public sector; requiring public sector institutions to operate more like those in the private sector; and encouraging private (individual/family) decision making. The increasing emphasis on competition and choice has also brought with it what Whitty calls a "hidden curriculum" of marketization. This is far less hidden now than when Whitty wrote in 2000: both conservative and New Labour general election manifestos for the May 2005 general election stressed competition and privatization (Labour Party, 2005) and the 2006 Education and Inspections Act has taken this further, in particular with its expansion of the Academies program, referred to below.

Schoolhouse Commercialism in the Unites States

Possibly the most consistent and thorough analysis of "schoolhouse commercialism" is that by Molnar in the United States (e.g., Molnar, 2004; Molnar et al., 2004). For example his latest (sixth) *No Student Left Unsold—Annual Report on Schoolhouse Commercialism Trends*—examines the following eight types of schoolhouse commercialism, and most are showing a year-on-year increase (Molnar et al., 2004).

Among those cited types of schoolhouse commercialism are corporate sponsorship of school programs and activities; exclusive agreements—agreements giving marketers exclusive rights to sell a product or a service on school or district grounds; incentive programs—the use of commercial products or services as rewards for achieving an academic goal; appropriation of space—the selling of naming rights or advertising space on school premises or property; corporately sponsored educational materials; through to actual privatization—the private ownership of publicly funded schools, private management of publicly funded schools, public charter schools, and private, for-profit school involvement in voucher programs.

It is likely that many of these developments will assume increasing salience in schools and colleges in England and Wales, especially with the anticipated growth in the number secondary schools (high schools in the United States) publicly funded but privately sponsored and controlled Academies.

Developing and Less Developed Countries

The growth of the private sector in schooling and in higher education is far more pronounced in developing and less developed countries. In a number of states, such as Pakistan and China,[9] governments simply request private companies to fill the gap of nonprovision publicly built and staffed/equipped schooling. In others, such as Haiti, Brazil, Thailand, public provision of schooling is of such poorer quality that the effect is the same—effective schooling is left to private companies—some not-for-profit (such as some religious schools), others very much for profit. Increasingly, for example, in Latin America, the profits from for-profit schools flow not only to national corporations, but also to U.S. chains and brands of schools (See Shugurensky and Davidson-Harden, 2003; Hill, 2005a, 2006a).

Part Three: Impacts of Neoliberal Education Policies

Impacts of Neoliberalization of Schooling on Workers' Pay, Conditions, and Securities

Globally, where neoliberalism has triumphed in education, common results have been increased casualization of academic labor, increased proletarianization, increased pay and conditions differentials within education sectors, cuts in the wages/salaries and in "the social wage" of state benefits and rights, cuts, increased intensification of labor, with larger classes, decreased autonomy for school and college teachers over curriculum and pedagogy, increased levels of surveillance, monitoring and report-writing, and accompanying increased levels of stress. There is also the curtailment of trade union rights and attacks on trade unions as organizations that defend and promote working-class interests.

Radical change, or restructuring, of an education institution either means fewer and/or different teachers, professional staff, and support workers. This means lay-offs, forced early retirements, or dismissals, as in the closure of institutions deemed ineffective or failing, and a radical change in what governments and schools and colleges themselves see as their mission. Instead of the liberal-humanist or social democratic ends of education, human capital is now the production focus of very many education systems and institutions.

Trade union rights and capacities are under attack in many countries through a number of means. Performance Related Pay (PRP) is being introduced in various countries, undermining the collective bargaining function of trade unions. The current trend to introduce performance-related pay in higher education in England and Wales, a trend that can be observed all over the world, is challenging established structure of collective bargaining. This is a form of individual pay bargaining. Globally, the World Bank insists (e.g., in Brazil) on the introduction of an additional remuneration for federal higher institution teaching personnel based on individual performance (Siqueira, 2005).

This is part of the deregulation agenda of neoliberal capital, to undermine nationally agreed pay and conditions, and to replace them withy locally negotiated agreements. Furthermore, the decentralization of control, the autonomy of education institutions, weakens union abilities to act. The transfer of negotiations from the national to the provincial or local level, such, in England and Wales, school level or university level, can lead to different pay scales. Unions oppose this because it strikes at the heart of professional equity, under which teachers having similar qualifications can expect the same pay and conditions at any education institution of the same level across the country. This weakens unions! Without strong unions, the pay and working conditions of average teachers/lecturers will further deteriorate, except for the few who receive performance-related pay enhancements or other "merit rewards."

Casualization, Temporary Contracts, and the Trend to a Part-Time Workforce

One way that teachers as workers are kept in line is through job insecurity. Staff on fixed-term contracts have the least job security in the sector, and usually have inferior terms and conditions to their permanent colleagues. Fixed-term contracts leave many staff feeling very exposed and undervalued; having difficulty finding things such as loans, mortgages, and other financial benefits and are discriminatory, as their use disproportionately affects women, black, and other minority groups of workers. They harm the possibility of career progression as individuals find themselves stuck on the lowest pay grades, on a succession of short-term, poorly funded projects that offer no room for staff development; mean staff coming to the end of contracts must inevitably spend time applying for funding or other posts (UCU, 2006: 161).

In England and Wales, as well as more part-time employment, the postincorporation/postdecentralization further education college sector has made increasingly greater use of temporary staff. Using the census returns to the funding council 1994–1995, NATFHE (National Association of Teahers in Further and Higher Education) estimated that 42 percent of staff employed for more than 15 hours per week had temporary contracts, this compared to a national average across all sectors of 9 percent (Hill, 2006a).

Not only is casual work increasing, so is part-time work. In the United States, for example, the percentage of part-time work within all faculty at community colleges rose from 22 percent in 1970 to 60 percent in 2001 (Longmate and Cosco, 2002: B14). In England and Wales, since further education colleges were changed from publicly controlled and funded colleges in 1993 into publicly funded independent colleges, there has been a decline in core funding. There are more part-time staff, less teaching hours for students and an increased use of temporary and agency staff. In 1995–1996, the funding council estimated that 55 percent of all college staff and 39 percent of teaching staff worked part time. This compares to a NATFHE estimate of 15 percent part-time working prior to incorporation, that is, decentralization (Hill, 2006a).

In universities in England and Wales there are more part-time staff, less teaching hours for students and an increased use of temporary and fixed-term contact staff. Nearly half of academic and academic-related staff are on fixed-term posts, a figure that rises to an astounding 93 percent for research-only staff (AUT, 2004b). And in the United States the share of faculty members who are tenured declined from 35 percent to 32 percent between 1992–1993 and 1998–1999 (Hill, 2006a). One reason is the increasing share of faculty that teaches part time since the majority of them are not tenured.

A further aspect of proletarianization is substitutability—and increasingly, nonqualified teachers are replacing qualified teachers in some countries. McLaren et al. (2004: 1) comment that "[i]n California more than 47,000 uncertified teachers are teaching in its public schools...in Baltimore...over one third of the school district's teachers do not hold full teaching credentials." Substitutability is one of the defining characteristics of deprofessionalization, or proletarianization.

Pay and Conditions

A key element of capital's plans for education is to cut its labor costs. For this, a deregulated labor market is essential—with schools and colleges able to set their own pay scales and sets of conditions—busting national trade union agreements, and, weakening union powers to protect their workforces. The growth of the private sector in education globally, as a form of extreme deregulation and decentralization, together with the deregulated decentralization of quasi-market schooling systems have impacted notably on education workers' pay and conditions.

For example, in Pakistan, private schools offer 35–80 percent less salary while job security and career prospectus are the missing elements in mostly short-term/temporary assignments. Pensions, medical, and other facilities and group insurance, and so on are not taken care of by most of the schools in private sector. In England and Wales, the neoliberalization of Schools and Further Education Colleges has taken a number of forms. Widespread contracting out of school meals, school cleaning, and other services to schools and colleges in the 1980s led to severe deterioration of pay and conditions for thousands of low-paid mainly women workers.

In higher education, too, pay levels have deteriorated. In England and Wales, university staff salaries have declined by 37 percent in comparison to the rest of the nation's workforce since 1981 (AUT, 2004a). And the current policy by government (being resisted by the AUT, and by its successor, the UCU (University and College Union) is to establish local pay bargaining.

Even the percentage of full-time teaching staff working in nontenure track position in the United States has increased in recent years, from 8 percent in 1987 to 18 percent in 1998. In other words, a growing number of institutions do not offer tenure. Lower status higher education institutions, community colleges, are least likely to offer tenure and public research universities the most likely, according to the study of the Education Department.

Finally, most part-time workers are hourly paid rather than on permanent fractional contracts and hence have no job security other than the one commonly agreed for the year ahead. They are second-class citizens in the academy.

Neoliberalism and Widening Class Inequalities Globally

There has been an increase in (gendered, raced, linguistically differentiated) social class inequalities in educational provision, attainment, and subsequent position in the labor market. For example, the movement to voucher and charter schools as well as other forms of privatized education such as chains of schools in United States (Molnar, 2004; Molnar et al., 2004) and, suggests Rikowski (2005a, 2005b) in future "federations of schools" (in the UK) have proven to be disproportionately beneficial to those segments of society who can afford to pay for better educational opportunities and experiences, leading to further social exclusion and polarization. (For example, see Whitty et al., 1998; Gillborn and Youdell, 2000.)

Hirtt has noted the apparently contradictory education policies of capital, "to adapt education to the needs of business and at the same time reduce state expenditure on education." He suggests that this contradiction is resolved by the polarization of the labor market, that from an economic point of view it is not necessary to provide high level education and of general knowledge, to all future workers: "it is now possible and even highly recommendable to have a more polarized education system...education should not try to transmit a broad common culture to the majority of future workers, but instead it should teach them some basic, general skills" (Hirtt, 2004: 446).

In brief, then, manual and service workers are to receive cheaper, inferior, transferable skills education, knowledge and elite workers to receive more expensive and more and internationally superior education—one manifestation of the hierarchicalization of schools and the end of the comprehensive ideal. Indeed, this is a form of educational triage—with basic skills training for millions of workers, more advanced education for supervision for middle class and in some countries the brightest of the working classes, and elite education for scions of the capitalist, and other sections of the ruling classes.

Reimers notes that

> the poor have less access to preschool, secondary, and tertiary education; they also attend schools of lower quality where they are socially segregated. Poor parents have fewer resources to support the education of their children, and they have less financial, cultural, and social capital to transmit. Only policies that explicitly address inequality, with a major redistributive purpose, therefore, could make education an equalizing force in social opportunity. (2000: 55)

The World Bank, Privatization, and Equity

The policies of the World Bank encourage this triage. While the World Bank nominally seeks equity, it also encourages the growth of an "educational private sector." This position is also emphasized by the IFC (International Finance Corporation) in IFC document entitled "Investing in Private Education: IFC's Strategic Directions." This outlines outline how fee-paying educational institutions, that is movement *away* from free education, might "improve" equity:

> Private education can indirectly benefit the lowest socioeconomic groups by attracting families who can afford some level of fee away from the public system, thereby increasing capacity and per student spending for the students who remain in the public system. Similarly, the emergence of private tertiary institutions allows governments to reduce funding in such institutions and instead to invest in lower levels of education, thus improving distributive efficiency. (IFC, 2001: 5)

There is a critical difference between conceptions of equity based on universal notions of access and those based on choice. The goals and principles of universal access, represented by international covenants such as the UN convention on economic, social, and cultural rights as well as in many Latin American charters, declarations, and educational reforms (cf. Schugurensky and Davidson-Harden, 2003) reflect a different notion of educational equity than that promoted by the IFC.

Increasing Inequalities: Polarized Schooling and Cherry-Picking

The liberalization of services such as schooling and further education is playing a very significant part in the increased and increasing inequalities *within states* marked by the intensification and exaggeration of (racialized and gendered) social class based and (in developing states rural/urban) differentiation in access and attainment. In sum, the poor have less access to preschool, secondary, and tertiary education; they also attend schools of lower quality where they are socially segregated.

One result of private schooling is that private schools "cream-skim" or "cherry-pick" wealthier families' children who are more equipped to succeed at school, with a corresponding burden on public system schools to absorb students with higher needs and challenges. As Hall notes, it also

> ...undermines the financial solidarity on which public services are based, undermines the political consensus needed to sustain public services, and draws resources away from those services into a consumer-oriented market. It may be exacerbated by cutting back of resources dedicated to public services, which reduces the quality of the public service and encourages those who can pay to buy themselves more resources from the private sector. (2003: 28)

Hall gives some examples—and effects—of the introduction or growth of private sector schools. Social Watch identified this process in Costa Rica: where quality public education has been a major factor in social equity and high living standards, a private school boom now draws better-off students away from public schools with declining resources. Thus, education has changed from being a mechanism for social mobility to becoming an instrument of status and exclusion; in Malaysia, where "two systems have emerged: higher quality private education for those who can afford it and poorer quality public education for those with low incomes" (ibid.: 26).

In South Africa, Educational International notes that the introduction of some charges for schooling "has created a two-tier system within public schooling: schools for the rich and schools for the poor" (2003b: 11). In Brazil, Gindin (in Hill, 2006a) reports that even though a greater proportion of the population than decades ago has access to education, the system is still highly segmented, that public (i.e., state) schools at the basic education level—which are free of charge—are for poor people and private schools are for the rich. The poor are systematically excluded from higher education. In Haiti, Education International reports that nearly 90 percent of Haitian schools are private, though parents have a choice between sending their children to a private school or to a state school. Some private schools are set up for the elite. Schools of the latter type are often run by religious institutions. State schools are completely free of charge, but nevertheless children attending them do pay a price: classes are grossly overcrowded and the education standards are mediocre (Education International, 2003a).

Gendered and Raced Social Class Inequalities

When schooling is not universal, compulsory, and free, then the resulting social bifurcation and reproduction, as well as being social class specific, is gender specific. The *2003/4 Global Monitoring Report on Education for All* by UNESCO (2004) condemns the imposition of school fees as "deterrents for poor families" that "force them to make the choice to reserve education opportunities only for boys." This point is also made in the UNICEF *State of the World's Children Report 2004* (Education International, 2004).

The education needs of indigenous peoples are also not met where schooling is not free, universal, and compulsory. Literacy rates and attendance rates of indigenous children are generally lower than the national averages throughout many states. In Bolivia, for example, indigenous children receive about three years less schooling than nonindigenous children. In the view of Education International, indigenous peoples' education requires specific attention from policymakers, educators, and indigenous communities (Education International, 2003b: 5).

Increasing Social, Economic, and Educational Inequalities in Britain

In assessing the impacts of neoliberal policies on education and on societies, I will conclude this chapter with a brief examination of one country, the UK.

In the UK, the wealth of the super-rich has doubled since Tony Blair came to power in 1997, according to a report on social inequality produced by the Office for National Statistics. Nearly 600,000 individuals in the top 1 percent of the UK wealth league owned assets worth £355 billion in 1996, the last full year of conservative rule. By 2002 that had increased to £797 billion. Part of the gain was due to rising national prosperity, but the top 1 percent also increased their share of national wealth from 20 percent to 23 percent in the first six years of the Labour government. Meanwhile the wealth of the poorest 50 percent of the population shrank from 10 percent in 1986 toward the end of the Thatcher government's second term to 7 percent in 1996 and 5 percent in 2002 (Kampfner, 2005).

And the working classes are paying more tax, with the richest groups paying less, in comparison to 1949 and to the late 1970s—that is, at the end of two periods of what might be termed "Old Labour," or social democratic governments (in ideological contradistinction to the primarily neoliberal policies of New Labour). As a percentage of income, middle, and high earners in Britain pay less tax in 2003 than at any time for 30 years. It is the poorest, the lowest paid, who are paying more. In comparison with the late 1970s, the "fat cats" (Johnson and Lynch, 2004) are now paying around half as much tax (income tax and insurance contribution rate). The fat cats are paying less income tax and national\insurance as a percentage of their earned income than in 1949. In contrast, the average tax rate for the low paid is roughly double that of the early 1970s—and nearly twice as much as in 1949. The subtitle for their article is "sponging off the poor." "As a percentage of income, middle and high earners pay less tax now than at any time in the past thirty years" (ibid.).

The above data concerns the UK—but it is replicated in essence in other advanced capitalist states. To take one example, in the United States, Korten (2004:17) highlights the immense increase in salaries taken by top executives since the early 1990s has shown that in the United States between 1990 and 1999 inflation increased by 27.5 percent, workers' pay by 32.3 percent, corporate profits by 116 percent, and, finally, the pay of chief executive officers by a staggering, kleptocratic, 535 percent.

Conclusion

In this chapter I have tried to indicate how neoliberal policies of marketization, commodification, and privatization of public services comprise an intensification of class war from above by the capitalist class against the working class.

This chapter has shown how the class war from above is being fought on many fronts. Three of these, described in this chapter, resulting from policy changes in the education sector result, are (1) widening social class educational inequalities; (2) attacks on the key working-class organizations—trade unions; and (3) worsening pay and conditions of education workers.

In this chapter, I argue these developments highlight Marxist analyses that social class is the essential and the salient structural cleavage, and form of exploitation, in the current period of neoliberal capitalist globalization, as indeed in all other capitalist periods. I also tried to advance a more structuralist reproductionist Marxist analysis, a classical Marxist analysis, as compared to the postmodern analyses, social movement analysis, and revisionist Left/Left liberal analyses, such as that of Apple.

Notes

1. As Kelsh and Hill (2006) note,

    By "revisionist left," we mean, following Rosa Luxemburg (1899/1970), those theorists who consider themselves to be "left" but who believe there is no alternative to capitalism, and thus do "not expect to see the contradictions of capitalism mature." Their theories consequently aim "to lessen, to attenuate, the capitalist contradictions"—in short, to "adjust" "the antagonism between capital and labor." As Luxemburg explained, the core aim of the revisionist left is the "bettering of the situation of the workers and...the conservation of the middle classes."

2. I have criticized other aspects of Apple's work, his espousal of relative autonomy theory, in Hill (2001, 2005b); Cole et al. (2001); Kelsh and Hill (2006). So has Farahmandpur (2004).
3. He continued his critique in Apple (2006a). Rikowski has made a thorough critique of this, Apple's latest article, in Rikowski (2006).
4. As Kelsh and Hill (2007) note,

    In the place of the Marxist theory of class, the revisionist left has installed a Weberian-derived notion of class as a tool of classification useful only to describe strata of people, as they appear at the level of culture and in terms of status derived from various possessions, economic, political, or cultural. Use of such classifications can be useful in exposing differentials. But for what purpose is such exposure useful? That is, who benefits from it?

Such differentials, because they are not understood from the vantage point of a binary concept of class, are used to fracture the working class by promoting anger, envy, guilt and blame among its various fractions. What is masked from workers, because the capitalist class and its agents work to augment ideology in place of knowledge, is that some workers are poor not because other workers are wealthy, but because the capitalist class exploits all workers, and then divides and hierarchizes them, according capitalist class needs for extracting ever more surplus value (profit). However, as a tool of categorization, such a concept of class cannot provide reliable knowledge to guide transformative praxis. It can provide indications and motivations for reformist measures, such as social democratic redistributive expenditure and policy programmes, but these are limited in nature. Ultimately, as we explain, such Weberian- derived classifications serve to occlude class consciousness and the class contradiction within Capital.

5. I argue this at length in Hill (2001, 2005a); Kelsh and Hill (2006); and in Cole et al. (2001).
6. The most obvious and profound consequence in Britain of a modification in the social composition of the workforce is the vast reduction in manual working class in line with the collapse of manufacturing industries such as steel, shipbuilding, and coal—the proletariat, and the substantial increase in the professional and managerial strata. In making this internal distinction it is important to note that the designation of "proletarian" identifies only that section of the working class who are *directly* involved in the production of surplus-value. Members of the proletariat are by definition part of the working class whilst most working-class people in Britain are no longer proletarians (Gordon, 1995: 36).

There are manifestly different layers, or strata among the working classes. Skilled workers (if in work, and particularly in full-time, long-term work), in general have a higher standard of living than semiskilled or unskilled, or unemployed workers. Their weekly and annual income is likely to be considerably higher. And their wealth is likely to be higher. They are more likely, for example, to have equity, or surplus-value, on an owner-occupied home. In contrast, poorer families, in poorer sections of the working class, may have no wealth whatsoever, and are far more likely to live in private rented accommodation or in rented council housing (what in the United States is termed, "the Projects").

Another important "internal" class distinction, and one which is inevitable as many workers no longer directly produce surplus-value, is the growth of a professional and managerial stratum. That is those who are "between capital and labour" in the sense that whilst being entirely dependent on capital, often in the shape of the national or local state, they exercise *supervisory functions* over the working class (Walker, 1979: 5; *Subversion,* 1998). Examples of this category would be social workers, teachers, lecturers in further and higher education, probation officers, employment service workers, and so forth. Many of them have a consciousness of status in which they place themselves above other, especially manual, sectors of the working class. (Indeed, the Ehrenreichs talk of a professional and managerial *class* [PMC] [Walker, 1979: 5].)

Some, of course, though not many, retain a working-class consciousness, family, social, and political affiliations while, *pace* casualization and proletarianization of many managerial and professional jobs, at the same time benefiting from superior income, conditions of work, and wealth.

For a debate on, and rebuttal of, the thesis that "class is dead," and/or that the working class has diminished to the point of political insignificance, see Callinicos and Harman (1987); Callinicos (1995); German (1996); Harman (2002); Hill et al. (2002); Kelsh and Hill (2006); Smith (2007).

7. For a Left description and analysis of these struggles see Livingstone (1987); Cliff and Gluckstein (1996); Hatton (1988); Taaffe and Mulhearn (1988); Harman/Socialist Worker (1993); Benn (1996); Ali (2005).

Tony Cliff, Donny Gluckstein, and Chris Harman are/were leading members of the Trotskyist Socialist Workers Party (which then became the major component of the Socialist Alliance around the 2001 general election in Britain and is currently the leading component of the RESPECT coalition. RESPECT won a number of council seats in the May 2005 local elections and one MP, George Galloway, in the 2005 general election.

Derek Hatton, Peter Taaffe, and Tony Mulhearn were leading members of "Militant'" (renamed The Socialist Party following its expulsion from the Labor Party in the late 1980s). It was, during the 1980s, one of the two most significant Trotskyist groups in Britain, having two of its members (and another five close sympathisers) elected as members of parliament, controlling dozens of Constituency Labour Parties, and controlling Liverpool City Council through the mid 1980s. Currently the Socialist Party has a handful of local council seats.

Tariq Ali was a leader of the third of the influential Trotskyite parties/groups over the past 40 years, the International Marxist Group (now the International Socialist Group).

Tony Benn has been the leading figure in democratic socialist (or hard) Left through the 1980s and 1990s. He remains in the Labour Party. Benn is "the grand old man" of British Hard Left politics. He left parliament as an MP in 2005, "to devote more time to politics."

8. Academies are publicly funded schools that are independent of the democratically elected local authorities (known as "school districts" in the United States). They have voluntary or private sector sponsors who control the school, or series of schools. The British government intends to have at least 200 academies established or in the pipeline by 2010 (DfES, 2006). Academies are outside LEA control. They can set their own pay and conditions, and change/"vary" the curriculum. The price of buying control of an academy, a state school (with its buildings, teachers, pupils, and curriculum) is between £2 and £2 and a half million. The government/taxpayer funds the £25 million building/rebuilding/start-up costs and hands control to whichever business person, or corporation, or religious group has stumped up the two-and-a-half million pounds purchase price.

In the discussion on the 2005 White Paper (i.e., the then proposed legislation) and the 2006 Education and Inspections Bill, a number of Old Labour MPs rebelled over "selling off state schools to assorted fringe Christian groups and second hand car salesmen" (see, e.g., Paton [2006]; *The Socialist* [2006];

Wintour [2006]). There is widespread, and sometimes successful, opposition to Academies, some of which is detailed at Socialist Teachers' Alliance (2007), and the website of the Anti-Academies Alliance (2007), and in Beckett (2007).
9. Zhou Ji (2005), China's Minister of Education typifies education policy in many developing countries on private schools in announcing

> Encouraging the development of private schools will be necessary to meet the country's growing educational demands. Private schools have played an important role in offering more chances to students and help ease the pressure on crowded public schools.

References

Agger, B. (1989) *Fast Capitalism: A Literary, Political and Sociological Analysis* (Dix Hills, NY: General Hall).
Ali, T. (2005) *Street-Fighting Years* (London: Verso).
Allman, P., G. Rikowski, and G. McLaren. (2005) After the Box People. In P. McLaren (ed.) *Capitalists and Conquerors: A Critical Pedagogy against Empire*, 135–166 (Boulder, CO: Rowman and Littlefield).
Anti-Academies Alliance. (2007) Available at http://www.antiacademies.org.uk/.
Apple, M. W. (1993) *Official Knowledge: Democratic Education in a Conservative Age* (New York: Routledge and Kegan Paul).
———. (1996) *Education and Cultural Politics* (New York: Teachers College Press).
———. (1999) *Power, Meaning, and Identity: Essays in Critical Educational Studies* (New York: Peter Lang).
———. (2001) *Educating the "Right" Way: Markets, Standards, God, and Inequality* (New York; London: Routledge/Falmer).
———. (2005a) Audit Cultures, Commodification, and Class and Race Strategies in Education. *Policy Futures in Education*, 3 (4): 379–399.
———. (2005b) Speech at the 2005 American Educational Research Association Annual Meeting, Montreal.
———. (2006a) *Educating the "Right" Way*, 2nd ed. (New York: Routledge).
———. (2006b) Review Essay: Rhetoric and Reality in Critical Educational Studies in the United States. *British Journal of Sociology of Education*, 27 (5): 679–687.
Association of University Teachers (AUT). (2004a) Higher Education Funding. Available at http://www.aut.org.uk/index.cfm?articleid=598.
———. (2004b) Security Alert II—Ending the Abuse of Fixed-Term Contracts. Available at http://www.aut.org.uk/index.cfm?articleid=904.
Baldoz, R., C. Koeber, and P. Kraf. (2001) Introduction. In R. Baldoz, C. Koeber, P. Kraf, *The Critical Study of Work: Labor, Technology and Global Production*, 3–16 (Philadelphia: Temple University Press). Available at http://www.temple.edu/tempress/chapters_1400/1546_ch1.pdf.
Beckett, F. (2007) *The Great City Academy Fraud* (London: Continuum).
Bellamy, R. (1997) The Intellectual as Social Critic, Antonio Gramsci and Michael Walzer. In A. Kemp-Welsh and J. Jennings (eds.) *Intellectuals in Politics: From the Dreyfus Affair to the Rushdie Affair*, 25–44 (London: Routledge).
Benn, T. (1996) *The Benn Diaries* (London: Arrow Books).
Beyer, L. and D. Liston. (1992) Discourse or Moral Action? A Critique of Postmodernism. *Educational Theory*, 42 (4): 371–393.

Callinicos, A. and C. Harman. (1987) *The Changing Working Class* (London: Bookmarks).
Centre for Public Services. (2003) *Mortgaging Our Children's Future* (Sheffield: Centre for Public Services). Available at http://www.centre.public.org.uk/publications/briefings/mortgaging-our-childrens-future-the-privatisat/.
Cliff, T. and D. Gluckstein. (1996) *The Labour Party: A Marxist History* (London: Bookmarks).
Cole, M. and D. Hill. (1995) Games of Despair and Rhetorics of Resistance: Postmodernism, Education and Reaction. *British Journal of Sociology of Education*, 16 (2): 165–182.
———. (2002) Resistance Postmodernism: Progressive Politics or Rhetorical Left Posturing? In D. Hill, P. McLaren, M. Cole, and G. Rikowski (eds.) *Marxism against Postmodernism in Educational Theory*, 89–107 (Lanham, MD: Lexington Books).
Cole, M., D. Hill, and G. Rikowski. (1997) Between Postmodern and Nowhere: The Predicament of the Postmodernist. *British Journal of Education Studies*, 45 (2): 187–200.
Cole, M., D. Hill, P. McLaren, and G. Rikowski. (2001) *Red Chalk: On Schooling, Capitalism and Politics* (Brighton: Institute for Education Policy Studies).
Dehal, I. (2006) *Still Aiming High*. Available at http://www.blss.portsmouth.sch.uk/training/ppt/DFES_ID_KD.ppt#13.
Department for Education and Skills (DfES). (2006) *A Short Guide to the Education and Inspections Act 2006* (London: Department for Education and Skills). 8 November. Available at http://www.dfes.gov.uk/publications/educationandinspectionsact/docs/Guide%20to%20the%20Education%20and%20Inspections%20Act.pdf.
Dumenil, G. and D. Levy. (2004) *Capital Resurgent: Roots of the Neoliberal Revolution* (London: Harvard University Press).
Ebert, T. (1996) *Ludic Feminism and After: Postmodernism, Desire, and Labor in Late Capitalism* (Ann Arbor, MI: University of Michigan Press).
Education International. (2003a) Privatisation: Government's Cop-Out Means Slow Death for Haiti. *Worlds of Education*, Vol. 6 (Brussels, Belgium: Education International).
———. (2003b) South Africa: Fighting for the Principle of State Provision of Public Services. *Worlds of Education*, Vol. 2 (Brussels, Belgium: Education International).
———. (2004) UNICEF State of the World's Children Report: 2004—Getting Girls into School Is Crucial. *Worlds of Education*, 7 (Brussels, Belgium: Education International).
Farahmandpur, R. (2004) Essay Review: A Marxist Critique of Michael Apple's Neo-Marxist approach to Education Reform. *Journal of Critical Education Policy Studies*, 2 (1). Available at http://www.jceps.com/index.php?pageID=article&articleID=24.
German, L. (1996) *A Question of Class* (London: Bookmarks).
Gillborn, D. and H. Mirza. (2000) *Educational Inequality; Mapping Race, Class and Gender—A Synthesis of Research Evidence* (London: Ofsted).
Gillborn, D. and D. Youdell. (2000) *Rationing Education: Policy, Practice, Reform and Equity* (Buckingham, UK: Open University Press).
Gimenez, M. (2001) Marxism and Class, Gender and Race: Rethinking the Trilogy. *Race, Gender & Class*, 8 (2): 23–33.

Gordon, F. (1995) Workers and Masses. *Open Polemic*, 36. 11 March. Contact P. O. Box 1169, London, W3 9OF.

Hall, D. (2003) *Public Services Work! Information, Insights and Ideas for the Future* (London: PSIRU [Public Services International Research Unit]). Available at http://www.world-psi.org/Content/ContentGroups/English7/Publications1/En_Public_Services_Work.pdf.

Hardt, M. and A. Negri. (2001) *Empire* (Cambridge, MA: Harvard University Press).

Harman, C. (1993) *In the Heat of the Struggle: 25 Years of Socialist Worker* (London: Bookmarks).

———. (1999) *A People's History of the World* (London: Bookmarks).

———. (2002) The Workers of the World. *International Socialism*, 96: 3–45.

Harvey, D. (2005) *A Brief History of Neoliberalism* (Oxford: Oxford University Press).

Hatton, D. (1988) *Inside Left, the Story So Far* (London: Bloomsbury Publishing).

Hill, D. (1999) Social Class. In D. Matheson and I. Grosvenor (eds.) *An Introduction to the Study of Education*, 73–86 (London: David Fulton).

———. (2001) State Theory and the Neo-liberal Reconstruction of Schooling and Teacher Education: A Structuralist Neo-Marxist Critique of Postmodernist, Quasi-Postmodernist, and Culturalist Neo-Marxist Theory. *The British Journal of Sociology of Education*, 22 (1): 137–157.

———. (2003) Global Neo-liberalism, the Deformation of Education and Resistance. *Journal for Critical Education Policy Studies*, 1 (1). Available at http://www.jceps.com/index.php?pageID=article&articleID=7.

———. (2004) Books, Banks and Bullets: Controlling Our Minds—The Global Project of Imperialistic and Militaristic Neo-liberalism and Its Effect on Education Policy. *Policy Futures*, 2 (3). Available at http://www.wwwords.co.uk/pdf/viewpdf.asp?j=pfie&vol=2&issue=3&year=2004&article=6_Hill_PFIE_2_3-4_web&id=86.155.100.248.

———. (2005a) Globalisation and Its Educational Discontents: Neoliberalisation and Its Impacts on Education Workers' Rights, Pay, and Conditions. *International Studies in the Sociology of Education*, 15 (3): 257–288.

———. (2005b) State Theory and the Neoliberal Reconstruction of Schooling and Teacher Education. In G. Fischman, P. McLaren, H. Sünker, and C. Lankshear, (eds.) *Critical Theories, Radical Pedagogies and Global Conflicts*, 23–51 (Boulder, CO: Rowman and Littlefield).

———. (2006a) Education Services Liberalization. In E. Rosskam (ed.) *Winners or Losers? Liberalizing Public Services* (Geneva: ILO).

———. (2006b) New Labour's Education Policy. In D. Kassem, E. Mufti, and J. Robinson (eds.) *Education Studies: Issues and Critical Perspectives*, 73–86 (Buckingham: Open University Press).

———. (2006c) Six Theses on Class, Global Capital and Resistance by Education and Other Cultural Workers. In O.-P. Moisio and J. Suoranta (eds.) *Education and the Spirit of Time*, 191–218 (Rotterdam, The Netherlands: Sense Publishers).

———. (2007a) Critical Teacher Education, New Labour in Britain, and the Global Project of Neoliberal Capital. *Policy Futures*, 5 (2). Available at http://www.wwwords.co.uk/pfie/content/pdfs/5/issue5_2.asp.

———. (2007b) Global Neo-liberalism, Inequality and Capital: Contemporary Education Policy in the USA. In E. Wayne Ross and R. Gibson (eds.) *Neoliberalism and Education Reform* (Cresskill, NJ: Hampton Press).

———. (ed.). (Forthcoming) *Contesting Neoliberal Education: Public Resistance and Collective Advance* (London: Routledge).
———. (ed.). (Forthcoming) *The Rich World and the Impoverishment of Education: Diminishing Democracy, Equity and Workers' Rights* (New York: Routledge).
Hill, D. and M. Cole. (2001) Social Class. In D. Hill and M. Cole (eds.) *Schooling and Equality: Fact, Concept and Policy*, 137–160 (London: Kogan Page).
Hill, D. and R. Kumar. (eds.). (Forthcoming) *Global Neoliberalism and Education and Its Consequences* (New York: Routledge).
Hill, D. and Rosskam, E. (eds.) (Forthcoming) *The Developing World and State Education: Neoliberal Depredation and Egalitarian Alternatives* (New York: Routledge).
Hill, D., S. Macrine, and D. Gabbard. (eds.). (Forthcoming) *Neo-liberalism, Education and the Politics of Inequality* (London: Routledge).
Hill, D., M. Sanders, and T. Hankin. (2002) Marxism, Class Analysis and Postmodernism. In D. Hill, P. McLaren, M. Cole, and G. Rikowski (eds.) *Marxism against Postmodernism in Educational Theory*, 159–194 (Lanham, MD: Lexington Books).
Hill, D., P. McLaren, M. Cole, and G. Rikowski. (2002) *Marxism against Postmodernism in Educational Theory* (Lanham, MD: Lexington Books).
Hirtt, N. (2004) Three Axes of Merchandisation. *European Educational Research Journal*, 3 (2): 442–453. Available at http://www.wwwords.co.uk/eerj/.
Hursh, D. (2002) Neoliberalism and the Control of Teachers, Students, and Learning: The Rise of Standards, Standardization, and Accountability. *Cultural Logic*, 4 (1). Available at http://www.eserver.org/clogic/4-1/hursh.html.
International Finance Corporation (IFC). (2001) *Investing in Private Education: IFC's Strategic Directions* (Washington, DC: IFC). Available at www.ifc.org/.../$FILE/Final%20Public%20Version%20Education%20Strategy%20Paper%202001.pdf.
Johnson, P. and F. Lynch. (2004) Sponging Off the Poor. *The Guardian*, 10 March. Available at http://www.guardian.co.uk/analysis/story/0,3604,1165918,00.html.
Kampfner, J. (2005) The Bling-Bling List. *New Statesman*, 7 March. Available at http://www.newstatesman.com/200503070004.
Kelsh, D. (2001) (D)evolutionary Socialism and the Containment of Class: For a Red Theory of Class. *The Red Critique: Marxist Theory Critique Pedagogy*, 1 (1): 9–13. Available at http://www.redcritique.org/spring2001/spring_2001.htm.
Kelsh, D. and D. Hill. (2006) The Culturalization of Class and the Occluding of Class Consciousness: The Knowledge Industry in/of Education. *Journal for Critical Education Policy Studies*, 4 (1). Available at http://www.jceps.com/index.php?pageID=article&articleID=59.
Korten, D. (2004) Sustainable Development: Conventional versus Emergent Alternative Wisdom. *Third World Traveller. Available at* http://www.thirdworldtraveler.com/Korten/Sustainable%20Develop_Korten.html.
Labour Party. (2005) *Labour: The Future for Britain*. Manifesto for the 2005 General Election (London: Labour Party).
Livingstone, K. (1987) *If Voting Changed Anything, They'd Abolish It* (London: Collins).
Longmate, J. and F. Cosco. (2002) Part-Time Instructors Deserve Equal Pay for Equal Work. *The Chronicle of Higher Education*, 48 (34): B14.
Marx, K. (1847) *Poverty of Philosophy*. Available at http://www.marxists.org/archive/marx/works/1847/poverty-philosophy/index.htm.

Marx, K. (1852/1999) The Eighteenth Brumaire of Louis Bonaparte. In The Works of Marx and Engles. Available at http://www.marxists.org/archive/marx/works/1852/18th-brumaire/ch07.html.

———. (1977) *Selected Writings*, ed. D. McLellan (Oxford: Oxford University Press).

Marx, K. and F. Engels. (1848/1985) *The Communist Manifesto* (New York: Penguin Books).

McLaren, P. and N. Jaramillo. (2006) Critical Pedagogy, Latino/a Education, and the Politics of Class Struggle. *Cultural Studies/ Critical Methodologies*, 6 (1): 73–93.

McLaren, P. and V. Scatamburlo d'Annibale. (2004) Class Dismissed? Historical Materialism and the Politics of "Difference." *Educational Philosophy and Theory*, 36 (2): 183–199.

McLaren. P., M. Cole, D. Hill, and G. Rikowski. (2001) Education, Struggle and the Left Today: An Interview with Three UK Marxist Educational Theorists: Mike Cole, Dave Hill and Glenn Rikowski by Peter McLaren. *International Journal of Education Reform*, 10 (2): 145–162.

McLaren, P., G. Martin, R. Farahmandpur, and N. Jaramillo. (2004) Teaching in and against Empire: Critical Pedagogy as Revolutionary Praxis. *Teacher Education Quarterly*, 31 (3): 131–153.

Molnar, A. (2004) *Giving Kids the Business: The Commercialisation of America's Schools*, 2nd ed. (Boulder, CO: Westview Pres).

Molnar, A., G. Wilson, and D. Allen. (2004) *Profiles of for Profit Education Management Companies, Sixth Annual Report, 2003–2004* (*Arizona:* Arizona State University Education Policy Studies Laboratory). Available at http://www.asu.edu/educ/epsl/CERU/Documents/EPSL-0402-101-CERU.pdf.

Mukhtar, A. (2008, forthcoming) Pakistan. In D. Hill and E. Rosskam (eds.) *The Developing World and State Education: Neoliberal Depredation and Egalitarian Alternatives*. (New York: Routledge).

Ngugi wa Thiong'o and Ngugi wa Mirii. (1985) *I Will Marry When I Want* (London: Heinemann).

Ngugi Ngugi Wa Thiong'o and Ngguggi, Moses Isegawa. (2005) *Petals of Blood* (London: Penguin).

Office for National Statistics. (2001) *Living in Britain 2000* (London: Stationery Office). Available at http://www.statistics.gov.uk/lib2000/index.html.

Pakulski, J. (1995) Social Movements and Class: The Decline of the Marxist Paradigm. In L. Maheu (ed.) *Social Movements and Social Classes*, 55–86 (London: Sage).

Pakulski, J. and M. Waters. (1996) *The Death of Class* (London: Sage).

Paton, G. (2006) Onward Christian Sponsors. *Times Educational Supplement*. 7 April.

Reimers, F. (2000) *Unequal Schools, Unequal Chances: The Challenges to Equal Opportunity in the Americas* (Cambridge, MA: Harvard University Press).

Rikowski, G. (2001) After the Manuscript Broke Off: Thoughts on Marx, Social Class and Education. A paper prepared for the British Sociological Association Education Study Group Meeting, King's College London, 23 June. Available at http://www.leeds.ac.uk/educol/documents/00001931.htm.

———. (2002a) Education, Capital and the Transhuman. In D. Hill, P. McLaren, M. Cole, and G. Rikowski (eds.) *Marxism against Postmodernism in Educational Theory*, 111–114 (Lanham, MD: Lexington Books).

———. (2002b) Fuel for the Living Fire: Labour-Power! In A. Dinerstein and M. Neary (eds.) *The Labour Debate: An Investigation into the Theory and Reality of Capitalist Work*, 179–202 (Aldershot: Ashgate).

———. (2003) Schools and the GATS Enigma. *Journal for Critical Education Policy Studies*, 1(1). Available at http://www.jceps.com/index.php? pageID=article&articleID=8.

———. (2005a) *The Education White Paper and the Marketisation and Capitalisation of the Secondary Schools System in England*, 24 October. Part I Available at http://journals.aol.co.uk/rikowskigr/Volumizer/entries/571. Part II Available at http://journals.aol.co.uk/rikowskigr/Volumizer/entries/572.

———. (2005b) *Habituation of the Nation: School Sponsors as Precursors to the Big Bang? The Volumizer*, 19 October. Available at http://journals.aol.co.uk/rikowskigr/Volumizer/entries/566.

———. (2005c) *Silence on the Wolves: What Is Absent in New Labour's Five Year Strategy for Education*. Occasional paper published by the Education Research Centre, University of Brighton, UK. *Falmer Papers in Education*, 1 (1).

———. (2006a) Education and Inspections Act 2006: First Thoughts. *The Volumizer*, 12 November. Available at http://journals.aol.co.uk/rikowskigr/Volumizer/entries/2006/11/12/education-and-inspections-act-2006-first-thoughts/1250.

———. (2006b) *In Retro Glide. Journal for Critical Education Policy Studies*, 5 (1). Available at http://www.jceps.com/index.php?pageID=article&articleID=81.

Rowthorn, R. E. (2001) Where Are the Advanced Economies Going? In G. M. Hodgson, M. Ito, and N. Yokokawa (eds.) *Capitalism in Evolution* (Cheltenham, UK: Edward Elgar Publishing).

Schemo, D. J. (2004) Education Study Finds Weakened Charter Results. Public School Students Often Do Better—Data Bode Ill for Bush's Philosophy. *San Francisco Chronicle (SFGate)*, 15 October. Available at http://www.sfgate.com/cgi-bin/article.cgi?file=/c/a/2004/08/17/MNGCT89CA51.DTL.

Schugurensky, D. and A. Davidson-Harden. (2003) From Cordoba to Washington: WTO/GATS and Latin American Education. *Globalization, Societies and Education*, 1 (3): 321–357.

Simon, R. (1982) *Gramsci's Political Thought* (London: Lawrence and Wishart).

Smith. M. (2007) The Shape of the Working Class. *International Socialism*, 113 (Winter): 49–70. Available at http:www.isj.org.uk/index.php4?id=293&issue=113.

Siqueira, A. C. (2005) The Regulation of Education through the WTO/GATS: Path to the Enhancement of Human Freedom? *Journal for Critical Education Policy Studies*, 3 (1). Available at http://www.jceps.com/index.php?pageID=article&articleID=41.

The Socialist. (2000) *World Education Report 2000* (Paris: UNESCO).

———. (2004) *Global Monitoring Report on Education for All* (Paris: UNESCO). Available at http://portal.unesco.org/education/ev.php?URL_ID=23023andURL_DO=DO_TOPICand URL_SECTION=201.

———. (2006) Education under Attack. 2–8 April. Available at http://www.socialistparty.org.uk/2006/429/index.html?id=pp5.htm.

Socialist Teachers' Alliance. (2006) Campaigns: Academies and Anti-privatisation. Available at http://www.socialist-teacher.org/campaigns.asp?expand=5.

Subversion. (23 June 1998) (Dept. 10, 1 Newton Street, Manchester, M1 1HW, UK).

Taaffe, P. and T. Mulhearn. (1988) *Liverpool: City That Dared to Fight* (Liverpool: Fortress Books).

Toynbee, P. (2002) After the Jubilation Must Come the Reckoning. *The Guardian*, 5 June. Available at http://www.guardian.co.uk/jubilee/story/0,,727456,00.html.

Toynbee, P. (2003) *Hard Work: Life in Low-Pay Britain* (London: Bloomsbury).

UNESCO. (2000) *World Education Report 2000* (Paris: UNESCO).

———. (2004) *Global Monitoring Report on Education for All* (Paris: UNESCO). Available at http://portal.unesco.org/education/ev.php?URL_ID=23023andURL_DO=DO_TOPICand URL_SECTION=201.

UNICEF. (2004) *State of the World's Children Report 2004* (New York: UNICEF). Available at http://www.unicef.org/sowc04/.

University and College Union (UCU). (2006) *"Further, Higher, Better." Submission to the Government's Second Comprehensive Spending Review. Section 30* (London: UCU). Available at http://www.ucu.org.uk/csrdocs/csrsection30.pdf.

Walker, P. (ed.). 1979 *Between Labour and Capital* (Brighton, UK: Harvester Press).

Whitty, G. (2000) *Privatisation and Marketisation in Education Policy*. Speech given at the National Union of Teachers (NUT) Conference on "Involving the Private Sector in Education: Value Added or High Risk?" London, 21 November 2000. Available at http://k1.ioe.ac.uk/directorate/NUTPres%20web%20version%20(2%2001).doc.

Whitty, G., S. Power, and D. Halpin. (1998) *Devolution and Choice in Education: The School, the State and the Market* (Buckingham, UK: Open University Press).

Wintour, P. (2006) Labour Party Members Voice Opposition to School Reforms. *The Guardian*. 13 March. Available at http://education.guardian.co.uk/policy/story/0,,1729546,00.html.

Wright, E. O. (1985) *Classes* (London: Verso).

———. (2005) *Approaches to Class Analysis* (Cambridge: Cambridge University Press).

Wright, E. O., U. Becker, J. Brenner, M. Burawoy, V. Burris, G. Carchedi et al. (eds.) (2001) *The Debate on Classes* (London: Verso).

Zhou, Ji. (2005) Interview. Cited in *Chinese Private Schools Get Legal Protection. PKU News*, 10 November. Available at http://ennews.pku.edu.cn/news.php?s=81472573.

Chapter 5

Neoliberalism and Education: A Marxist Critique of New Labour's Five-Year Strategy for Education

Mike Cole

Introduction

In this chapter I begin by outlining some defining features of global neoliberalism. I go on to address the issue of democracy in the age of global neoliberalism. After this, I look at neoliberalism, GATS (General Agrement on Trade Services), and world poverty. I then consider neo-liberalism and global education. In this context, I differentiate between the *Capitalist Agenda for Education,* and the *Capitalist Agenda in Education.* I conclude by examining the specific case of Britain, focusing on New Labour's *Five-Year Strategy for Children and Learners* and argue that both agendas are at work in the contemporary British Education System, but in different spheres.

Neoliberalism and GATS

Martinez and García (2000) have identified five defining features of the global phenomenon of neoliberalism:

1. *The Rule of the Market*
    - the liberation of "free" or private enterprise from any bonds imposed by the state no matter how much social damage this causes;
    - greater openness to international trade and investment;
    - the reduction of wages by deunionizing workers and eliminating workers' rights;

- an end to price controls;
- total freedom of movement for capital, goods, and services.

2. *Cutting Public Expenditure*

    - less spending on social services such as education and health care;
    - reducing the safety net for the poor;
    - reducing expenditure on maintenance, for example, of roads, bridges, and water supply.

3. *Deregulation: Reducing Government Regulation of Everything That Could Diminish Profits*

    - less protection of the environment;
    - lesser concerns with job safety.

4. *Privatization: Selling State-Owned Enterprises, Goods and Services to Private Investors, such as*

    - banks;
    - key industries;
    - railroads;
    - toll highways;
    - electricity;
    - schools;
    - hospitals;
    - fresh water.

5. *Eliminating the Concept of "The Public Good" or "Community"*

    - replacing it with "individual responsibility";
    - pressuring the poorest people in a society to find solutions by themselves to their lack of health care, education, and social security.[1]

Global neoliberalism was given a major boost in 1994, with the signing of the GATS at the World Trade Organisation (WTO).[2] The aim of this agreement, which came into force in January 1995, is to remove any restrictions and internal government regulations in the area of service delivery that are considered to be "barriers to trade." The list of services of the GATS includes 12 types, subdivided in many others:

1. Business (accounting, computer science and related subjects, legal, marketing and correlated, medical and dental services, architecture, etc.);
2. Communication (telecommunication, mail, audiovisual, radio, motion picture, etc.);
3. Construction and related engineering services;
4. Distribution (franchising, retail and wholesale, etc.);
5. Education (primary, secondary, higher, adult education, and others);

6. Environmental (sewage, sanitation, disposal, etc.);
7. Financial (insurances, banking, leasing, asset management, etc.);
8. Health and related social services (hospital, other human health services, social, etc.);
9. Tourism and travel related (hotel, restaurant, travel agencies, etc.);
10. Recreational, Cultural, and Sporting (news agency, libraries, archives, museums, theater, sports, etc.);
11. Transports (maritime, aerial, railway, railroad, passenger, freight, maintenance and repair, towing, pipelines, warehouses, etc.); and
12. Other services not mentioned in any other place. (WTO, 2003, cited in de Siqueira, 2005)

Democracy in the Age of Global Neoliberalism

Since February 2000, negotiations are underway in the WTO to expand and "fine-tune" the GATS. As GATSWatch has pointed out, these negotiations have aroused concern worldwide. A growing number of local governments, trade unions, NGOs, parliaments, and developing country governments are criticizing the GATS negotiations and call for a halt on the negotiations. Their main points of critique are

- negative impacts on universal access to basic services such as health care, education, water and transport;
- fundamental conflict between freeing up trade in services and the right of governments and communities to regulate companies in areas such as tourism, retail, telecommunications and broadcasting;
- absence of a comprehensive assessment of the impacts of GATS-style liberalisation before further negotiations continue;
- a one-sided deal. GATS is primarily about expanding opportunities for large multinational companies. (GATSWatch)

In response to criticisms such as these that governments are losing their sovereignty to multinationals, the WTO (2005: 2–3) poses the question

> Why should any Government, let alone 140 Governments, agree to allow themselves to be forced, or force each other, to surrender or compromise powers which are important to them and to all of us?

This view that governments are not to be coerced by capital is based on the WTO claim that the "ultimate aim of Government is to promote human welfare in the broadest sense" (ibid.: 2). David Geoffrey Smith (2003) has provided a trenchant critique of the notion that contemporary government aims have anything remotely to do with human welfare, and has exposed the sham of contemporary "democracy" in the age of global neoliberalism. He argues that special circumstances, such as "the condition of contemporary North American culture" require the creation of new language and new terminology. He has coined the phrase "enfraudening the public sphere" to describe "not just simple or single acts of deception, cheating or misrepresentation" (which may be described as "defrauding"), but rather "a more generalized active conditioning of the public sphere through systemized lying, deception and misrepresentation" (488–489). While Smith's focus is North America, his arguments apply equally to other "democracies."

Smith argues that the "new politics" is the "opposite" of democracy, that it is marked by "unilateralism" and "monological decision-making," by "bullying, both domestically and internationally" (ibid: 498).[3] If it is the case that democracy is no longer a valid term to describe modern states, in its imposition of GATS, the WTO is a steadfast ally to proneoliberal governments in foreclosing democratic resistance to their policies. As GATSWatch puts it,

> under GATS, ... it becomes practically impossible for citizens of a country to reclaim basic public services once they have been liberalized, or to introduce new regulations on social or environmental grounds. Moreover, the WTO positively welcomes this anti-democratic aspect of GATS. In its own question and answer introduction to the Agreement, the WTO Secretariat recommends GATS to pro-liberalization governments for the political assistance it can bring them in "overcoming domestic resistance to change." (GATSWatch, cited in Hill, 2005a: 4).

The WTO claims that the GATS is committed to promoting "the interests of all participants on a mutually advantageous basis" (WTO, 2005: 2). In reality, the WTO is one of the most untransparent and undemocratic global institutions (Sardar and Davies, 2002: 72, cited in Beckmann and Cooper, 2004: 2), largely due to the tendency for decisions to be made in miniministerial gatherings of a select group of rich OECD (Organization of Economic Cooperation and Development) member countries, which are dominated by the United States and the European Union (Rady, 2002, cited in Beckmann and Cooper, 2004: 2).

This is not to say, of course, that GATS is not resisted. As mentioned above, there is opposition from a number of constituencies, and with some degree of success. In 2003, for example, European anti-GATS campaigners scored a victory when European Trade Commissioner Pascal Lamy announced that the European Commission would not further commit Europe's health and education sectors to the free market rules of GATS. In the same year, the fifth ministerial meeting in Cancun came to a dramatic end without agreement on 14 September, leaving negotiations in a deadlock. On the agenda were a range of issues, including agriculture, services, e-commerce, and the environment. The European Union had continued to push for an expansion of the WTO's powers to include new issues (investment, competition policy, government procurement, and trade facilitation), until the eleventh hour, insisting that these issues get dealt with first before the development issues that should have been at the top of the agenda.

As the WDM points out,

> despite the Ministerial Meeting's abrupt ending, the outcome is not a negative one. It was the only option for the developing countries, as no deal is better than a bad deal at this stage. Developing countries refused to be pushed into a corner and have proven they are a force to be reckoned with at the WTO negotiating table.

In addition, the Bolkestein directive, which sought to develop the EU Commission's Lisbon Declaration of 2000 to further liberalize the EU, particularly with respect to services, was withdrawn by the European Commission on 2 February 2005 (McLaughlin, 2005, cited in Hill, 2005a: 6). Under this directive, almost all services would have been exposed to market-based competition. In essence, and for the most part, the directive would have "leveled down" regulatory standards and pay, rather than bringing all services operators up to best standards.

As part of the ongoing hegemonic struggle, the new EU Internal Market Commissioner Charlie McCreevy (appointed in November 2004) "insisted that he was 'committed' to reducing legislation on the services industry during his five-year term" (McLaughlin, 2005, cited in Hill, 2005a: 6). Negotiations are now underway to extend the 1994 agreement and to create even tighter control over 160 service sectors (WDM, 2005).

Neoliberalism, the WTO, and World Poverty

The reality of global neoliberalism is that an estimated 800 million go hungry every day (Vidal, 2005: 14), while 20 million people die each year from starvation and malnutrition (Alcoff and Mendieta, 2000: 2). More than 1 billion people in developing countries live in slums; 27 million adults are slaves; and 245 million children have to go to work (Vidal, 2005: 14). It does not take much insight to work out that if multinational corporations are seeking to make a profit out of health, education, and water those without purchasing power are likely to lose out. For example, recent water privatization in Puerto Rico has meant that poor communities have gone without water, while an unlimited supply has been available in U.S. military bases and tourist areas (WDM, 2005: 2).

Developing countries make pledges to address these issues. However, as Kevin Watkins of Oxfam puts it, "[i]ndustrialised countries...have collectively reneged on every commitment made" (*Guardian*, 12 November 2001: 22). In fact, organizations such as the WTO, the World Bank, and the IMF are constitutionally destined to fail in any attempt at addressing the marginalization of "the developing world." The WTO can only set maximum standards for global trade, rather than the minimum standards that might restrain big corporations, while the World Bank and the IMF, entirely controlled by the creditor nations, exist to police the poor world's debt on their behalf. Rather than recognize these inherent defects, their backers blame the poor countries themselves. Peter Sutherland, former head of the WTO, has asserted that it is "indisputable that the real problem with the economies that have failed [is] their own domestic governments," while Maria Cattui, who runs the International Chamber of Commerce, insisted that the "fault lies most of all at home with the countries concerned" (Monbiot, 2001: 17).

Neoliberalism and Global Education

Hatcher (2001: 1) has identified three agendas for neoliberal capital with respect to schooling in Britain. He describes them as the "business agenda for *what* the school system should produce; an agenda for *how* it should do it; and an agenda for what business itself should do *within* the school system, i.e. make profit." Hill (e.g., 2004b, 2005b, 2004b, emphasis in original) has renamed the first and third of Hatcher's agendas as capital's agenda *for* education and capital's agenda *in* education, and applied them globally. For the purposes of this chapter, I will deal with these two agendas only.

The Capitalist Agenda for Education

This agenda relates to the role of education in producing the kind of workforce that is currently required by global capitalist enterprises. It is thus about making profits *indirectly*. In economic theory, this agenda is connected to *human capital theory*. In mainstream labor economics, human capital theory uses a restricted account based on skills and knowledge: create workers who are flexible and meet the requirements of capitalist enterprises at any given time. Marxists have long argued that personality traits and attitudes should be added to skills and knowledge (e.g., Bowles and Gintis, 1976; Frith, 1980; Rikowski, 2000, 2005). The Capitalist Agenda *for* Education is thus about creating the kind of workers who will "fit in" with capital's needs. In practical terms, the Capitalist Agenda *for* Education means involving the private sector in the running of schools to ensure that government and institutional aims for education correspond to market needs. Not only governments regulate this process, but so also do (relatively) new state apparatuses. In the case of Britain, for example, there is the Office for Standards in Education (Ofsted), policing schools and teacher education, the Teacher Training Agency (TTA) regulating teacher education and the Qualifications and Curriculum Authority (QCA) as a general overseer. As Hatcher (2006a: 600) puts it, control by teachers and Local Education Authorities (LEAs) has been displaced by two new categories of agents: Ofsted, the TTA, and the QCA on the one hand, and private companies, on the other. Their role "is to discipline and transform the old institutional sites of power" (ibid.).

The Capitalist Agenda in Education

As Hill (2005b) argues agenda relates to the role of education in providing profits for capitalists *directly*. It centers on setting business "free" in education, extracting profits from privately controlled/owned schools (ibid.: 260). This involves privatizing either the schools themselves or privatizing services to schools. Privatizing schools is a particular feature of "developing countries" (e.g., Argentina El Salvador, Nicaragua, and Pakistan) (Hill, 2005a, 2005b). Hatcher (in personal correspondence) has described this as "the educational dimension of imperialism." Developed countries, in general, do not want to privatize their own school systems, relying instead on forms of public-private partnerships. However, "peripheral services" (catering, security, reprographics, and consultancy fees) are privatized for profit. In addition, student fees or loans are run for profit by private corporations rather than by the local or national state.

Neoliberalism and Education in Britain

Since the election of the New Labour government in 1997, there has been a qualitative extension of the role of the private sector in the schools system in England. Indeed, in a 1998 background note on education services, the WTO and its Council for Trade in Services (CTS) expressed praise for the British government for having promoted "greater market responsiveness" and an "increasing openness to alternative financing mechanisms" (cited in Rikowski, 2001: 28). As Hatcher (2005: 2) points out, "almost every major government policy initiative has relied on private companies to translate it into practice." Citing Smithers (2004) gives the example of a five-year contract worth £177 million given by the government to *Capita* to manage the delivery of the National Primary and Key Stage 3 Strategies (2005: 2). Hatcher (2005: 3) also lists the chronological privatization of LEA functions: the handing over of some functions, such as supply teachers and school inspections permanently to the private sector; the contracting out of entire LEA provision to private companies, which results when LEAs are designated as "failing" by Ofsted; and the current dominant model of "public-private partnership" between LEAs and private companies.

The *Five-Year Strategy for Children and Learners*: For-Profit or Not-for-Profit?

The most obvious example of New Labor's intention to privatize the whole education system is the *Five-Year Strategy for Children and Learners* (DfES 2004).

Foundation Schools

One of the main features of the *Five-Year Strategy* is the expansion of secondary Independent Specialist Schools (or Foundation Schools), so that all secondary schools will eventually achieve this status. The highest performing ones will be able to opt out of LEA control. These schools, massively expanded by the New Labour government, now make up 50 percent of all secondary schools in the state sector (about 500). They are subject to constant monitoring, and can select 10 percent of their intake based on ability. They must teach the National Curriculum, but can focus on specialist subjects, such as arts or sciences. In order to achieve specialist status, a school has to submit detailed development plans that show increased exam results, attendance, and improved behavior year by year (Smith L., 2004: 1). In addition, they need to be able to raise £50,000 from business.

Another feature of the *Five-Year Strategy* are Academies. These are new state secondary schools, which in general replace schools deemed to be "failing." They are controlled by external sponsors, including business entrepreneurs. Sponsors are required to pay 20 percent of the capital costs of the school, approximately £2 million, and the government provides funding of about £25 million. The sponsor has decisive control over the school. Academies are funded directly by government and set up under private school legislation. This means that they fall outside the legislative framework that governs other state-maintained schools. Ownership of the land and buildings of the existing state school which the Academy replaces, previously the property of the local council, is transferred to the new Academy. Academies are not subject to LEA control and have complete freedom, for example, to devise the curriculum. The sponsor can appoint a majority of the school governing body and thus has control over staffing. Seventeen Academies have opened so far. The government's aim is 200 Academies by 2010 (Hatcher, 2005: 4–5).

However, whereas the 2002 Education Act (DfES, 2002) makes it clear that schools can be run for profit, the *Five-Year Strategy* makes only two references to running schools for profit, both excluding it (Hatcher, 2006a: 603). One reference is to foundation schools: "[a]s now, foundations will not be permitted to run school budgets for profit' (DfES, 2004, Chapter. 4 para. 24, cited in Hatcher, 2006b: 603). The other reference is to Academies. As Hatcher concludes:

> academies are not profit-making investments. On the contrary sponsors are required to put money into the school, not take it out.

The *Five Year Strategy for Children and Learners* (DfES, 2004) states that Academies are promoted and managed by independent sponsors, including philanthropic individuals, educational trusts, faith sponsors and companies on a nonprofit basis. (DfES, 2004, Chapter 4 para. 35, cited in Hatcher, 2006a: 603).

Hatcher points outs that the *Five-Year Strategy* also refers to failing schools being dealt with forcefully, but the *Strategy* does not make any mention of handing them over to private companies (ibid.). He suggests,

> Presumably a major factor in government thinking is the risk of substantial popular and political opposition to the outsourcing of state schools for profit. According to a survey commissioned by UNISON, 89% of the public agreed

that "public services should be run by the government or local authorities, rather than by private companies." (UNISON, 2005; Hatcher, 2006a: 603–604)

Given that there is no evidence that investment in schools in Britain is more profitable for big business than investing elsewhere, big business is less concerned with making profits directly from schools (the *Capitalist Agenda in Education*), and more interested in the production of human capital in state-run schools (*the Capitalist Agenda for Education*). As the Business and Industry Advisory Committee to the OECD (BIAC, 2004) (representing big capital) put it,

> Initial education is essential to a healthy economy. It contributes to competitiveness, enhanced productivity, and the capacity for innovation. Companies depend upon a labour pool that is flexible, technologically literate and work-ready, and have a vital interest in schools turning out young people equipped to take on the ever more technologically sophisticated and knowledge intensive jobs. (cited in Hatcher, 2006a: 607)

As Rikowski (2005: 73) points out, the *Five-Year Strategy* also argued for bringing employers closer into the design of education and training programs for 14–19-year-old students, and attempted to summarize what employers' educational needs were regarding young people:

> We will build employers much more closely into the process of designing and delivering education and training. Employers tell us they need young people above all to have well developed skills in communicating (including writing well), using numbers and using ICT. But they need much more than this. To succeed in the contemporary work environment, young people must be able to handle uncertainty and respond positively to change, to create and implement new ideas, to have the capacity to solve problems and make sound decisions on the basis of evidence, and to be self-reliant and motivated. We must ensure that our offer to young people gives them the right opportunities to develop these skills, and we need the help of employers to make this a reality.

As he goes on, apart from the references to ICT, this summary of employers' educational needs could have been written 20 or even 30 years ago. Rikowski (2005) cites Simon Frith's (1980) arguments about how employers perceived the relationship between the labor process and schooling and how their priority for schools was to develop personality traits and work attitudes (the *Capitalist Agenda for Education*).[4]

Specialist Schools and Academies as a Trojan Horse?

If the *Capitalist Agenda for Education* is the principal form of business involvement in the reshaping of schools in Britain in the opening decade of the twenty-first Century, that is sponsorship of specialist schools and

Academies on a nonprofit basis, this begs the question as to whether putting money into the school system, not taking it out, is a Trojan horse for a future business takeover of schools for profit (the *Capitalist Agenda in Education*). This case is argued, for example, by Crouch (2003) and Rikowski (2003, 2005). It also surfaced at the 2005 National Union of Teachers Conference. As Rikowski puts it, "within the schools system as a whole schools will increasingly be forming long-term social and economic relationship with companies" (2005: 30). This habituates schools to the business environment and

> ...if this habituation process develops according to plan and without effective opposition then that may tempt New Labour to go much further and faster regarding the politically more sensitive issue of outsourcing of educational services on a for-profit basis. (Ibid.: 31)

Hatcher (2005: 8) argues against Rikowski on two grounds: first, with regard to specialist secondary schools, which account for the overwhelming majority of business sponsorships, in most cases they do not have *long-term* relationships with their sponsors; second, Rikowski's argument fails to explain the qualitative leap from businesses making donations to businesses extracting profits. Hatcher (2006a: 608) states that "[a]n analysis of the companies and business entrepreneurs sponsoring Academies and specialist schools confirms the view, I would argue, that they have no interest in taking over and running them for profit." After a detailed empirical analysis of such companies, which appears to back up this claim, he argues that the "fundamental task facing government in terms of the school system is a radical transformation of its culture in order to meet the government's human capital objectives" (ibid.: 614)—that the functions of business sponsorship, therefore, are twofold—innovatory school management (including weakening the unions), and shaping the future workforce (the *Capitalist Agenda for Education*). This is the explicit aim of a new model of Academies being pioneered in Manchester and Birmingham, where the local councils are brokering authority-wide groups of vocational Academies, sponsored by major local employers with curricula geared to local labor market needs, as the hubs of their area plans for implementing the government's 14–19 agenda (Hatcher, 2006b). While the "possible future scenario of the business takeover of schools for profit is real, and a legitimate concern," it is not for Hatcher "the principal lens" through which we should analyze "the actual developments taking place today" (ibid.: 614–615). Hatcher (2006a: 615) concludes that the role of business and other interests as key agents takes two forms:

> the colonisation of government education policy implementation and delivery of national and local educational services by private companies for profit; and the reshaping of schools through non-profit sponsorship, which in the case of the Academies amounts to a non-profit form of privatisation.

In the context of the central themes underpinning this chapter this means that the capitalist agendas *in* and *for* are at work in the contemporary British education system, but in different spheres.

Conclusion

In this chapter, I began by looking at some of the defining features of the global phenomenon of neoliberalism. I then looked at how GATS has exacerbated its effects. I then addressed neoliberalism and world poverty. In the final part of the chapter I looked at neoliberalism and global education, focusing on two major capitalist agendas, the capitalist agenda *for* education, and the capitalist agenda *in* education. I concluded by examining two differing perspectives from *within* the Marxist tradition on New Labour's *Five-Year Strategy*. Only time will tell which of these differing interpretations is more accurate.

Notes

I would like to thank Richard Hatcher, Dave Hill, and Terry James for their helpful comments on an earlier draft of this chapter.

1. "Neoliberalism" *in the economic sense* as is described here has its origins in nineteenth-century liberalism, which was based on the ideas of Adam Smith. Smith and others advocated no government intervention in economic matters. Economic liberalism prevailed throughout the nineteenth and early twentieth centuries. It was proceeded by welfare economics based on the ideas of John Maynard Keynes. Keynes believed in mass government spending to create full employment, and to create jobs in times of depression. Keynesian economics had a great influence on President Roosevelt's "New Deal" in the United States, and on the mixed economy (part state/part private ownership), a major feature in Britain from the end of the World War II until the 1970s. Liberalism's "neo" or new incarnation arose as a result of shrinking profit rates that encouraged capitalists, with the connivance of sympathetic governments, to launch a world offensive–to seek unfettered profit making around the globe to compensate for these shrinking profits. With respect to economics, the current dominant ruling faction in the United States (George W. Bush and his followers) epitomizes neoliberal thinking. With respect to other spheres of life—culture, morality, etc.—the faction has been described as "neoconservative" (or "neocon") and, taking cognizance of its strong right-wing Christian leanings, as "theoconservative" or "theocon." "Liberal" *in the political sense*, on the other hand, is used in Britain to describe "middle-of-the-road" politics. In the United States, it refers to those whose politics is considered to be "on the Left" (for an analysis of various political perspectives, see Hill, 2001).
2. Although in this chapter, I concentrate on the WTO and GATs, there are other organizations involved in promoting neoliberalism worldwide, in particular the World Bank and the International Monetary Fund (IMF).
3. Citing Weatherford (1990), Smith's solution to this enfraudening process is a rethinking of liberal democracy (Smith, 2003: 499) and a "return to the theory of democracy Thomas Paine learned, not from the Greeks or the French,

but from the Iroquois on the banks of the Delaware river" (ibid.: 500). This theory of democracy, Smith continues, relates to "consensus making [which]...arises from 'sitting together' until that truth is found which can be held in common" (ibid.). I would argue that this is a utopian vision in the context of current U.S. imperialism and global neoliberal capitalism, a context critically explored by Smith in his paper. Elsewhere (Cole, 2005) I have pointed out that while Marxists, of course, acknowledge and abhor current gross transgressions of democracy, it is their argument that bourgeois democracy is *always* a numbers game, *always* distorts the truth, and *always* involves manipulation by politicians and by the media.

4. In February, 2005, the government published a *White Paper* (DfES, 2005) that deals specifically with the 14–19 age range. This White Paper paved the way for two distinct routes for 14–19-year-olds, an academic one for the minority and training for work for the majority, and provided further evidence of the *Capitalist Agenda for Education*. The *14–19 White Paper* together with the *Five-Year Strategy* "bury any last hopes that comprehensive education can survive under New Labour" (Hatcher, 2005: 1).

References

Alcoff, L. M. and E. Mendieta. (2000) Introduction. In L. M. Alcoff and E. Mendieta (eds.) *Thinking from the Underside of History* (Oxford: Rowman and Littlefield).

Beckmann A. and C. Cooper. (2004) Globalisation, the New Managerialism and Education: Rethinking the Purpose of Education in Britain. *Journal of Critical Education Policy Studies*, 2 (2). Available at http://www.jceps.com/print.php?articleID=31 (accessed 29 March 2005).

Bowles, S. and H. Gintis. (1976) *Schooling in Capitalist America: Educational Reform and the Contradictions of Economic Life* (London: Routledge and Kegan Paul).

Cole, M. (2005) Transmodernism, Marxism and Social Change: Some Implications for Teacher Education. *Policy Futures in Education*, 3 (1): 90–105. Available on http://www.wwwordsco.uk/pfie/content/pdfs/3/issue3_1.asp#9 (accessed 8 August 2006).

Crouch, C. (2003) *Commercialisation or Citizenship* (London: Fabian Society).

Department for Education and Science (DfES). (2002) *Education Act 2002* (London: HMSO).

———. (2004) *Five Year Strategy for Children and Learners* (London: HMSO).

———. (2005) *14–19 Education and Skills* (London: HMSO).

Frith, S. (1980) Education, Training and the Labour Process. In M. Cole (ed.) *Blind Alley: Youth in a Crisis of Capital*, 25–44 (Ormskirk: G. W. and A. Hesketh).

GATSWatch. (n.d.). Available at http://www.gatswatch.org/ (accessed 31 March 2005).

Hatcher, R. (2001) *The Business of Education: How Business Agendas Drive Labour's Policies for Schools* (Stafford: Socialist Education Association).

———. (2005) The 14 Plus Youth Training Scheme. *Socialist Teacher*, 72 (Spring): 24–25.

———. (2006a) Privatisation and Sponsorship: The Re-agenting of the School System in England. *Journal of Educational Policy*, 21 (5): 599–619.

———. (2006b) Academies, Building Schools for the Future and the 14–19 Vocational Agenda. Talk given at Anti-Academies Alliance Conference, London, 25 November 2006. Available at www.socialist-teacher.org.

Hill, D. (2001) Equality, Ideology and Education Policy. In D. Hill and M. Cole (eds.) *Schooling and Equality: Fact, Concept and Policy* (London: Kogan Page).

———. (2004a) Books, Banks and Bullets: Controlling Our Minds—The Global Project of Imperialistic and Militaristic Neo-liberalism and Its Effect on Education Policy. *Policy Futures*, 2 (3). Available at http://www.triangle.co.uk/pfie/.

———. (2004b) Educational Perversion and Global Neo-liberalism: A Marxist Critique.*Cultural Logic: An Electronic Journal of Marxist Theory and Practice.* Available at http://eserver.org/clogic/2004/2004.html.

———. (2005a) *The Global, Regional and National Levers of Neo-liberalisation*, Brighton: IEPS (Institute for Education Policy Studies). Available at http://www.ieps.org.uk.cwc.net

———. (2005b) Globalisation and Its Educational Discontents: Neo-liberalisation and Its Impacts on Education Workers' Rights Pay and Conditions. *International Studies in the Sociology of Education*, 15 (3): 257–288.

Martinez, E. and A. García (2000) What Is "Neo-liberalism" a Brief Definition. *Economy*, 101. Available at http://www.globalexchange.org/campaigns/econ101/neoliberalDefined.html (accessed 29 March 2005).

Monbiot, G. (2001) Tinkering with Poverty. *The Guardian*, 20 November, 17.

Rady, F. (2002) The "Green Room" Syndrome, *Al-Ahram Weekly*. Available at http://weekly.ahram.org.eg/2002/613/in4.htm (accessed 29 March 2005).

Rikowski, G. (2000) *That Other Great Class of Commodities: Repositioning Marxist Educational Theory*, BERA conference paper, Cardiff University, 7–10 September. Available at http://www.leeds.ac.uk/educol/documents/00001624.htm (accessed 29 March 2007).

———. (2001) *The Battle in Seattle: Its Significance for Education* (London: Tufnell Press).

———. (2003) The Business Takeover of Schools. *Mediactive: Ideas Knowledge Culture*, 1: 91–108.

———. (2005) *Silence of the Wolves*. Education Research Centre, University of Brighton Occasional Paper (Brighton: University of Brighton).

Sardar, Z. and M. W. Davies. (2002) *Why Do People Hate America?* (Cambridge: Icon Books).

Siqueira, A. C. (2005) The Regulation of Education through the WTO/GATS *Journal for Critical Education Policy Studies*, 3 (1) (March). Available at http://www.jceps.com/?pageID=article&articleID=41.

Smith, D. G. (2003) On Enfraudening the Public Sphere, the Futility of Empire and the Future of Knowledge after "America." *Policy Futures in Education*, 1 (2): 488–503.

Smith, L. (2004) Britain: Government Outlines Plans to Dismantle State Education. World Socialist Web Site (WSWS). Available at http://www.wsws.org/articles/2004/aug2004/educ-a11.shtml (accessed 29 March 2005).

Vidal, J. (2005) Global Poverty Targeted as 100,000 Gather in Brazil. *The Guardian*, 26 January.

Weatherford, J. (1990) *Indian Givers; How the Indians of the Americas Transformed the World* (New York: Fawcett Books).

World Development Movement (WDM). (2005) *Stop the GATSastrophe!* Available at http://www.wdm.org.uk/campaigns/GATS.htm (accessed 29 March 2005).

World Development Movement. (n.d.) *The WTO in Cancun*. Available at http://www.wdm.org.uk/campaigns/cancun03/cancun.htm.
World Trade Organisation (WTO). (2005) *GATS: Fact and Fiction*. Available at http://www.wto.org/english/tratop_e/serv_e/gats_factfiction_e.htm (accessed 3 April 2005).

Part III

Discourse, Postmodernism, and Poststructuralism

Chapter 6

Indecision, Social Justice, and Social Change: A Dialogue on Marxism, Postmodernism, and Education

Elizabeth Atkinson and Mike Cole

Introduction

This chapter comprises a lightly edited record of a dialogue between Elizabeth Atkinson and Mike Cole. It was recorded during the MERD seminar (October, 2002) chaired by Glenn Rikowski (GR), in which they explored different approaches to critical thinking, analysis, and practices. As in previous exchanges between them, they express challenging alternatives as well as some significant elements of complementarity around the issues and grounds on which Marxism and postmodernism may engage in relation to education, political action, and social justice. The format had Elizabeth and Mike in dialogue as well as addressing questions from the floor. Mike led off with an opening statement followed by Elizabeth's responses and discussion.

Glenn Rikowski

We have two speakers who are going to engage in a dialogue, then open the discussion up to the floor. First of all we have Elizabeth Atkinson, from the University of Sunderland. Elizabeth is one of the UK's leading postmodern thinkers and has written a range of critiques of New Labour's education policy as well as critiques of the way that research is conducted and the ethics of research, and has published widely in journals such as *International Studies in Sociology of Education* and the *British Journal of Sociology of Education*. Then we have Mike Cole, from the University of Brighton, who is a senior lecturer and Research and Publications Mentor in the School of Education at the University of Brighton. He is a Labour Union activist, and is a Marxist educational theorist and one of the founder members of the Hillcole Group of Radical Left Educators. His publications include *Education, Equality and Human Rights* (2000) and *Promoting Equality in Secondary Schools* (1999). His latest book is *Marxism against Postmodernism in Educational Theory* (2002), coedited with Peter McLaren, Dave Hill, and myself.

Mike Cole

I have three main topics; first of all, the *ordeal of the undecidable or the need for theory*; second, *did Marx and does Marxism have a theory of social justice;* third, I am going to look at *social change*. Here is a summary of the *Abstract* from Cole (2003a) that deals in more depth with these issues (see also 2003b).

This paper looks at attempts in Britain and the United States to argue the case, within educational theory, that postmodernism and poststructuralism can be forces for social change and social justice. Concentrating on the work of Elizabeth Atkinson, but also looking at some of the work of Patti Lather and Judith Baxter, I argue that such claims are illusory. Nothwithstanding some tensions within Marxism, in respect of its relationship with social justice, I make the case that Marxism remains the most viable option in the pursuit of social change and social justice.

The Ordeal of the Undecidable

This is the development of a paper that first appeared in the journal *The School Field* (Cole, 2001) to which Elizabeth replied. Elizabeth challenges the view that "it is essential to chose one theoretical perspective or course of action over another." This relates very much to one of Derrida's key concepts (1992), which he calls the *ordeal of the undecidable*, with its "obligations to openness, passage and nonmastery" where "questions are constantly moving" and where "one cannot define, finish, close" (Lather, 2001: 184).

Patti Lather is someone who has influenced Elizabeth's work, and this concept relates in turn to one of Lather's concepts: "a praxis of not being so sure" (ibid.: 184), "a praxis in excess of binary or dialectical logic" (ibid.: 189) a "post-dialectical praxis" that is about:

> ontological stammering, concepts with a lower ontological weight, a praxis without guaranteed subjects or objects, oriented towards the as yet incompletely thinkable conditions and potentials of given arrangements. (ibid.: 189)

I think that that sums up a very central tenet of postmodernism.

One of Marxism's defining characteristics, on the other hand, is a recognition of "guaranteed subjects or objects," for example, the working class and surplus-value. For Lather, however, nothing is certain or decided. Citing Derrida, Lather asserts that undecidability is "a constant ethical-political reminder" "that moral and political responsibility can only occur in the not knowing, the not being sure" (2001: 187).

So this is a central tenet of postmodernism: nothing can be sure; there's no overarching theory to explain society. Marxists, on the other hand, believe firmly that we *need* theory both to understand, and importantly, as several people have pointed out this morning, to change it. So that's the first thing the tension between the postmodernists' belief in the ordeal of the undecidable and Marxists' insistence that we do need a theory to understand society and to change it.

Truth and Social Justice

Paula Allman is in the audience. In one of her books, Paula (1999) has developed a fourfold conception of truth, which maybe she can talk about afterwards. Since in this chapter, I am arguing that only Marxism has a theory of social justice that is realizable, it needs to be pointed out that there is a great controversy within Marxism, as to whether Marx had a theory of social justice, and I refer to some of the works on that, but just for the purposes of this brief exposition, I would say that I agree with Alex Callinicos, who gives several clear examples of Marx's inherent belief in some universal principle of justice: "the burning anger with which Marx describes the condition of the working class"; "the ostensibly egalitarian 'needs principle,'" and I know this is sexist language, "from each according to his abilities, to each according to his needs!" in the *Critique of the Gotha Programme* (Marx, 1970); his description of capitalist exploitation as the "the theft of alien labour time" in the *Grundrisse* (Marx, 1973), which, since Marx makes it clear that this does not violate capitalist property laws, must, as Callinicos points out, imply an appeal to some transhistorical principle of justice (Callinicos, 2000: 28); his moral position on the collective ownership of land in *Capital,* (Volume 3) (Marx, 1966: 29) and his implication, again from his *Critique of the Gotha Programme (82)* that treating unequals equally is unjust.

The point to make is that whatever Marx's relationship to the concept of justice, the important point for Marxists is that Marx's vision of a socialist society allows us to look beyond the multiple injustices of capitalist society. Norman Geras (1989) is someone who has written a lot about whether Marx had a theory of social justice. To quote from him to finish off this section, "the largest paradox here is that Marx, despite everything, displayed a greater commitment to the creation of a just society than many more overtly interested in analysis of what justice is" (267).

So whether Marx had a theory of social justice or not, and I believe he did, what is more important is that his works, his many writings and the writings of Marxists, move us in the direction of a socially just society.

Social Change

Elizabeth has said that postmodernism "does not have and could not have a 'single' project for social justice" (2002: 75). My response to that was "socialism then, if not social change, is thus ruled out in a stroke." Elizabeth responded to that by saying "well maybe socialism with a capital 'S.'" So I would very much welcome hearing what socialism with a capital "S" is, as opposed to socialism as a general concept of running the world. We also had differences of views about the importance of networking. I agree with Elizabeth that networking is important, but it doesn't replace, I think, mass action, and, again, I'd be interested in Elizabeth's views on whether there is a postmodern view on the current (2002) firefighters' action and the possibility of mass industrial action. How would a postmodernist respond to that or analyze it? It's certainly something that's going to cause a lot more disruption than I would say is the case with networking.

Finally (Cole, 2003b), I look at a recent paper by Judith Baxter (2002). She promotes the concept of what she calls feminist poststructural analysis. I know that poststructuralism is not identical to postmodersnism, and I've dealt with that in an endnote (Cole, 2003b: 496–497). Like postmodernists in general she tries to argue that poststructuralism has 'potentially transformative possibilities" (ibid.). She gives a number of reasons for this, but ends up by saying, and hence the main title of Cole (2003b): "Might it be in the practice rather than the theory that feminist poststructuralist analysis fails to succeed?" This, I think, is my main point: the practical implications of postmodernism and poststructuralism are very limited. I won't do the conclusion (but see ibid.: 494–496), but I think you've got the gist of what I'm saying.

Elizabeth Atkinson

Mike has referred to Judith Baxter's question, "Might it be in the practice that it fails to succeed?" and Judith Baxter asks her own question of herself, so she is critiquing her own possibilities for applying poststructuralist critique or poststructuralist research approaches to real-life situations. I would like to start by saying "Yes, of course it might be in the practice that it fails to succeed." In *any* critical perspective on any human situation, it might be in the practice that it fails to succeed, and no one of us is exempt from the possibility that when we try to put our philosophy or our critical perspective into practice, we may fail to succeed. I don't think that we can level that at postmodernism and therefore imply that this makes postmodernism invalid or untenable. So *yes*, we are all open to the possibility of failure when we apply what we think to what we do. I have a slight problem of course, sitting here, because I am, as I can tell from this morning's discussion of certain postmodernists at whom you were all invited to sling mud, that I may be here as a token postmodernist, so I don't feel very comfortable in that position. However, as Mike has reminded me, it is also very uncomfortable being in the position of token Marxist, having had mud slung at him—so I hope it's not going to be too dirty! I also have a problem because I can only represent my own perspective in relation to this; I cannot stand here for Patti Lather or Judith Baxter, so although I will try and represent postmodernism to some extent as whole (1) it doesn't exist as a whole and (2) it wouldn't be fair for one person to try and do that anyway.

I think it would probably be helpful to try and summarize ways in which postmodern thinking can be characterized. It has so many different explanations and people think of it in so many different ways, that at least it might help to identify some features that are associated with postmodern thinking, and I'm deliberately resisting the term definition here, as no such thing exists as a definition of postmodernism. Some characteristics of postmodernism could be described in the following ways (this list is in my paper concerning "the responsible anarchist" (Atkinson, 2002). Postmodernism can be characterized by

- resistance toward certainty and resolution;
- rejection of fixed notions of reality, knowledge, or method;
- acceptance of complexity, of lack of clarity, and of multiplicity;
- acknowledgment of subjectivity, contradiction and irony;
- irreverence for traditions of philosophy or morality;
- deliberate intent to unsettle assumptions and presuppositions;
- refusal to accept boundaries or hierarchies in ways of thinking;
- disruption of binaries that define things as either/or.

This is what I see as being "the ordeal of undecidability." I like the idea of not being so sure, and I like the idea of uncertainty as a productive way forward in critiquing current social and educational realities, including current social and educational policy. "Not being so sure" is a way of resisting the certainty of "what works," "best practice," "key skills," "standards," "improvement" and all of those aspects of rhetoric in which we become embroiled as soon as we become part of the education system. Just not being so sure is a starting point for resisting that kind of certainty that washes over us and creates its own kind of hegemony.

So what I want to do is, first of all, to highlight what I see as the key differences between our perspectives, because I think that's very important, and I want to go on to say what I think the possibilities for postmodern thinking are as well as trying to answer the specific points that you've raised, Mike, and to look at some ways forward that maybe we can all make use of, and I do appreciate what Ken (Martin) said about drawing on what's valuable in Marxism and postmodernism and feminism and a whole range of different critical-theoretical perspectives in order to *make a difference*. And I believe in making a difference as much as you all do, even if I might be saying it from a different theoretical perspective.

So, differences between myself and Mike (and I summarize these in the paper in *The School Field* [Atkinson, 2001]). First of all, Mike, you talk about Marxist solutions to social injustice, and what I'm talking about is alternative possibilities to the inertia of the status quo. I'm talking about a range of possibilities, not about one single solution, and of course that's one of the key differences between us. You say it's important to theorize the world because we need to provide a way of understanding it; postmodernism doesn't provide *a* way of understanding the world, it provides ways of looking and seeing and interpreting and constructing, not *an* answer to a problem. So that's obviously one of the biggest differences between our two perspectives. I'm saying that we need to look at a range of alternative possibilities, not to weigh them up, but to see how complex the reality of these things is; how very complex the intersections of different discourses in any situation can be, and what we can make of those intersections.

Next, Mike, you talk about the state as a complex of institutions acting in the interests of capitalism. I'm more interested, from a postmodernist perspective, not only in the state operating to produce and reproduce capitalism, but also to produce and reproduce *all* of the structures within which we are enmeshed; all of the grids that we create with our language. Elizabeth St. Pierre (2001) says that we construct grids and hierarchies with our social institutions, our educational institutions, which we need to try and unpack, to make sense of, to try and deconsteruct, to see *how* we've enmeshed ourselves in them. So the decentered system of networks which, from your perspective, is mostly to do with the production and reproduction of capitalism, is about a whole range of social and political institutions from a postmodern perspective.

The third difference between us, I think, is that whereas your implication is that nothing is constructed after a postmodernist has had a go at it—there is nothing left, nothing can exist after deconstruction—I argue at some length in the paper in the *British Journal of Sociology of Education* (Atkinson, 2002), following Judith Butler (1992), that *deconstruction* is not the same as *destruction;* that to deconstruct is to take apart and look at; it is to reconsider the possible meanings of something; it is to open up new possibilities of meaning. And again, that is in order to understand *complexities* and their intersections, not to say, "It's a free market place—OK, you decide on that meaning and I'll decide on another one." For me, postmodernism is not about relativism; it's a different thing altogether.

And finally, the fourth difference between us is Mike says, "Postmodernism is clearly capable of asking questions, but by its own admission it has no answers." I don't admit, I *claim* it has no answers: postmodernism is not trying to provide you with an answer as to why society is as it is. It is trying to ask more questions, and I find that the asking of more questions is the thing that drives me forward in change every day, day on day, in my daily life with my students, with the people I work with in my own institution, with the thinking and writing that I do. It is asking the questions, it is *not being so sure*, that makes a difference.

So, ways forward. I think postmodernism can offer challenges that we need to take up. It can challenge rationalist certainty with the uncertainties that don't allow us to accept as commonsense and obvious and taken for granted the kind of things that are thrown at us all the time in social and educational policies, for example. It can challenge the idea of objectivity, the idea that we can go out and find something out without having to ask any further questions: it continually asks questions. It can challenge the sort of old-fashioned positivist validity which allows us perhaps to reach conclusions and to make assumptions too easily. And it can also challenge the rationalist claims to truth—and I'm not contradicting Paula Allman's analysis (which you've described in detail in your paper, Mike) of four types of truth. What postmodern theorists are saying is that there is an end to the idea of an innocent truth, the idea that there is some sort of truth which will save the world, some sort of truth that overrides all else. Jane Flax (1992) talks about the end of innocence and the end of the possibility of *explaining* by virtue of an overarching truth.

Finally, I think it does make a difference when we talk about things and I think it does make a difference when we think about things. I think what Helen [Colley] says about mentoring makes a difference to how mentoring is going to happen, and it was lovely to hear Dave [Hill] say, "Yes, it's the spread of ideas that makes people feel differently." And that in itself is a sort of mass movement. I think that when people see the way that multiple injustices and oppressions are rethought through the voices of those marginalized, others whose voices are *heard* through postmodern thinking, then we have a different way of looking at those injustices. It's not just a simple answer in relation to one specific perspective, it's a way of seeing how all these voices make a difference in being heard themselves. And I think that, when you get organizations like, for example, the newly formed Center for Anti-Oppressive Education, which has just opened in the United States and is bringing together poststructuralist, postcolonial, and queer theory perspectives to challenge the assumptions about normalcy, to challenge the normalization of the way that we think, then that is living proof of the way that rethinking, thinking otherwise, troubling the certainties around what we take for granted, is actually showing itself as influencing the realities that do change within the everyday social world.

I'll stop there. I haven't addressed networking versus mass action, and I haven't addressed Socialism with a capital S, but maybe people would like to say things first, and I might come back to that.

Discussion

Q: Two questions for Elizabeth: How do you see reality? Is there a multiplicity of realities, or is there one reality? And the second question is, what is the postmodern theory of history?

EA: Two very interesting questions! On reality, of course you can get into the reductionist "a brick is a brick is a brick" question, can't you: why can't the postmodernists see that this is a real thing, and if a postmodern pilot flies a plane, it won't go, and all that sort of stuff. But I don't think that's what postmodern theorizing is about. I see reality as multiple social constructions, which is very close to many of the other critical perspectives on the social construction of reality which don't necessarily have a postmodern label on them, but it is a question of not being able to say, "This is this thing and unchangeable and stable," but to say that it is constructed in our creation of it and our response to it. So we are constructing both jointly and individually our understanding of something. I think if you apply that idea of reality to things like educational institutions, which become reified through social construction, then I see the postmodern perspective on that as being something which deconstructs those assumptions about what makes that thing real.

History: postmodernism is accused of being ahistorical, and of not dealing with history. I don't think I could provide you with an immediate answer as to what is a postmodern view of history, other than to say that one of the things that postmodernism is also interested in, along with Marxism, is unpacking the way in which we tend to see history through the lens of the present, and that's been taken up particularly by Helen (Colley). And I notice that in your paper, Helen, when you were talking about viewing the past through the lens of the present and creating a new kind of image of the past, you were actually using terminology from postmodernism, such as the *simulacrum*—Baudrillard's term for the identical copy of the original which never existed. And I find that combination of your feminist Marxist perspective with that use of a postmodernist perspective completely tenable: it doesn't create a problem for me. So I guess a quick answer would be, not allowing us to assume that the way that we look at history is set or stable or fixed, and unpacking the way that we interpret past events through the lens of the present.

Q: Two questions for Elizabeth: firstly, you talk about the characteristics of postmodernism and you talk about uncertainty. It seems to me that when one is uncertain, one is inactive. When you deconstruct and you have a look at things, and then you put it back together again, you actually end up with nothing new. So I'm quite interested in how you would put thinking into action. And secondly, can you describe how you see the difference between postmodernism and relativism?

MC: Could we tie that question to your attitude to the [firefighters'] strike action? Do postmodernists join unions and go on strike and stand on picket lines?

EA: Yes, right. Postmodernism and action: to me, uncertainty does not in any way result in inaction. Actually, I was talking to Julia this lunchtime about confessions and comings-out and epiphanies, and the tales we tell of how marvellous it's been—and this sounds ridiculously confessional, but as a result of thinking and writing and working within this particular critical field, I have become more active in relation to issues to do with identity, diversity, and social injustice than I ever have been before. And it actually has been as a direct result of rethinking the assumptions I make about identity or about social groupings or about the way society works. If you're asking, does that mean I stand on picket lines, yes it does. Does it mean I join demonstrations against racist attacks on asylum seekers? Yes it does—and to the extent that I found someone photographing my house last week, I think as a direct result of having been on a demonstration.

Now you might say, "Oh well, she's gone over to the other side; she's become a socialist—she can't do that," but it's not a *problem* for me. I don't find a contradiction there. I don't have to put away my postmodern perspective in order to go and stand on a picket line or join a demonstration. On the contrary, I was tending to put away aspects of my identity in order to do one thing instead of another, prior to having what I felt was a more useful critical understanding of the multiplicity within identity. So I couldn't possibly say what the postmodernists of the world would make of the firefighters' struggle, but I know that for me there isn't an issue about supporting aspects around social justice—where I'm still doing the same thing. I haven't stopped deconstructing what's going on—it doesn't stop me from doing that. And the same with Socialism with a capital S: what I was saying to Mike in my response to his paper was, "No, postmodernism might rule out Socialism with a capital S, because postmodernism isn't Socialism—you wouldn't expect them to be the same thing." But it doesn't rule out the possibility of a socialist interpretation of society... which has useful ways forward, along with a lot of other ways. So socialism as a perspective is not ruled out.

Q: And just adding on to that, in fact Rachael, in her paper this morning, talking about Kurdish women, made it very clear also that trauma doesn't produce inactivity either, and I interpret undecidability and trauma as being along very similar lines.

Q: I'm deeply concerned about both the theory and the practice of postmodernism, and I'm talking right now not only as a social scientist but as a political activist, which I consider myself to have been for the last 20 years, working in different contexts in North America and Europe. And when I try to understand postmodernism, and put it into the practice of understanding the world, I see such a big juncture, and I feel that it has helped me. What it has helped me to think about and rethink is the deeper understanding of Marxism, and I appreciate that sense that I've got from postmodernism.

Let me explain how it's helped me to do that: we had in the discussion this morning the question about bringing in more analysis of the philosophical understanding of Marxism in order to explain economic relations and the understanding of Capitalism that Marxism is all about. And I think that what postmodernism helped me as a Marxist to pay more attention to was exactly that philosophical understanding of Marxism, which I think, especially in Western Marxist academia, has been totally ignored. And that is the *dialectical*—and I emphasize the *historical materialist* dialectical understanding of Marxism, not the mechanical, not the idealist understanding of dialectical relations, but the dialectical historical materialist understanding of ontology and epistemology which is at the roots of all the debates that we are engendering. But we don't go to that level of understanding it. I think what Marxism helps us to do is to see that dialectical relationship. There is a contradiction, a tension, in terms of human beings and knowledge.

Of course, what postmodernism is doing is focusing on the ontology and forgetting about the epistemology, which is what we are talking about here—that is, consciousness. The individual, no matter how many multiple identities that individual has, does not turn into a thing unless it is turned into the epistemology of that individual through consciousness and what we've labelled, through Paula Allman's work, as revolutionary consciousness. I think if we want to talk about where postmodernism is taking us, and where Marxism isn't going to take us, it is really deeply understanding this relationship between ontology and epistemology, and bringing the philosophical understanding of Marxism into this debate. Because I think that even the debate on the understanding of race, gender, class... and all sorts of other diversities that we're all working with... and I know that, no matter how many times we talk to a Moslem woman and appreciate her identity and confirm her identity and try to accommodate her, try to make everything accessible to her, the bottom line is that, structurally, we are living in a society that doesn't go beyond that. At the end, it's imperialist, it's colonial—and it's understanding that it's all sorts of other structural issues and relations that overlay issues of identity of this particular woman. And it is to that understanding that we can pay attention.

And this is the only place in postmodernism that I would say made me go back to a philosophical understanding of Marxism, in paying more attention to the particularities. But I'm not going to stop there: it's not going to help me to stay and spend a whole day talking about all these particularities. What am I going to do with it? How many times are we going to antiracist demonstrations? What are we doing? Where has the antiracism movement been able to converge with the labor unions, with women, with environmental issues—if you name it, we have it. But what are we doing with that?...And I think what you say, Elizabeth, makes sense, but in textbooks, in discussion, in academia. In the real world of the lives of people that I'm dealing with, including Kurdish women living in Britain today, including asylum seekers, it doesn't make sense. It makes sense for the moment about that individual and then nothing more beyond that.

MC: Can I just briefly say that the European Social Forum in Florence, which I am going to as a union delegate at my own expense, is trying to do that very thing: to connect antiracism, the antiwar movement, antiglobalization, with the possibility of a different future in a very practical way.

Q: A question for Mike. How do the Marxist educational theorists see the contexts of the changes in education today?

MC: From when to when?

Q: The writers on educational change in the last 20 years see the change, marketization etc., as being a response to the end of the neo-Keynesian consensus etc. etc. They have a very narrow context for the change in education. How do the Marxists see the contexts?

MC: Well, I think you've answered the question. The crisis in the Welfare State, the onset of Thatcherism, of Reagonomics, the way in which Thatcher used the collapse of the Soviet Union to try and make a false equation, in my view, that this collapse means the collapse of socialism. And, as we all know, this is what brought Blair into power: the notion of a Labour politician who could put all that "socialist old Labour stuff" behind him and move in the context of the "free market." I don't think one can give a Marxist analysis of the last 20 years in a couple of minutes.

Q: Have you got any positive stories, Mike?

MC: Positive stories, about?

Q: About the last 20 years from a Marxist perspective

MC: Yes, I think we are living at a very hopeful time. I think that while people aren't openly advocating Marxism and socialism (well some of us are, but not in terms of a mass movement), but with the antiglobalization movement, I think that there is a realization among masses of people that there must be another way to run the world than free market globalized capitalist economics. And people are aware that societies are not democratic, that there is something a bit odd about George Bush who is managing to get masses of Americans to be convinced that Sadam Hussein is about to drop a nuclear bomb around the world, and to cover up the notion of oil which we all here know about I am sure (and it has come to light in the press), but we are still bombarded with this notion that America who has been involved in how many attacks on independent nations since the Second World War, that America is somehow standing up for "the free world," is, for Marxists, a nonsense. Is that what you were asking me?

Q: I think one of the things we can talk about Marxists being involved in is things like the Anti Nazi League—if you look at the difference between here and France, where the National Front are very strong activists and...of course it's the case that events in Stoke recently appear to show that the BNP are very strong, but I think Marxism and Marxist groups are centrally involved in that (the Anti Nazi League) and that has made a difference to the context in which we work in education and in society. So I think we absolutely do not have to be...and I think one of the problems with the whole notion of identity politics throughout the debate with postmodernism is precisely that the authenticity of what you are talking about is basically your own selfauthenticity. So you can talk about progressive antiracism, you can talk about progressive antisexism, but if all you have is your own authenticity, self-defined, then how do we deal with fascism? How do we deal with racism? In the sense that these are, surely, other multiple identities that people may or may not have? And I think that links back to the problem that postmodernists have in actually coming to terms with actually getting practically involved in the world, rather than writing books. And I think Alex Calinicos in terms of that, and in terms of quoting Habermas, is absolutely right: we should not be at all defensive about that.

Q: But Elizabeth said she didn't have a problem with that. She's already said that she does get involved in lots of antiracist...

MC: Could we ask Elizabeth *why* she gets involved in picket lines and supports strikes? It may be for different reasons from why a Marxist does.

EA: It may be...

Q: And also, how you make judgments about whether your strategies which are developing are more likely to succeed or less likely to succeed or whatever and maybe Mike can respond to this as well.

MC: Yes, sure.

EA: I want to also bring into that the question I didn't answer—I'm sorry—about the difference between postmodernism and relativism, because I think that is part of the answer, and also the earlier comment about ontology and epistemology. So, trying to do all those things at once, when I stand on a picket line or in a demonstration, for a start I don't agree that postmodernism stops short of epistemology. What I think you end up with is a different sort of epistemology from what you might end up with with Marxism. So what you're saying is that you are denied a sort of identity which allows you to move forward if you adopt a postmodern perspective, and what I'm saying is no, it doesn't deny you that sort of possibility, because the fact of recognizing the intersection of a multiplicity of identities within myself does not stop me from making judgments about how those identities interact with the everyday world. And when you say that it's only a question of individual authority for one's own identity, it doesn't feel like that to me; it's not a question of, "OK, so here I am, I'm Elizabeth, I'm a woman, I'm a lesbian, I'm a postmodernist, I'm a feminist, I'm a white middle-class intellectual snob, and I am therefore going to do this, this, this and this," it's that all of those things are all intersecting within me *and* it's like a web that connects me with an awful lot of different people, who are also intersecting with an awful lot of other different people. So that I'm part of a much bigger network, which is where networking partly comes in.

And it's not a question of saying, "Oh well, so if I stood in this box then it would look like this, and if I stood in that box it would look like that, therefore I will decide to do this or that," which is how a relativist position might look. It is a question of recognizing that not only are all these things operating within me at the one time, but also that they are sometimes mutually contradictory, and that that is a fact of life—it is something which happens. So my authoritarian perspective on education, which I was born with and grew up with through my own experience, is at battle all the time with my libertarian philosophy, which in itself is at battle with other aspects of my identity. And the postmodern perspective allows for that to be the case, while at the same time allowing me to decide that there are things I wish to do which will identify me with these different aspects of my identity. So I will stand in different places according to those different aspects. Again, the contradictions are there, so that one of the things I might be fighting for might at times be in contradiction to another thing I might be fighting for, but I think that's probably the case for a lot of us a lot of the time: we're full of hugely contradictory perspectives.

MC: As I suspected, the reason Marxists would support strikes are very different. Marxists would argue that there is a fundamental contradiction between capital and labor, and a strike is one of the ways in which this contradiction comes to the surface. Strikes and workers' movements have historically been the only way workers have made gains. Marxists would see most strikes as essentially struggles over surplus-value. On picket lines and on strike workers become politicized, and we can see a move maybe towards workers being a *class-for-itself* as well as a *class-in-itself*. That the working class is a class-in-itself is, for Marxists, an objective fact. This is very different from Elizabeth's reasons. I think we both agree that we would support workers for reasons of social justice. Beyond that, I think our analyses differ.

Q: I'm absolutely intrigued to know how writing books or inhabiting the academy isn't practical or isn't part of the real world. For me, that's one of the old dichotomies which surely shouldn't actually exist. Writing a book or a poem, or making a film or a play or any of those things—and this discussion—to my mind are incredibly practical things to do. And I find extraordinary the notion that somehow writing books is separate from the dirty, imbricated, ordinary world that other people live in.

Q: I'm just going back to the framework I was talking about earlier. To me, the postmodern political position seems to be the absolute individualizing, the absolute atomizing of self, and I'm also aware that I'm speaking academically, in the middle class. And if I want to indulge in taking my identity apart and thinking about which thing is better for me as an absolute individual every day, I may have the privilege to be able to do that, depending upon how things go politically in the world, for a while, but... actually when I first came out, I didn't have that luxury, because we were still engaged in a struggle over sexual identity 10 or 15 years ago, where I couldn't be out. So I was forced to be part of a larger group that was making claims about group identity... that we needed to fight an oppression. And... even is recognizing how that struggle against oppression is connected to other struggles against oppression, so they made it this larger collective thing, so I may right now at this moment have the luxury of being able to say well I can sort of do whatever, but that's not historically going to always be the case, so I think this position is sort of.... And in terms of using it for political organizing, it simply hasn't been true. It's about, it's absolutely individualized, any type of personal... is.

...I think it's fine for middle-class people, isn't it—you say quite rightly that... you're an individual, but what about the poor people that are dying and starving in Afghanistan, you know, that was horrific, wasn't it, when people were bombing them and people were starving to death because the aid workers...had to come out. Millions of people died...]and you need a Marxist analysis, as far as I'm concerned, to understand why this horrific situation exists.

GR: ...I think this one could go on for another hour or so, or longer. It's been tremendous. We're going to have to wind it up because we're running overtime. Yes, just one more...

Q: Can I just say, I think we need a postmodern perspective in order to understand George Bush, you know, the simulacrum of the original that never existed...

Q: In the forthcoming book, the first words here are that postmodernism has become the orthodoxy in educational theory, and...postmodernism is the enemy of Marxism according to this, so...do you think really that postmodernism has become a kind of orthodoxy? How do you feel about that?

EA: I know there's no time, but I can say one thing, that the only group of theorists who see postmodernism as being the orthodoxy, I think, is the group of Marxists here today. Nowhere else is postmodernism recognized in education as being a force that's worth struggling against, so it's an interesting perspective!

References

Allman, P. (1999) *Revolutionary Social Transformation: Democratic Hopes, Political Possibilities and Critical Education* (Westport, CT: Bergin and Garvey).

Atkinson, E. (2001) A Response to Mike Cole's "Educational Postmodernism, Social Justice and Societal Change: An Incompatible Ménage-a-trois." *The School Field: International Journal of Theory and Research in Education*, 12 (1 and 2): 87–94.

———. (2002) The Responsible Anarchist: Postmodernism and Social Change. *British Journal of Sociology of Education*, 23 (1): 73–87.

Baxter, J. (2002) A Juggling Act: A Feminist Post-structuralist Analysis of Girls' and Boys' Talk in the Secondary Classroom. *Gender and Education*, 14 (1): 5–19.

Butler, J. (1992) Contingent Foundations: Feminism and the Question of "Postmodernism." In J. Butler and J. Scott (eds.) *Feminists Theorize the Political*, 3–21 (New York: Routledge).

Callinicos, A. (2000) *Equality* (Oxford: Polity).

Cole, M. (ed.). (2000/2006) *Education, Equality and Human Rights: Issues of Gender, "Race," Sexuality, Special Needs and Social Class* (London: Routledge/Falmer).

———. (2001) Educational Postmodernism, Social Justice and Social Change: An Incompatible Ménage-à-trois. *The School Field*, 12 (1/2): 69–85.

———. (2003a) Might It Be in the Practice That It Fails to Succeed? A Marxist Critique of Claims for Postmodernism and Poststructuralism as Forces for Social Change and Social Justice. *British Journal of Sociology of Education*, 24 (4): 487–500.

———. (2003b) Global Capital, Postmodern/Poststructural Deconstruction and Social Change: A Marxist Critique. Paper presented at a conference entitled *The Work of*

Karl Marx and Challenges for the XXI Century, Institute of Philosophy of the Ministry of Science, Technology and Environment of Havana, Cuba. Available at http://www.nodo50.org/cubasigloXXI/congreso/cole_05abr03.pdf).
Derrida, J. (1992) Force of Law: The "Mystical Foundation of Authority." In D. Cornell, M. Rosenfeld, and D. G. Carlson (eds.) *Deconstruction and the Possibility of Justice*, 3–67 (London: Routledge).
Flax, J. (1992) The End of Innocence. In J. Butler and J. Scott (eds.) *Feminists Theorize the Political*, 445–463 (New York: Routledge).
Geras, N. (1989) The Controversy about Marx and Justice. In A. Callinicos (ed.) *Marxist Theory* (Oxford: Oxford University Press).
Hill, D. and M. Cole. (eds.). (1999) *Promoting Equality in Secondary Schools* (London: Cassell).
Hill, D., P. McClaren, M. Cole. and G. Rikowski. (2002) *Marxism against Postmodernism in Educational Theory* (Lanham, MD: Lexington Books).
Lather, P. (2001) Ten Years Later, Yet Again: Critical Pedagogy and Its Complicities. In K. Weiler (ed.) *Feminist Engagements: Reading, Resisting and Revisioning Male Theorists in Education and Cultural Studies*, 183–195 (London: Routledge).
Marx, K. (1966) *Capital*, Vol. 3 (Moscow: Progress Publishers).
———. (1970) Critique of the Gotha Programme. *Marx/Engels Selected Works*, Vol. 3 (Moscow: Progress Publishers).
———. (1973) *Grundrisse* (London: New Left Review, Allen Lane).
St. Pierre, E. (2001) The Intelligibility of Postmodern Educational Research. Paper presented at the annual meeting of the American Educational Research Association, Seattle, 10–14 April.

Chapter 7

Textual Strategies of Representation and Legitimation in New Labour Policy Discourse

Jane Mulderrig

Introduction

This chapter presents a critical discourse analysis of education policy texts issued under the New Labour government. The analysis focuses on the discourse representation of key educational actors, as well as the discourse strategies by which policy decisions are legitimated. In order to interpret the sociological significance of the findings, the data is interpreted in relation to its wider socioeconomic context. It is postulated that a broadly instrumental rationality underlies the representation of educational actors. This is theorized as an indicator of a general shift toward the commodification of education that stems in part from a subordination of social to economic policy. Furthermore, the texts build an inclusive and vague social identity for the government, which is shown to play a significant role in constructing an apparent consensus over neoliberal policy statements.

Since the early 1990s, and in particular following the Lisbon Strategy, EC policy initiatives have increasingly placed education at the forefront of strategies for achieving a successful knowledge-based economy and learning society (Brine, 2006). Increasingly, investment in learning was seen as a key political mechanism for achieving both economic growth and social cohesion. At the UK level this progressive policy orientation was clearly illustrated during the 1997 electoral campaign, when New Labour placed education at the vanguard of its modernization agenda. Addressing primary education first, most notably with its National Literacy and Numeracy

Strategies, the government then moved on to secondary education, the subject of the policies examined in this chapter. In its own words, its mission is "[o]pening secondary education to a new era of engagement with the worlds of enterprise, higher education and civic responsibility."[1] With its emphasis on forging new links between sectors, this strategic vision challenges the traditional scope and content of educational practices. In so doing, it also challenges the social identities and relations that help (re)produce those practices. The aim of this chapter is therefore to examine the textual dynamics by which the new educational identities and relations of a knowledge economy and learning society are constructed in policy discourse. These textual patterns are critically examined with respect to their role in negotiating the distribution of power in education.

The study employed a method of discourse analysis developed in order to conduct systematic—and thus replicable—critical discourse analysis. The method combines critical discourse analysis, with its commitment to sociologically grounded textual analysis, with corpus linguistic computer software tools (Scott, 1997). Among the advantages of this approach is the ability of such software to generate and reveal patterns of textual prominence in the data, which are then amenable to more detailed qualitative analysis. The analyst can, for example, generate lists of the words in the data that cooccur most frequently (termed "collocates"), or whose frequency of occurrence is unusually high given the genre (termed "keywords"). A further advantage of this approach is that it allows one to carry out detailed critical textual analysis on potentially large bodies of data, thus facilitating historical analyses (e.g., Mulderrig, 2006) for further details on the application of this method to a corpus of policy documents spanning 30 years).

This chapter examines the reform proposals set out in just two consultation documents, analyzing the representation of the three most textually salient participants in the data: the government, teachers, and students. The texts analyzed in this chapter are the 2001 White Paper *Schools: Achieving Success* (DfES, 2001), and the 2002 Green Paper *14–19: Extending Opportunities, Raising Standards* (ibid.). They were digitized to form a small corpus of 58,739 words. Although not a longitudinal study, occasional comparison was made between this corpus and the White Paper *Secondary Education for All: A New Drive* (Ministry of Education, 1958) published under Prime Minister Harold Macmillan's conservative government (1957–1963). General patterns in the data were investigated using corpus linguistic software (see Mulderrig, 2006, for a fuller account of this method). A detailed critical discourse analysis of these preliminary search results is presented in the section "The Representation of Social Actors." The findings are interpreted in relation to their wider political economic context, which I outline below in the sections "New Labour in Context" and "Political and Economic Agendas in Education Policy." I begin with some arguments about the unique insights textual analysis can bring to critical policy research.

Theoretical Connections

The chapter presents a critical discourse analysis of the language of policy. A key tenet of this approach is that the origins and social effects of discourse can only be understood by examining the range of social practices and human relations with which it shares a dialectical relation. It is this social embeddedness of discourse that determines which discourses[2] will be taken up in a given policy text, and which are likely to become naturalized and accepted in various social contexts. Thus, for instance, if we are to understand the postulated increase in education policy texts of commercial values and discourses, we must also recognize the changes in governance structures that allow representatives from the commercial sector an unprecedented voice in policy-making procedures through a burgeoning of advisory and interventionist powers. Equally however, sociological analyses of education policy that ignore discourse, risk overlooking its important role in shaping, enacting and legitimizing policy. As Ball (1990) puts it, both control and content of policy are significant; both the structural mechanisms and the discourse.

In understanding the relationship of education to its wider socioeconomic context, I adopt a dialectical approach, viewing educational change in terms of its dynamic relationship with other elements of the capitalist social formation, rather than seeing it as simply responding to external economic, social, and political forces. A critical analysis of educational discourse can illustrate this dynamic at work, highlighting its inherently unstable nature as it struggles to bring together the contradictory logics of the economic and the extra-economic. This form of analysis makes the important theoretical contribution of thematizing and problematizing the complex and increasingly salient role played by discourse in the uneven and often contradictory construction of the knowledge economy and knowledge society.

This brings me to my view of education policy discourse, which is the basis of my argument for using discourse analysis as a research tool. Education policy discourse construes education as an object of governance. It circumscribes the educationally possible within the parameters of what is seen to be politically and economically possible. In order to arrive at political solutions, one must first define the nature of the problem and the wider context in which it is situated. Thus, for example, contemporary political strategies, like that of the "learning society," are formulated and presented as a response to the "demands" of a fast-changing, knowledge-driven, global economy. Such representations of the realities and exigencies of the "new world order" can be seen as "imaginaries" (Jessop, 2002) that help reduce the infinite complexity of the social world by selectively representing it in order to render it amenable to intervention. In the case of education policy therefore they help set the parameters of what is seen to be feasible, desirable, and thinkable for education. I would further argue that these necessarily reductive representations contribute to the hegemonic strategy of presenting the rapid spread of commodification to new domains of society as inevitable and natural, thereby assisting the neoliberal project of removing obstacles (like the welfare state) to new capitalist transformations (Fairclough, 2003).

However, no political program carries guarantees of success. Indeed it is precisely because of this fact, and the concomitant need for those programs to be legitimated, that discourse analysis can play such a valuable role in political analysis. It can add a missing dimension to social research that accepts the role of legitimation strategies in political processes, and yet fails to attend to how those strategies are realized. Textual analysis does precisely this, thus methodologically enhancing our understanding of this sociological issue.

New Labour in Context

In assessing the specificity of New Labour's education policy, there is a danger of isolating it spatiotemporally, that is, New Labour's politics are partly shaped by both historical and geopolitical parameters. The former concerns the legacy of Thatcherism, while the latter relates to ongoing changes in both global capitalism and the role of the nation state. Put simply, can New Labour's approach to education be seen simply as a continuation of Thatcherite policies?

Following Hay (1996, 1999), I see New Labour as operating within the conditions of a post-Thatcherite political consensus.[3] The Thatcher period entailed a major state restructuring project in response to the perceived crisis in capitalism and its mode of regulation (Jessop, 2002). This involved a break with many of the social and economic policies of the postwar Keynesian welfare state; in particular through policies of privatization and marketization, alongside welfare retrenchment. In the case of education, this period created the context for New Labour to effect structural and ideological transformations that align education more closely with its economic function. As Dale (1989) has it, it was during this period that the "vocabularies of motives," that is, the Discourses that articulate the goals and values of education, were changed, thereby redefining what education is, and what it is for. This discursive shift in the educational debate toward its economic function was a necessary legitimatory tool in the concomitant structural changes that entailed funding cuts and new forms of organization and regimes of evaluation. In effect it paved the way for further "modernization" programs by a reinvented Labour party that placed economic competitiveness at the center of its political agenda. Hill (1999, 2001) argues that New Labour education policy is in consonance with its overall political ideology. This self-termed Third Way ideology can be seen as neoliberalism[4] with a Discourse of social justice (Fairclough, 2000). In essence, it entails opening up education to business values, interests, principles, methods of management, and funding.

In addition to the structural and discursive legacy of Thatcherism, New Labour education policy should also be understood as an aspect of the government's relationship to more global political and economic forces. Globalization sits in a complex relationship with the modernization of public services; it is simultaneously the set of political and economic processes modernization helps construct, and the Discourse by which it is legitimated. The theme of globalization and the imperative it creates for economic competitiveness is a core rationale running through New Labour's educational initiatives.

Recent education policy research has identified the increasingly international convergence in education policy. Hatcher and Hirtt (1999) argue that this is in fact a response to explicit calls from influential economic and political organizations (OECD [Organisation for Economic Cooperation and Development], EC [European Commission], ERT [European Round Table]) for rapid educational reform to meet the needs of the new globalized, knowledge-based economy. However, to call education policy solely a *response* seems to endorse the government's own legitimatory rhetoric that constructs globalization as an inexorable force of change to which nations and individuals must be prepared to adapt, and which obfuscates the realities of the capitalist system whose intrinsic instability demands adaptability and flexibility from its workforce. Moreover, in this context of education policy convergence and services-trading (notably under GATS [General Agreement on Trade and Services]), education itself becomes an important tool in the normalization of neoliberal politicoeconomic strategies on a global scale (Robertson and Dale, 2006). The power of the rhetoric of globalization lies precisely in its self-representation as an abstract challenge to be met, rather than the agent-driven processes of capitalist development. Policies are represented as simply meeting the challenges of a contemporary world, thus serving general interests, rather than as contributing to capitalism's ongoing globalized construction, and thus in fact serving particular interests.

Political and Economic Agendas in Education Policy

The education policy imperatives that arise out of this increasingly internationalized process are manifold and subject to adaptation within each nation state. However, in the case of Britain, they come under three broad and interrelated projects, each of which can be understood in terms of its relationship to the development of capitalism. They are one, creating a business agenda in and for education; two, making education a principal agent in the construction of the workfare state; and three, creating the lifelong learning society.

Creating a Business Agenda in and for Education

Creating a business agenda in and for education involves a complex of structural and content-based transformations. The structural aspects entail processes termed marketization (i.e., creating an educational market through interschools competition) and managerialization (modelling the administration and running of schools on techniques employed in commercial organizations).

Creating a business agenda in terms of educational content can be seen in moves toward a more vocationally relevant curriculum, in particular the skills and dispositions appropriate to the continuance of a technologically driven knowledge economy. Ball (1990) theorizes these processes in terms of a redefinition of the meaning of education's autonomy. He states that under the postwar educational settlement, education was relatively autonomous from the sphere of production, but has now been subordinated to the logic of commodity circulation, giving rise to a new definition of autonomy for individual schools within the sphere of production. Thus, through interschool competition for funding and pupils, tighter controls over teaching (or "delivery") practices, and a more outcome-oriented curriculum, the functional role of education has penetrated the content and form of schooling. One consequence of the new managerial logic in educational organization is an intensified codification and regulation of teachers' working practices, alongside an increased emphasis on standards, targets, quality, and delivery. Dale (1989) sees this as a removal of teachers' professional autonomy, or judgment. This means moreover, a significant role for discourse in inculcating the right attitudes and values; the hegemonic construction of a new consensus on the nature of teaching and education.

Making Education a Principal Agent in the Construction of the Workfare State

The two broad agendas for education, constructing the postwelfare (or "workfare") state and creating the lifelong learning society, are closely interrelated. Education plays a newly significant role in an integrated social policy aimed at supporting the economy and reducing the welfare burden on the government. The move from Keynesian policies means that welfare is no longer a state-run economic system, but a set of practices designed to bring about a fundamental change of culture founded on self-reliance, enterprise, and lifelong learning. Changing attitudes and conceptions of citizenship and equality are thus central to the workfare system. In Blair's "stakeholder society," the emphasis is placed on individual endeavor and responsibility, in which the government is cast as an "enabler," rather than a guarantor of citizens' rights. This entails redefining fundamental concepts on which social conformity and consensus depend. Citizenship rights, like the right to welfare, are linked to individual responsibilities and the personal investment of hard work (Ellison, 1997). Indeed, the economic metaphors of "stakeholding" and "investment" illustrate the instrumental, exchange-value logic that underpins the mechanisms to achieve New Labour's goal of social justice; they illustrate its Third Way discourse that claims "social justice and economic competitiveness should not be treated as though they were distinct and separate from one another" (Giddens, 2002: 79).

Creating the Lifelong Learning Society

Education policy forms part of a wider social policy aimed at creating the "learning society," in which education and training are subsumed under "learning" that is "lifelong." The ongoing accumulation, credentializing and upgrading of skills, which is constructed as one of the key objectives for both pupils and teachers in New Labour education policy, supports the progressive development of the knowledge economy and its managerial infrastructure. Moreover, the textual representations of educational roles and relations in policy, linking success (and by implication, failure) with individual commitment and aspirations, potentially acts as a powerful form of social control. Not only does it establish a practice of lifelong learning and individual adaptability with which to occupy and appease the unemployed, it also constitutes a form of self-regulation in which the individual is responsible for and invests, through learning, in her own success. The coercive force comes not from the government, which is construed as a facilitator, but from the implicit laws of the market. The lifelong learning policy is often described as a response to the instability in the labour market and the demands of the economy for rapid technological development by creating a highly skilled, motivated, and adaptive learning society. However, rather than being a "response" to the globalized economic system, I would argue that this learning policy constitutes a key ideological mechanism in actively constructing and legitimizing globalization and our roles in it.

In the remaining sections I set out a necessarily schematic illustration[5] of how the key social actors represented in the data are construed through New Labour's educational discourse. Drawing on Fairclough's critical discourse analytical method (1992, 2003; and Chouliaraki and Fairclough, 1999), I assess the contribution made by this discourse to the wider sociopolitical processes outlined above: in short, to the ongoing legitimation and constitution of the current globalizing, knowledge-based phase of capitalism, and the forms of citizenship, or "consumership," that this entails.

The Representation of Social Actors

The overall picture of representation in the data is that of a nonauthoritarian government cast as a collaborator and facilitator, rather than a coercive force. Teachers are relatively passivated, yet self-monitoring; hierarchized and in competition, yet expected to collaborate, with one another. Students are both "pupils" who are classified, competitive, and committed to individual success; and "young people" who, in later school years, are vocationally and attitudinally prepared to be entrepreneurial, responsible and flexible citizens, committed to lifelong learning.

The "Enabling" Government and Strategies of Legitimation

One of the most immediately striking differences between the 1958 Macmillan government policy text (cited above) and those of New Labour is the way in which the government is textually represented. In the older text, it is with the words *the government* which in terms of frequency ranks third after *schools* and *children*, whereas in the New Labour text this term is barely used, favoring instead the collective pronoun *we* that is the highest frequency grammatical participant (while not used once in the 1958 text, it occurs 905 times in NL—1.5 percent of the total word count). In other words, in the New Labour corpus, the government is statistically the most dominant participant. The use of the pronoun *we* rather than *the government* may signal what Fairclough (1992) calls the "democratization" of discourse, of which one aspect is a tendency toward more informal language and the removal of explicit textual markers of power asymmetries. Thus *the government*, with its authoritarian tone, may have been removed in favor of *we* in order to create a style more consonant with New Labour's claims to "participatory democracy." However, as Fairclough observes, democratized discourse can in fact be simply a means of disguising these power asymmetries, rather than removing them. Moreover, because *we* potentially includes the reader[6] (which at times in the text it does), it removes space for dialogue and oppositional voices one would expect of a consultation paper, thus having a dedemocratizing effect.

We is both the Subject (the entity performing the actions) and the Theme (the first element in a sentence) and occurs in the texts far more frequently than any of the other participants. The stylistic effect of this grammatical pattern is the construction of a predominantly government-centered text, where it is acting upon processes (mainly structural and organizational changes) and upon people (mainly by facilitating their actions). The most frequent verbs expressing the government's actions are *make, support, ensure, continue to, develop, ask, encourage, legislate, consider, provide,* and *introduce*. It was said earlier that one of the most noticeable features of the representation of the government is its facilitating role. This is realized grammatically through what Van Leeuwen (1999) terms "Managed Actions." Thus, in examining the agency of textual participants, he distinguishes between, (1) the agents of actual actions (e.g., *teachers have raised standards*), and (2), the instigators or "Managers" of "Managed Actions" (e.g., *the government has enabled teachers to raise standards*). This analytical term he uses as a sociological alternative to Halliday's (1994) grammatical term "Causatives," since it captures the sense that certain actors are managing, instigating, or in some other sense controlling others' actions.

Adopting Van Leeuwen's analytical scheme, we can see that *we* figures either as an agent or a Manager. The other participants (schools, teachers, students), on the other hand, are Managed Actors, collocating with processes like *enable* or *support*. They are thereby stylistically backgrounded, whereas the government is foregrounded as the principal social actor who directly or indirectly (through its policies) makes others' actions possible. A concordance of the verb *enable* (which occurs 40 times) reveals its role in constructing 2 key agendas in education: a competitive market and "fast-tracking" the most successful, alongside the generalized principles of individual responsibility, autonomy, and self-governance that help create the workfare society. The most frequent beneficiaries of *enable* are schools, teachers, and pupils. Schools are *enabled* to raise standards through innovation, specialization, sharing *best practice* between schools, partnership with other educational *providers* to extend *opportunity*, and *tailoring* educational programs to *individual needs*. Schools are thus encouraged to diversify and operate according to the principles of expertise-sharing, partnership, and market competition that lead to success in the commercial sector.

I would argue that this textual patterning of Manager-Managed relations between the government and other participants is a linguistic realization of a political phenomenon in which new modes of governance involve greater networked steerage and management practices that encourage innovation and autonomy alongside, of course, regulation in the form of target-setting and auditing practices. This move toward facilitating competitiveness and innovation can be seen in the educational texts, although the autonomy assigned to schools is circumscribed by targets and standards of practice. The following extract illustrates what one might term this type of "Schumpterian"[7] innovation-oriented competitiveness discourse, in which the government is cast as an enabler:

> We also want to enable successful and popular schools to expand more easily. [...] Within this framework, we want to deregulate to increase flexibility where possible, to reduce burdens, enable schools to innovate and find new ways to raise standards. (DfES, 2001: 43)

Raising *standards* and *opportunity for all* are presented throughout the corpus as the main objectives for all participants in education. They draw respectively on a traditional right-wing Discourse, in which a thematic focus on standards in education is articulated within a complex of nationalist Discourses, and a broadly social democratic Discourse of *equality of opportunity*. They simultaneously evoke popular fears about failure, and moral concerns with social justice. The weaving together of apparently contradictory right- and left-wing views illustrates the way in which New Labour rhetoric redefines social justice as the widening of opportunity to enter into competition. Indeed, this discursive reinterpretation of equality in meritocratic terms can be seen as central to a wider project of redefining socialism in the postwelfare era. Equality and justice are thus redefined as the right to succeed in an open competition.

This project of "updating" socialism is explicitly argued for in Giddens' recent apologia for New Labour. While rightly eschewing the unfettered competition of neoliberalism, he argues for meritocracy (here worded "fluidity") as a means to equality: "Fluidity is morally as well as economically desirable, since talented individuals have the chance to live up to their potential" (Giddens, 2002: 39). The claim ignores the mutual exclusivity of genuinely redistributive and just social policies, and a competitive market system that, while it continues to operate, will progressively neutralize any policies aimed at achieving social equality, since competition depends upon *relative* gain in a system where individuals operate against, rather than with, each other. The argument for meritocratic social democracy is, of course, premised on the economic advantages that talented individuals can bring to the country. Thus, the economy becomes the driving force behind social policy. But what about the untalented individuals? As in Giddens' tract, in the New Labour texts, they are not represented, while *pupils* are frequently modified with the phrase *gifted and talented*. Thus, redefining equality also involves classifying students in competitive terms, and reconstructing the learner as a strategic entrepreneur, investing in his or her education in order to reap future rewards in the labour market. And its legitimation in the text relies heavily on a discourse of globalization.

Where the use of *we* includes the reader (i.e., the nation as a whole), it claims a consensus on the desirability of equipping our children with the skills and dispositions necessary *if we are to participate successfully in the global economy*. Note the presupposition[8] here that what is at issue is not our participation in it, but whether we win. The inevitability of our taking an active role in global capitalism is thus constructed textually through such a presupposition. Moreover, this *reality of the world we live in* poses not a threat to the social and intellectual integrity of the education system, but rather *challenges* us to redefine education and pupils in human capital terms: *if young people are to fulfil their economic and social potential*. Their juxtaposition as equal modifiers in this noun phrase constructs a parity of worth between *economic* and *social*.

A related aspect of the use of *we* is to absorb responsibility for demands made on education into a vague linguistic agent who could be the government or the nation as a whole. Thus, for example, deontic modals that express an imperative such as *must, need to have to* occur most frequently with ambivalent instances of *we*, where the referent of the pronoun is unclear.

While most of the arguments made here stem from analysis of general patterns in the texts, critical discourse analysis also involves much close qualitative analysis, linking together different levels of theoretical abstraction. To illustrate, the following extract from the 2001 White Paper is analyzed with respect to its role in legitimating the business agenda in and for education, by drawing on a neoliberal discourse of globalization:

> To prosper in the 21st century competitive global economy, Britain must transform the knowledge and skills of its population. Every child, whatever their circumstances, requires an education that equips them for work and prepares them to succeed in the wider economy and society. We must harness to the full the commitment of teachers, parents, employers, the voluntary sector, and government—national and local—for our educational mission. (DfES, 2001: 5).

The first sentence rests on the assumption that there *is* a global economy already in existence and that it is somehow just "out there" demanding a response. In this sense it can be seen as a fragment of a neoliberal discourse of globalization that claims its inevitability, and the need for countries to be internationally competitive. The main proposition of this sentence (what

Britain must do) is an "imaginary"; out of all possibilities, it selects education-dependent practices as the strategy for dealing with this global state of affairs, thereby constructing, through education, a knowledge-based economy. This discourse of globalization is here recontextualized in an educational discourse in order to legitimate reforms. The basis of this legitimacy is the premise that competitiveness is both necessary and good in a context of globalization. The power of the rhetoric of globalization lies precisely in its representation as an abstract challenge to be met, rather than the agent-driven processes of capitalist development. The legitimacy of these processes and concomitant political strategies is, in turn, partly constructed through educational discourse; by texturing together educational and neoliberal discourses, a codependency and apparent compatibility between educational and economic agendas is constructed.

In terms of the wording in this extract, the force of the imperative is strengthened through modality: "Britain must," "we must," as well as in the choice of words: "every child requires"; to "transform" (not "change," since something more radical and urgent is required), "we must harness to the full" (not "we need"), "our educational mission" (not "aims" or "plans"—something much more radical and grand). These lexical choices contribute to the style of the text by which the government constructs its identity. I would argue this is partly modelled on a corporate identity of dynamism, energy, commitment, and concern that it projects to the public.

Returning to the use of the pronoun *we*, its use in the above extract is apparently inclusive, but shows an interesting and contradictory slippage. The government here construes itself as one of the collaborators in the joint project of education whose commitment is to be harnessed (which contradicts its pervasive structural mechanisms of control). However, the aim of this is to achieve "*our educational mission.*" This must surely be the government's educational mission, yet it is not an agent of this process, but rather one of the actors whose commitment is to be harnessed. Since this is contradictory, the only alternative interpretation is that the government is making a claim that all agree with their mission and it is thus shared. Given that the imperative for this mission stems from an "inevitable" global economy, and the government is construed in equal terms with other social actors forced to respond to the challenges it poses, this textual organization also contributes to the neoliberal agenda of representing unfettered global competition as an agentless and inexorable force of change. It is worth noting here that one aim of doing quantitative statistical analysis in addition to close textual analysis is to determine the social significance of patterns of representation like this.

Teachers: Hierarchized, Collaborative, Regulated

Let us turn briefly to the representation of teachers. They are enabled by the government to perform two main types of activities: skills-updating (usually worded as professional development) and raising standards. This appears to confirm the logic of lifelong learning policies that knowledge is seen as a perishable product, requiring constant upgrading. Teachers are also hierarchically classified with modifiers like *expert, advanced skills, outstanding, fast track,* and *excellent*. These teachers are involved in the training of other teachers, in which they share their expertise and specialism.

The texts also represent teachers' *willingness* to engage in *professional development* that can be seen as serving both a legitimatory and regulatory function. The legitimation stems from the implicit consensus among teachers that helps represent policy as meeting strongly felt needs rather than enforcing new constraints on practice. The regulatory function stems from the networking of a set of new practices and roles represented in the texts, of which *professional development* forms one part. It intersects with other practices in forming an implicit web of responsibility and accountability. These include meeting *targets* and *standards,* appraisals, and performance-related pay. Added to this is the creation of a hierarchy of expertise within the profession, which illustrates the core rationality of the market, in which competition is used to drive up standards. Herein lies a basic contradiction in the policy: a competitive market among teachers undermines the teamwork and collaboration that is also represented as forming an essential part of teachers' practice. By creating a breed of expert teachers to advise others, this effectively removes the *trust in teachers' professional judgment* the government claims it is committed to.

Students: Lifelong Learning Entrepreneurs

The contradiction between competitive self-advancement, and cooperation and inclusion, through the data, is perhaps most clearly illustrated in the representation of students, primarily worded as either *pupils* or *young people.* The latter wording, unsurprisingly, is mostly used to refer to older students in the 14–19 age range. However, there is an interesting difference in the way they are represented. *Pupils* are more frequently classified according to age, ethnicity, and ability. They more frequently occur within a discourse of individualism in which they are supported in their individual programs of learning and accumulation of skills. They are also represented as value-adding commodities linked to national standards of attainment, as well as entering into relations of consumption in which knowledge is a commodity to be acquired (the verbs that collocate with *skills,* for example, express relations of possession). This instrumental logic is most explicitly illustrated in the following extract concerning the science curriculum: *This will engage pupils with contemporary scientific issues and focus on their role as users and consumers of science.* Science is actually a diverse set of practices designed to further our understanding of the world and our relationship with it. Yet, here it is being constructed as a commodity to be used or bought. The potential effect of the proposed science curriculum is thus to alienate pupils from the intrinsic value of their own and others' learning, and thereby to view others as obstacles or coentrepreneurs, rather than the means, to their intellectual development.

Let us turn finally to *young people*: as well as being involved in vocational training, their most statistically significant textual environment is that which helps construct discourses of lifelong learning, and of social inclusion. The project of the learning society requires strategies that inculcate in individuals a commitment to continual learning and self-improvement. Social inclusion policies complement this by ensuring wider access to educational opportunities underpinned by financial assistance and incentives for those at greatest risk of "disaffection." The texts thus represent an education system that must *motivate, include, raise expectations, meet individual needs, and aspirations,* as well as prepare *young people* to be *responsible citizens*. Both the functional and socializing roles of education are most clearly encapsulated in the following statement:

> [education must] meet the needs and aspirations of all young people, so that they are motivated to make a commitment to lifelong learning and to become socially responsible citizens and workers; broaden the skills acquired by all young people to improve their employability, bridge the skills gap identified by employers, and overcome social exclusion. (DfES, 2004)

The statement textures together particular interdependencies and equivalencies: between citizenship and working; between individual responsibility and work; between effort (commitment) and reward (employability and qualifications); between education and the needs of employers; and between social justice and education. In effect, this places education at the forefront of constructing the postwelfare society in which individuals are afforded rights in the shape of education and training, in return for their commitment, effort, and responsibility to others. Moreover, it illustrates a shifting conception of citizenship forged in practices of consumption, and oriented to what Rose (1999) calls the "enterprise of the self." Within this paradigm, education functions as a form of strategic investment in one's own future capital. This is a necessarily commodifying move, and one at odds with education as part of the intellectual commons wherein the more knowledge is freely shared, the more is produced. Just the opposite is true of an educational market where the value of a commodity lies in its scarcity; its unavailability. Viewed thus, widening access in the spirit of social inclusion and recasting education as an investment become tension-riven educational strategies for constructing the inclusive workfare society.

Conclusion

I have tried to illustrate how the representation of social actors in New Labour education policy discourse helps legitimate its educational reforms by drawing on a neoliberal discourse of globalization. Moreover, in construing the right sort of social identities, values, relations, and practices for education, the discourse also contributes to the ongoing and contingent processes of shaping the knowledge-based economy and the workfare competition state as its emergent regulatory regime. This project entails an unprecedented role for education in helping construct and inculcate the dispositions necessary to the learning society. In the fight for a more egalitarian and genuinely empowering education system free from the exigencies and biased interests of capital, I would suggest that a critical analysis of discourse can play an important role in countering the strategies of inculcation and legitimation necessary to any political program that seeks increasingly to reduce education to its economic function, and to use it to recast citizenship rights as the right to work.

Notes

1. DfES (2001: 2).
2. I am using the term "discourse" in its most general sense of textual and visual forms of semiosis. This is to be distinguished from the capitalized form "Discourse," which refers to a constellation of meanings that together constitute a representation of some aspect of the world from a particular point of view. When used in this sense, the term is either in plural form, or it is qualified by an expression classifying the Discourse-type in question. For example, "managerial Discourse" or "the Discourse of globalization."
3. This refers to broad cross-party agreement on the form and role of the British state. It implies a broad set of agreed parameters circumscribing the policy options generally regarded as feasible by politicians and civil servants. By adopting this political analysis, one can distinguish between what Hay (1996) terms "state-shaping" governments like those of Attlee and Thatcher; and "state-accommodating" governments like New Labour, that is, "state-accommodating" governments that tend to adopt policies that do not substantially challenge the structures of the state inherited from the previous government, although they may make important "adjustments" to them, both consolidating and managing the new regime. For instance, in the case of Major, there was a "deradicalization" of some Thatcherite policies, such as moving away from the destruction of welfare, toward the subordination of welfare to workfare.
4. An important assumption underlying the analysis in this chapter is that in fact they *are* distinct; that social justice in the educational arena is mutually exclusive with a narrow focus on education's role in generating economic competitiveness.
5. For a fuller account of the findings of this textual analysis, see Mulderrig (2003).
6. It is possible to distinguish two usages of *we*, termed "inclusive" and "exclusive." The former includes the addressees of the text; the latter refers only the speakers and the group to which they belong (i.e., the institution of government). Which form is being used may be more or less explicit in the text. For instance, the referents are clear in phrases such as "we, the nation" or "we, the government"; other times it may be ambivalent' 90 percent of occurrences of *we* are used in the exclusive sense.
7. Following Jessop (2002), I use this term to refer to a particular conception of economic competitiveness that is tendentially emerging as the dominant model replacing the Keynesian economics of the postwar era. In essence, the

economist Joseph Schumpeter saw competitiveness as depending on developing individual and collective capacities to engage in permanent innovation in both economic and extra-economic spheres.

8. In the sense used here, a presupposition is that information assumed by the speaker, as opposed to that information at the center of the speaker's communicative interest.

References

Ball, S. (1990) *Politics and Policy Making in Education: Explorations in Policy Sociology* (London: Routledge).
Brine, Jacky. (2006) Lifelong Learning and the Knowledge Economy: Those That Know and Those That Do Not—The Discourse of the European Union. *British Educational Research Journal*, 32 (5): 649–665.
Chouliaraki, L. and N. Fairclough. (1999) *Discourse in Late Modernity* (Edinburgh: Edinburgh University Press).
Dale, R. (1989) *The State and Education Policy* (Milton Keynes: Open University Press).
Department for Education and Skills (DfES). (2001) *Schools-Achieving Success*. White Paper (London: HMSO).
———. (2002) Green Paper *14–19: Extending Opportunities, Raising Standards*. Green Paper (London: HMSO).
Ellison, N. (1997) From Welfare State to Post-welfare Society? Labour's Social Policy in Historical and Contemporary Perspective. In B. Brivati and T. Bale (eds.) *New Labour in Power: Precedents and Prospects*, 34–65 (London: Routledge).
Fairclough, N. (1992) *Discourse and Social Change* (Cambridge, UK: Polity Press).
———. (2000) *New Labour, New Language?* (London: Routledge).
———. (2003) *Analysing Discourse: Textual Analysis for Social Research* (London: Routledge).
Giddens, A. (2002) *Where Now for New Labour?* (Cambridge, UK: Polity Press).
Halliday, M. (1994) *An Introduction to Functional Grammar* (London: Hodder Arnold).
Hatcher, R. and N. Hirtt. (1999) The Business Agenda behind Labour's Education Policy. In M. Allen, C. Benn, C. Chitty, M. Cole, R. Hatcher, N. Hirtt, and G. Rickowski (eds.) *Business, Business, Business: New Labour's Education Policy*, 12–23 (London: Tufnell Press).
Hay, C. (1996) *Re-stating Social and Political Change* (Buckingham: Open University Press).
———. (1999) *The Political Economy of New Labour: Labouring under False Pretences?* (Manchester: Manchester University Press).
Hill, D. (1999) *New Labour and Education: Policy, Ideology and the Third Way*. A Hillcole Paper (London: Tufnell Press).
———. (2001) *The Third Way in Britain: New Labour's Neo-liberal Education Policy*. Paper presented at Congress Marx International III, Université de Paris X Nanterre, Sorbonne.
Jessop, B. (2002) *The Future of the Capitalist State* (Cambridge, UK: Polity Press).
Ministry of Education. (1958) *Secondary Education for All—A New Drive*. Cmnd. 604 (London: HMSO).
Mulderrig, J. (2003) Consuming Education: A Critical Discourse Analysis of Social Actors in New Labour's Education Policy. *Journal of Critical Education Policy Studies*, 1 (1). Available at http://www.jceps.com/?pageID=article&articleID=2.

Mulderrig, J. (2006) *The Governance of Education: A Corpus-Based Critical Discourse Analysis of UK Education Policy Texts 1972–2005*. Unpublished doctoral thesis, University of Lancaster.

Robertson, S. and R. Dale. (2006) Changing Geographies of Power in Education: The Politics of Rescaling and Its Contradictions. In D. Kassem, E. Mufti, and J. Robinson (eds.) *Education Studies: Issues and Critical Perspectives* (Buckinghamshire: Open University Press). Available at http://www.genie-tn.net.

Rose, N. (1999) *Powers of Freedom: Reframing Political Thought* (Cambridge: Cambridge University Press).

Scott, M. (1997) *Wordsmith Tools* (Oxford: Oxford University Press). Available at http://www.lexically.net/wordsmith.

Van Leeuwen, T. (1999) Discourses of Unemployment in New Labour Britain. In R. Wodak and L. Christoph (eds.) *Challenges in a Changing World: Issues in Critical Discourse Analysis*, 87–100 (Wien: Passagen-Verlag).

Chapter 8

Marx, Education, and the Possibilities of a Fairer World: Reviving Radical Political Economy through Foucault

Mark Olssen and Michael A. Peters

> *It is clear, even if one admits that Marx will disappear for now, that he will reappear one day. What I wish for... is not so much the defalsification and restitution of a true Marx but the unburdening and liberation of Marx in relation to party dogma, which has constrained it, touted it, and brandished it for so long.*
>
> —(Foucault, 1998: 458)

Introduction[1]

Marxism, we are told by politicians and the popular press, is dead. The Left, as a historical movement tied to the labor movement, is frozen over, caught between the collapse of actually existing communism in Eastern Europe and the triumph of global market forces. Union membership in the traditional industrial economy in the UK is dwindling as multinationals relocate offshore; even insurance, information, banking, and call-center jobs of the "new economy" are increasingly outsourced to India and other emergent economies literate in information and computing technology and English. China has joined the World Trade Organization (WTO) and committed itself to a postsocialist market economy. At a time of an intensification of inequalities between regions and, perhaps more significantly, between North and South—between the developed world and the developing world—the Left in Britain, the United States, and most of Europe seems ideologically gutted by the Third Way preoccupation with the social market and with citizenship "responsibilities" rather than with traditional concerns of equality and advancing rights. The best offer on hand seems to be a *socialization* of the market and an acknowledgment of its moral limits. Neoliberalism, in the age of privatization, reduces the state's role more and more to one of regulation, rather than provision or funding of public services. The U.S.–UK neoliberal model of globalization has dominated the world economy and world politics for the past 20 years, defining the present crisis of fundamentalisms and restyling imperialism as a new age of barbarism. In this age, American-style democracy is exported alongside the ideology of "free trade." Yet many Americans have shifted their view since the Vietnam War on whether the United States is a force for good in the world or an imperialist power, and this is so despite Bush's recent election victory. Even the philosophers of 1968 have given way to a new breed of fashion-conscious savants, who now turn their attention to extolling the virtues of liberal individualism or sneer at the last great generation of Left-Nietzscheans, such as Foucault and Derrida.

The Left has certainly been marginalized and even in the home of European socialism it seems confused and crisis-ridden. Europe itself is fighting to establish a new identity, reshaping its territory through enlargement and integration, and desperately competing with the U.S. juggernaut of global power and the rising stars of East Asia—not only China, but also Japan, Taiwan, Hong Kong, Singapore, and Korea—which seem destined to develop a trading bloc at least as powerful as that of the United States and the European Union. The traditional Left, wedded to the rise of the industrial working class, some observers have remarked, is also tied to its demise. Is the Left history? Has it simply become an academic form of analysis or does it have the seeds to reconfigure itself as an organizing force once again?

In terms of emancipatory futures there are all sorts of oppressions to overcome; many of these oppressions have intensified in the neoliberal era. The question that Steve Brier (1999) asks is

> How do we position ourselves as a movement in relation to all the particular forms of oppression experienced by specific communities and people, defined by race, gender, nationality, sexual orientation, etc., especially at a time when no unified working-class movement exists that encompasses these communities and fights to eradicate the special injustices they face?

The question of unity becomes paramount. Against identity politics and certain forms of postmodernism we need to inquire what is the unifying principle? Is it the concept of "class" or even an overlapping set of concepts? Brier was writing at a time that had not yet seen the neoconservative hegemony in the White House or its consolidation after the reelection of Bush for a second term. In this environment of voter behavior and corporate corruption it is difficult to see the flourishing of social democracy even though the White House wants to export American-style democracy to the world as part of its neoconservative agenda. In these circumstances is it really enough to talk of "beyond left and right" as the future of radical politics as Tony Giddens

(1994) has done? Or does Alex Callinicos's (2003) *Anti-capitalist Manifesto* define a way forward?

These are weighty questions that do not admit easy answers. But it is clear that even in this environment of world politics there are new lines of struggle emerging that coalesce with the old articles of faith. There *are* expressions of new forms of socialism that revolve around the international labor movement and invoke new imperialism struggles based on the movements of indigenous and radicalized peoples. There are active social movements, perhaps less coherent but every bit as powerful as older class-based movements, such as the anticapitalism, antiglobalization movements, women's and feminist movements, and environmental movements. These new expressions do require engagement and retheorizing by the Left. One obvious challenge for Marxism and the Left more generally is its engagement with Islam and the enslavement of women.

There is also a host of struggles around the socialization of the market and a question of whether this can be pursued successfully at the level beyond the state. Indeed, as many theorists have asserted, the future of the Left is tied up with the future of world democracy and with the development of Left media cultures and centers. Part of the success of the Right has been its ability to privatize thinking and media, moving beyond the academy to set up dozens of new think tanks, private consultancies, and media centers that propagate partisan "news" or lobby and influence government departments at the highest levels.

Marx's Radical Political Economy

This chapter seeks to ask to what extent Foucault can provide a different vision of radical politics to that of Marx, and then to assess the implications of this for work in education, politics, and ethics. Central to Marx's model of political economy was a particular materialist inversion of the Hegelian dialectic, giving rise in Western Marxism to a particular formulation of base and superstructure.

In his Preface to *A Contribution to the Critique of Political Economy* (1971), (hereafter *CPE*), written in 1859, Karl Marx, as he says, examines "the system of bourgeois economy" and he gives a biographical account of how he first came to realize the position he adopts on the issue of the relationship of the economy to the cultural, educational, legal, and political domains of society—a fundamental and stunning insight whose force has not diminished even though the sophistication of structuralist and poststructuralist arguments against the rudiments of a base/superstructure model now have to be accepted. He indicates that he had started thinking about these issues in 1842 and returned to a fresh examination of Hegelian philosophy of law in order to rethink the origins of legal relations and political forms. As his massive bibliographical studies across several languages led him to conclude legal relations and political forms can not be understood by themselves or as a product of the development of the human mind but only in "the material conditions of life," the totality of which Hegel called "civil society" and whose "anatomy" Marx argued must be sought in political economy.

In that remarkable work that took him over 17 years to bring to maturity, Marx addresses the question of the method of political economy and is clearly influenced in his construction not only by the history of political economy and especially the major figures of the eighteenth century (especially Smith and Ricardo) but also Charles Darwin from whom he takes a newly scientized view of historical evolution. Marx's view is tantamount to a form of *historical naturalism*, which assumes laws of historical development. Marx calls it a "materialist account of history" and Engels shortened it to "historical materialism." Thus, he argues,

> Bourgeois society is the most advanced and complex historical organisation of production. The categories which express its relations, and an understanding of its structure, therefore, provide an insight into the structure and the relations of production of all formerly existing social formations the ruins and component elements of which were used in the creation of bourgeois society.[2]

Clearly his model here is Darwin as well as Hegel. This is confirmed when later in the text he argues, "The anatomy of man is a key to the anatomy of the ape." Marx claimed that all history should be thought of as the history of class struggles over surplus-value. Engles described "being determines consciousness" as the "law of evolution in human history" equating it with Darwin's "law of evolution in organic nature." The *Origin of Species*[3] was published in 1859 and Marx read it in 1860. Marx believed that Darwin's book "contains the basis in natural history for our views." In 1861 in a letter to Ferdinand Lassalle Marx wrote,

> Darwin's book is very important and serves me as a model for the class struggle in history... Despite all deficiencies, not only is the death-blow dealt for the first time here to "teleology" in the natural sciences, but its rational basis is empirically explained. (Marx and Engles, 1965: 123)

There is an oft quoted story that Marx sought permission to dedicate *Das Kapital* to Charles Darwin who declined the offer but this now seems suspect.[4]

Naturalism is the tendency to look upon the material universe as the only reality and to reduce all laws to uniformities in nature. To this end it denies the dualism of spirit and matter regarding the social and cultural as manifestations of matter that are governed by its laws. Naturalism, as Quine has suggested, is the position that there is no higher tribunal for truth than natural science itself; scientific method alone must judge the claims of science and there is no room for metaphysics or first philosophy. Naturalism is in this sense derived from materialism or pragmatism. Historical materialism explains changes in human history through material factors, for Marx, economic and technological. Where Marx is both a historical materialist and naturalist, Foucault is the former (although as we shall see he places no particular priority on the economic) but also firmly antinaturalist when it comes to the market as we will see in more detail in later sections. Naturalism, like empiricism and older forms of materialism that seek to represent the real or nature outside of discourse and independent of historicity, fails to adequately recognize the contingent dimension of knowledge. For Foucault, knowledge, like the human subject, is always already social, and attempts to establish a foundation in nature to anchor knowledge or the operations of institutions independently of history are not possible.

For Foucault, also, the Marxist conception of relations between economy and superstructures were problematic. In the Marxist conception of historical materialism, educational, legal, political institutions, as well as ideologies and discourses are represented as part of the superstructure of society that is split from material practices of the economic foundation or base, and are determined by it. In the same way, the mental operations of consciousness are represented as determined by the material base of society. As Marx (1971: 20–21) expresses the point,

> In the social production which men carry on they enter into definite relations that are indispensable and independent of their will; these relations of production correspond to a definite stage of development of their material powers of production. The sum total of these relations of production constitutes the economic structure of society—the real foundation, on which rise legal and political superstructures and to which correspond definite forms of social consciousness. The mode of production in material life determines the general character of the social, political and spiritual processes of life. It is not the consciousness of men that determines their existence, but on the contrary, their social existence determines their consciousness.

In the twentieth century one of the central issues addressed by Western Marxists has been an attempted resolution and reconceptualization of the nature of the relation between the economic base and the cultural superstructure of society. In the classical Marxist model both the character of a society's culture and institutions, as well as the direction set for its future development are determined by the nature of the economic base, which can be defined as the mode of production at a certain stage of development (Williams, 1980: 33). The simplest nature of this relation, as Williams tells us, was one of "the reflection, the imitation, or the reproduction of the reality of the base in the superstructure in a more or less direct way" (ibid.); that is, a relation in which the economic base and specifically the forces of production constituted the ultimate *cause* to which the social, legal, and political framework of the society can be traced back.

In the attempt to reformulate Marxism in the twentieth century the economic determinist conception is challenged by those who see Marxism as granting rather more "independence" or "autonomy" to the superstructures of society. Hence a "dialectical" notion of the relation was stressed suggesting a relation of reciprocal influence. It was argued that, although the base *conditions* and *affects* the superstructure, it is in turn *conditioned* and *affected* by it. In all cases, however, in order to remain as Marxists, the ultimate priority of the economic base as the causal determinant of the social character of a society was safeguarded by maintaining that the economic factor is "determining in the last instance." Hence, it was maintained that the superstructure had only a "relative autonomy," and the theory of "relative autonomy," as a shorthand designation of the base-superstructure relation, became a central concept of twentieth-century Marxism.

All of the Marxist studies on different aspects of education reflected the problematic determinism of the relations between the economy and the various cultural and ideological aspects of the society. If we consider the application of a particular case of Marx's base-superstructure analysis to education one important theme has been an understanding of the schooling system as a production system related to capitalism. The best known contemporary form of this kind of application is Bowles and Gintis' (1976) correspondence theory that hypothesises a set of correspondences between work and education at all levels: a subservient workforce, an acceptance of hierarchy, and motivated by external rewards.[5] In their later work they explain their original position as

> schools prepare people for adult work rules, by socializing people to function well, and without complaint, in the hierarchical structure of the modern corporation. Schools accomplish this by what we called the *correspondence principle*, namely, by structuring social interactions and individual rewards to replicate the environment of the workplace. (2001: 1)

In this later work they endorse the correspondence principle as more or less correct although they also mention shortcomings of the original work. Criticisms of structural overdetermination of the lives of working-class kids has been explored by Paul Willis (1977) in *Learning to Labour* and by many others at the Centre for Contemporary Cultural Studies at the University of Birmingham who demonstrate that working-class kids choose to fail through the development of a counterculture.

Foucault's Radical Political Economy

Political economy has a much longer tradition than as used by Marx. The Greeks considered it as pertaining to the management of the household of the state, as did Jean Jacques Rousseau and Adam Smith.[6]

Foucault's sense of the concept must be seen in relation to his way of conceptualizing social structure as well as his opposition to Marxism. In one sense, his own conception can be viewed as a form of "new political economy."[7] Foucault opposed both the determinism of the base-superstructure model as well as the Hegelian monistic conception of society and the Hegelian/Darwinian conception of progressive evolution of history through the unfolding of the dialectic to the communist utopia. Rather than seek to explain all phenomena in relation to a single centre, Foucault is interested rather to advance a polymorphous conception of determination in order to reveal the play of dependencies in the social and historical process. Hence, in opposition to the themes of totalizing history as found in Hegel, Foucault (1978: 10) substitutes what he calls a "differentiated analysis":

> Nothing, you see, is more foreign to me than the quest for a sovereign, unique and constraining form. I do not seek to detect, starting from diverse signs, the unitary spirit of an epoch, the general form of its consciousness: something like a Weltanschauung... I have studied, one after another, ensembles of discourse; I have characterised them; I have defined the play of rules, of transformations, of thresholds, of remanences. I have established and I have described their clusters of relations. Whenever I have deemed it necessary I have allowed systems to proliferate.

In advocating pluralism in place of monism, Foucault believed Marxism to reflect theoretical rules inherited from its time of origins. As he says (2001b: 269),

> Marx's economic discourse comes under the rules of formation of the scientific discourses that were peculiar to the nineteenth century... Marxist economics—through its basic concepts and the general rules of its discourse—belongs to a type of discursive formation that was defined around the time of Ricardo.

A central element of Foucault's critique of Marxism relates to the notion of "totalization." Essentially, for Foucault, Marxism was not just a "deterministic" but a "deductivistic" approach. That is, it directs attention not just to the primacy of the economy but it seeks to explain the parts of a culture as explicable and decodable parts of the whole totality or system represented as a closed system. Marxism, claims Foucault, seeks to ascertain "the principle of cohesion or the code that unlocks the system explaining the elements by deduction" (Thompson, 1986: 106). This was the approach of Marx took from Hegel, which seeks to analyze history and society in terms of "totality," where the parts are an "expression" of the whole—hence the notion of an "expressive totality."

The dissociation between Marxism, and Foucault's own position became more apparent after Foucault's turn to genealogy and Nietzsche at the close of the 1960s. With his growing interest in genealogy, Foucault became more concerned with power and history, and the historical constitution of knowledge. In this process, there was however, no integrative principle or essence, and history was not periodized according to economic stages. If the genealogist studies history "he finds that there is 'something altogether different' behind things: not a timeless and essential secret, but the secret that they have no essence or that their essence was fabricated in a piecemeal fashion from alien forms" (Foucault, 1977b: 142).

Foucault's objection to elements of Marxism explicitly reflects his Nietzschean heritage and his belief that certain aspects of Marxism distorted the liberatory potential of the discourse.

> The interest in Nietzsche and Bataille was not a way of distancing ourselves from Marxism or communism—it was the only path towards what we expected from communism. (Foucault, 2001b: 249)

It was in terms of the philosophy of difference and Nietzsche's conception of multiplicities through a rejection of Platonic hierarchies that Foucault enunciates a theory of discursive formations and rejects Marxist and Hegelian conceptions of history. The utilization of Nietzsche signalled a rupture from Marxism in relation to a series of interrelated conceptual, theoretical, and methodological precepts, including power, knowledge and truth, the subject, and the nature of historical change and determination.

Nietzsche focussed on power in an altogether different way to Marx. In "Prison Talk," Foucault (1980a: 47) states,

> It was Nietzsche who specified the power relation as the general focus, shall we say, of philosophical discourse—whereas for Marx it was the productive relation. Nietzsche is the philosopher of power, a philosopher who managed to think of power without having to confine himself within a political theory in order to do so.

Power, for Nietzsche, was conceived as a relation of forces within an analytics of power/knowledge/truth, which became important for Foucault to understand in the later 1960s after the publication of *The Archaeology of Knowledge* and his growing friendship with the Parisian Nietzschean Gilles Deleuze. Foucault accredits Nietzsche as the source of his interest in the question of truth and its relation to power. As he states, in "Truth and Power" (1980b: 133), "The political question ... is not, error, illusion, alienated consciousness or ideology, it is truth itself. Hence, the importance of Nietzsche." Nietzsche's importance to Foucault can be seen as "correcting" the Marxism developed after Marx, especially in relation to the linkage between power-knowledge-truth, and the functioning of knowledge as an instrument of power. As Alan Schrift (1995: 40) notes, Nietzsche's influence drew attention away from "substances, subjects and things, and focussed attention instead on the *relations between* these substantives" (italics in original). In a related way, Foucault "draws our attention away from the substantive notion of power and directs our attention instead to the multifarious ways that power operates through the social order" (ibid.). For Nietzsche, such relations were relations of forces. Foucault thus focussed on new relations as the relations of forces that existed and interacted within social systems as social practices. These were forces of repression and production that characterized the disciplinary society; forces that enable and block; subjugate and realize, and normalize and resist. In this model, power is not a thing, but a process, a becoming.

Foucault rejects Marxist models of a determining economic base and a determined superstructure as well as refinements based on conceptions of totality by Marx's twentieth-century successors. Foucault is not interested in accounting for the practices of the social structure solely in terms of a model of economic determination. Although, like Althusser, he utilizes a model of complex and multiple causation and determination within the social structure, the specific elements and mechanisms of such processes, as elaborated by Foucault, differ in important, indeed crucial, respects. In Foucault's conception of social structure, explaining the relations between discursive formations and nondiscursive domains (institutions, political events, economic practices, and processes) is recognized as the ultimate objective. As he formulates it in *The Archaeology of Knowledge* (1972), for archaeology, in comparison to Marxism,

> The *rapprochements* are not intended to uncover great cultural continuities, nor to isolate mechanisms of causality...nor does it seek to rediscover what is expressed in them...it tries to determine how the rules of formation that govern it...may be linked to non-discursive systems: it seeks to define specific forms of articulation. (162)

Unlike Marxists, he sees no one set of factors as necessarily directing human destiny. Rather, the forms of articulation and determination may differ in relation to the relative importance of different nondiscursive (material) factors in terms of both place and time. In the shift from a purely archeological to a genealogical mode of enquiry, Foucault's concern with the relation between discursive and nondiscursive domains is given a more historical and dynamic formulation, although, the concern with synchronic analysis is not abandoned. Throughout, however, as Mark Poster (1984: 39–40) explains, Foucault's central aim is to provide a version of critical theory in which the economic base is not the totalizing center of the social formation, whereby Hegel's evolutionary model of history is replaced by Nietzsche's concept of genealogy, and where causes and connections to an imputed center or foundation are rejected in favor of exposing the contingency and transitory nature of existing social practices. In Poster's view, this presents us with a crucial decision. In comparing Foucault and Althusser, he maintains that "the theoretical choice offered by these two theorists is dramatic and urgent. In my opinion Foucault's position in the present context is more valuable as an interpretative strategy...Foucault's position opens up critical theory more than Althusser's both to the changing social formation and to the social locations where contestation actually occurs" (ibid.). While having a generally historicized view of the nature and development of knowledge, Foucault rejects the possibility of any "absolute" or "transcendental" conception of truth "outside of history" as well as of any conception of "objective" or "necessary" interests that could provide a necessary "Archimedean point" to ground either knowledge, morality, or politics. Read in this way, historical materialism does not prioritize the economy in any necessary or universal sense, but is about the systematic character of society and how it might change. It is about the processes of change internal to social systems. It holds that societies are to varying extents integrated systematically through their material practices and discursive coherences, and break down and change as the component elements of the system change.[8]

Governmentality Studies

Based upon this general social ontology Foucault utilizes the notion of "governmentality." as the basis of his "new" conception of political economy. The working premise of governmentality studies is based on Foucault's insight and analysis of the modern regime of power in which power characteristically operates *internally* or subjectively in terms of a logic of "self-improvement" that demands the freedom of the individual. Governmentality is the key concept that links Foucault genealogy of the subject with his interest in political rationalities, that is, between the government of the state and the government of the self, and in so doing "solves" the problem of agency (liberal political economy) versus structure (Marxist political economy).

Foucault's overriding interest was not in "knowledge as ideology," as Marxists would have it, where bourgeois knowledge, say, modern liberal economics was seen as false knowledge or bad science. Nor was he interested in "knowledge as theory" as classical liberalism has constructed disinterested knowledge, based on inherited distinctions from the Greeks, including Platonic epistemology and endorsed by the Kantian separation of schema/content that distinguishes the analytic enterprise. Rather Foucault examined *practices* of knowledge produced through the relations of power.[9] He examined how these practices, then, were used to augment and refine the efficacy and instrumentality of power in its exercise over both individuals and populations, and also in large measure helped to shape the constitution of subjectivity. Fundamental to his governmentality studies was the understanding that Western society professed to be based on principles of liberty and the Rule of Law and said to derive the legitimation of the state from political philosophies that elucidated these very principles. Yet as a matter of historical fact, Western society employed technologies of power that operated on forms of disciplinary order or were based on biopolitical techniques that bypassed the law and its freedoms altogether. As Colin Gordon (2001: xxvi) puts it so starkly, Foucault embraced Nietzsche as the thinker "who transforms Western philosophy by rejecting its founding disjunction of power and knowledge as myth." By this he means that the rationalities of Western politics, from the time of the Greeks, had incorporated techniques of power specific to Western practices of government, first, in the expert knowledges of the Greek tyrant and, second, in the concept of pastoral power that characterized ecclesiastical government.

It is in this vein that Foucault examines government as a practice and problematic that first emerges in the sixteenth century and is characterized by the insertion of economy into political practice. Foucault (2001c: 201) explores the problem of government as it "explodes in the sixteenth century" after the collapse of feudalism and the establishment of new territorial states. Government emerges at this time as a general problem dispersed across quite different questions: Foucault mentions specifically the Stoic revival that focussed on the government of oneself; the government of souls elaborated in Catholic and Protestant pastoral doctrine; the government of children and the problematic of pedagogy; and, finally the government of the state by the prince. Through the reception of Machiavelli's *The Prince* in the sixteenth century and its rediscovery in the nineteenth century, there emerges a literature that sought to replace the power of the prince with the art of government understood in terms of the government of the family, based on the central concept of "economy." The introduction of economy into political practice is for Foucault the essential issue in the establishment of the art of government. As he points out, the problem is still posed for Rousseau, in the mid-eighteenth century, in the same terms—the government of the state is modelled on the management by the head of the family over his family, household, and its assets.

It is in the late sixteenth century, then, that the art of government receives its first formulation as "reason of state" that emphasizes a specific rationality intrinsic to the nature of the state, based on principles no longer philosophical and transcendent, or theological and divine, but rather centered on the *problem of population*. This became a science of government conceived of outside the juridical framework of sovereignty characteristic of the feudal territory and firmly focused on the problem of population based on the modern concept that enabled "the creation of new orders of knowledge, new objects of intervention, new forms of subjectivity and....new state forms" (Curtis, 2002: 2). It is this political-statistical concept of population that provided the means by which the government of the state came to involve individualization and totalization, and, thus, married Christian pastoral care with sovereign political authority. The new rationality of reason of state focussed on the couplet *population-wealth* as an object of rule, providing conditions for the emergence of political economy as a form of analysis. Foucault investigated the techniques of police science and a new biopolitics,

> which tends to treat the "population" as a mass of living and co-existing beings, which evidence biological traits and particular kinds of pathologies and which, in consequence, give rise to specific knowledges and techniques. (1989b: 106, cited in Curtis, 2002)

As Foucault (2001d) comments in "The Political Technology of Individuals," the "rise and development of our modern political rationality" as reason of state, that is, as a specific rationality intrinsic to the state, is formulated through "a new relation between politics as a practice and as knowledge" (407), involving specific political knowledge or "political arithmetic" (statistics); "new relationships between politics and history," such that political knowledge helped to strengthen the state and at the same time ushered in an era of politics based on "an irreducible multiplicity of states struggling and competing in a limited history" (409); and, finally, a new relationship between the individual and the state, where "the individual becomes pertinent for the state insofar as he can do something for the strength of the state" (409). In analyzing the works of von Justi, Foucault infers that the true object of the police becomes, at the end of the eighteenth century, the population; or, in other words, the state has essentially to take care of men as a population. It wields its power over living beings, and its politics, therefore has to be a biopolitics (416).

Foucault's lectures on governmentality were first delivered in a course he gave at the Collège de France, entitled *Sécurité, Territoire, Population*, during the 1977–1978 academic year. While the essays "Governmentality" and "Questions of Method" were published in 1978 and 1980, respectively, and translated into English in the collection *The Foucault Effect: Studies in Governmentality* (Burchell et al., 1991), it is only very recently that the course itself has been transcribed from original tapes and published for the first time (Foucault, 2004b), along with the sequel *Naissance de la biopolitique: Cours au Collège de France, 1978–1979* (Foucault, 2004a).[10] The governmentality literature in English, roughly speaking, dates from the 1991 collection and has now grown quite substantially (see, e.g., Miller and Rose, 1990; Barry et al., 1996; Dean, 1999; Rose, 1999).[11] As a number of scholars have pointed out Foucault relied on a group of researchers to help him in his endeavors: François Ewald, Pasquale Pasquino, Daniel Defert, Giovanna Procacci, Jacques Donzelot, on governmentality; François Ewald, Catherine Mevel, Éliane Allo, Nathanie Coppinger and Pasquale Pasquino, François Delaporte and Anne-Marie Moulin, on the birth of biopolitics. These researchers working with Foucault in the late 1970s constitute the first generation of governmentality studies scholars and many have gone on to publish significant works too numerous to list here. In the field of education as yet not a great deal has focussed specifically on governmentality.[12]

Gordon (2001: xxiii) indicates three shifts that took place in Foucault's thinking: a shift from a focus on "specialized practices and knowledges of the individual person" "to the exercise of political sovereignty exercised by the state over an entire population"; the study of government as a *practice* informed and enabled by a specific rationality or succession of different rationalities; and the understanding that liberalism, by contrast with socialism, possessed a distinctive concept and rationale for the activity of governing. Liberalism and neoliberalism, then, for Foucault represented distinctive innovations in the history of governmental rationality. In his governmentality studies Foucault focussed on the introduction of economy into the practice of politics and in a turn to the contemporary scene studied two examples: German liberalism during the period 1948–1962, with an emphasis on the Ordoliberalism of the Freiburg School, and American neoliberalism of the Chicago School. Foucault's critical reading of German neoliberalism and the emergence of the "social market" has significance not only for understanding the historical development of an economic constitution and formulation of "social policy" (and the role of education policy within it), but also the development of the European social model, more generally, and the continued relevance for Third Way politics of the social market economy.

Neoliberalism and the Birth of Biopolitics

Naissance de la biopolitique (Foucault, 2004a) consists of 13 lectures delivered by Foucault at the Collège de France (10 January–4 April 1979). It is helpful to see this course in the series of 13 courses he gave from 1970 to 1984. The first five courses reflected his early work on knowledge in the human sciences, concerning punishment, penal and psychiatric institutions: "La Volonté de savoir" (1970–1971), "Théories et Institutions pénales" (1971–1972), "La Société punitive" (1972–1973), "Le Pouvoir psychiatrique" (1973–1974), "Les Anormaux" (1974–1975). The remaining eight courses focussed squarely on governmentality studies, with a clear emphasis also on the problematic (and hermeneutics) of the subject and the relation between subjectivity and truth: "It faut défendre la société" (1975–1976), "Securité, Territoire, Population" (1977–1978), "Naissance de la biopolitique" (1978–1979), "Du gouvernement des vivants" (1979–1980), "Subjectivité et Vérité" (1980–1981), "L'Herméneutique du subjet" (1981–1982), "Le Gouvernement de soi et des autres" (1982–1983), "Le Gouvernement de soi et des autres: le courage de la verite" (1983–1984). Even from this list of courses, it becomes readily apparent that the question of government concerns Foucault for the last decade of his life and that for his governmentality studies, politics was inseparable in its modern forms both from biology—biopower and the government of the living—and truth and subjectivity. It is important to note that these same concerns in one form or another enter into Foucault's formulations in *Naissance de la biopolitique*.[13]

The *Ordoliberalen*[14] comprised a group of jurists and economists in the years 1928–1930 who published in the yearbook *Ordo*. Amongst their numbers were included William Röpke, Walter Eucken, Franz Böhm, Alexander Rüstow, Alfred Müller-Armack, and others. Preaching the slogan that "inequality is equal for all" they devised a social market economy influencing the shaping of West German economic policy as it developed after the war. Foucault refers to these *Ordoliberalen* as the "Freiberg School" who had some affinities (of time and place) with the Frankfurt School but were of a very different political persuasion. While they held that Nazism was a consequence of the absence of liberalism, they did not see liberalism as a doctrine based upon the natural freedom of the individual that will develop by itself of its own volition. In fact, for the Freiberg School the market economy was not an autonomous, or naturally self-regulating entity at all. As a consequence, their conception of the market and of the role of competition, says Foucault, is radically antinaturalistic. Rather than the market being a natural arena which the state must refrain from interfering with, it is rather constituted and kept going by the state's political machine. Similarly, competition is not a natural fact that emerges spontaneously from human social intercourse, as a result of human nature, but must be engineered by the state.

As a consequence of this, the traditional distinction between a sphere of natural liberty and a sphere of government intervention no longer holds, for the market order and competition are engineered by the practices of government. Both the state and the market are on this conception artificial and both presuppose each other. In Foucault's view such a conception means that the principle of laissez-faire, which can be traced back to a distinction between culture (the artificial state) and nature (the self-regulating market), no longer holds. For the *Ordoliberalen*, the history of capitalism is an institutional, nonnatural, history. Capitalism is a particular contingent apparatus by which economic processes and institutional frameworks are articulated. Not only is there no "logic of capital" in this model, but the *Ordoliberalen* held that the dysfunctions of capitalism could only be corrected by political-institutional interventions they saw as contingent historical phenomena. What this means, says Foucault, is that the *Ordoliberalen* support the active creation of the social conditions for an effective competitive market order. Education thus becomes pivotal in this constructivism. Not only must government block and prevent anticompetitive practices, but it must fine-tune and actively promote competition in both the economy and in areas where the market mechanism is traditionally least prone to operate. One policy to this effect was to "universalise the entrepreneurial form" (Lemke, 2001: 195) through the promotion of an enterprise culture, premised, as Foucault put it in a lecture given on 14 February 1979, on "equal inequality for all" (ibid.). The goal here was to increase competitive forms throughout society so that social and work relations in general assume the market form, that is, exhibit competition, obey laws of supply and demand. In the writings of Rüstow, this was called "vital policy" (*Vitalpolitik*) that described policies geared to reconstructing the moral and cultural order to promote and reward entrepreneurial behavior, opposing bureaucratic initiatives that stifle the market mechanism. To achieve such goals, the *Ordoliberalen* also advocated the redefining of law and of juridical institutions so that they could function to correct the market mechanism and discipline nonentrepreneurial behavior within an institutional structure in accordance with, and supported by, the law. In this sense, the *Ordoliberalen* were not simply antinaturalist, but constructivist.

In his analysis of neoliberalism, Foucault also directs his attention to the Chicago School of Human Capital theorists in America, focussing particularly on the works of Gary Becker. These neoliberals also opposed state interventionism when it was bureaucratic and supported it when it fostered and protected economic liberty. For Human Capital theorists the concern was the uncontrolled growth of the bureaucratic apparatus as a threat to the freedom of the individual. Foucault sees the major distinction between the German and U.S. neoliberals existing in the fact that in the U.S. neoliberalism was much less a political crusade as it was in Germany or France, for in the United States the critique was centrally directed against state interventionism and aimed to challenge the growth of the state apparatus. In his lecture of 28 March 1979, Foucault discusses Hayek and von Mises (whom he labels as the "intermediaries of US neoliberalism"), Simons, Schultz, Stigler, and Gary Becker, whom he says is the most radical exponent in the United States. The U.S. neoliberals saw the *Ordoliberalen* as representing the political as being above and outside the market but constantly intervening to correct its bureaucratic dislocations. From their viewpoint, they wanted to extend the market across into the social arena and political arenas, thus collapsing the distinction between the economic, social, and political in what constitutes a marketization of the state.

As Foucault sees HCT (Human Capital Theory), it is concerned with the problem of labor in economic theory. While classical political economy claimed that the production of goods depended upon real estate, capital, and labor, neoliberals held that only real estate and capital are treated appropriately by the classical theory, and that labor needs greater illumination as an active, rather than as a passive, factor in production. In this sense neoliberals concurred with Marx that classical political economy had forgotten labor and thereby they misrepresent the process of production. In order to correct this deficiency, neoliberals theorize the role and importance of labor in terms of a model of human capital. In essence their theory starts with the human individual in terms of a classification of skills, knowledge, and ability. Although, unlike other forms of capital, it cannot be separated from the individual who owns these resources, they nevertheless constitute resources that can be sold in a market. Becker distinguishes two central aspects to such human capital: (1) inborn, physical, and genetic dispositions, and (2) education, nutrition, training, and emotional health. In this model, each person is now an autonomous entrepreneur responsible ontologically for their own selves and their own progress and position. Individuals have full responsibility over their investment decisions and must aim to produce a surplus-value. As Foucault puts it in his 14 March 1979 lecture, noting the educational implication, they are "entrepreneurs of themselves."

Graham Burchell (1996: 23–24) has noted the core distinction between classical and neoliberalism. Whereas for liberalism the basis of government conduct is in terms of "natural, private-interest-motivated conduct of free, market exchanging individuals," for neoliberalism "the rational principle for regulating and limiting governmental activity must be determined by reference to artificially arranged or contrived forms of free, entrepreneurial and competitive conduct of economic-rational individuals." This means that for neoliberal perspectives, the end goals of freedom, choice, consumer sovereignty, competition, and individual initiative, as well as those of compliance and obedience, must be constructions of the state acting now in its positive role through the development of the techniques of auditing, accounting, and management. It is these techniques, as Barry, Osborne, and Rose (1996: 14) put it,

> [that] enable the marketplace for services to be established as "autonomous" from central control. Neo-liberalism, in these terms, involves less a retreat from governmental "intervention" than a re-inscription of the techniques and forms of expertise required for the exercise of government.

Notwithstanding this rather crucial difference between the two forms of liberalism, the common element expresses a distinctive concern. For both classical liberalism as well as neoliberalism, what defines this concern is a common orientation concerning "the limits of government in relation to the market" (Burchell, 1996: 22).

In addition to a common priority concerning the scope of the market, both classical liberalism and neoliberalism share common views concerning the nature of the individual, as rational self-interested subjects. In this perspective the individual is presented as a rational optimiser and the best judge of his/her own interests and needs. Being rational was to follow a systematic program of action underpinned and structured according to rules. The rules were rendered coherent and permissible in relation to the "interests" of the individual.

In summary, then, central to neoliberals such as the *Ordoliberalen* and Public Choice theorists, the state actively constructs the market. Far from existing within a protected and limited space, market relations now extend to cover all forms of voluntary behavior amongst individuals. Rather than absenting itself from interfering in the private or market spheres of society, Foucauldian political economy points out that in the global economic era neoliberalism becomes a new authoritarian discourse of state management and control. Rather than being a form of political bureaucracy, which Weber (1921) saw as the supreme form of modernist rationality, neoliberalism constitutes a new and more advanced technology of control. It is both a substantive political doctrine of control and a self-driving technology of operations. It incorporates both more flexible and more devolved governmental steering mechanisms than does bureaucracy. If, for Weber, bureaucracy constituted large scale organization comprising a hierarchy of offices and lines of control, enabling efficiency, predictability, calculability, and technical control, then neoliberalism, while incorporating these factors, goes beyond them to enable an extension of control in more devolved forms and in more flexible systems. This enables the function of control to be differentiated from the function of operations, or to use Osborne and Gaebler's (1992) metaphor, "steering" from "rowing." It points to a more effective means of social engineering and control than classical bureaucracy, scientific management, or the Fordist assembly line. Its overall rationale is to measure the costs of, and place a monetary or market-value on, all forms of human activity in order to render it controllable. It extends the market mechanism from the economic to the political to the social. Market exchanges now encapsulate all forms of voluntary behavior amongst individuals.[15]

Toward a Possible Foucauldian Politics

If Foucault is critical of neoliberalism as being a new form of superstructural sociology, in many ways highlighting a new operating model of capitalism in a global era, his model of political economy also supports, and has affinities to, a particular approach to economics and politics that can be represented to tie in closely with "regulation school" approaches developed by writers such as Michel Aglietta, Hughes Bertrand, Robert Boyer, Alan Lipietz, and Jacques Mistral. Aglietta has commented directly on Foucault's contribution in his conversations with François Dosse (1997a: 291) where he describes Foucault's importance as being "because he raised questions about institutions and gave answers." Furthermore, as Dosse explains, Aglietta was especially influenced by Foucault's

> concern for micropowers, his shift from the centre to the peripheral, his pluralization of a polymorphous power that corresponded to the regulationists' desire to reach intermediary institutional bodies. Moreover, Foucault had made it possible to take some distance from "the fundamental conception of Marxism" and to understand that this smooth growth curve depended on a system of conciliation and a concentration of interests. Until then, the antagonoism between capatalists and workers was considered irreconcilable. (1997: 291)

The "regulationists" rendered structuralism dynamic and bought microstructures and human beings back into the orbit of the analysis. As well as incorporating much from the tradition of Marxism, they were also influenced by Keynesian economics through the consideration of real demand, and by arguing for a consideration of money as an institution, and work as a relationship rather than a market. Robert Boyer (1986) and Alain Lipietz (1983, 1995) also accepted a "broad church" conception of the regulationist approach, distancing themselves from the more specific variations that also developed (see Dosse, 1997: 290–294).

Central to the regulation school approach was its rejection of the market order as a self-regulating entity, and its "openness to social and historical elements" (ibid.: 292). This presents the future as an always existing constellation of dangers and has enormous implications for ethics and education. As with Foucault there was an appropriation of some features of Marxism (especially Althusserianism), a conception of holism/particularism; an appreciation that the laws governing economic tendencies are historically contingent, and a concern for institutional forms of power as they arose from divergent conflicts or from market processes. The emphasis on historicity meant that there was no recognition of predetermined universal categories or systems such as forces of production, in preference for a recognition of the historical variability of other economic institutions, such as money or markets. What resulted was a reinterpretation of economic phenomena in terms of dynamic schemas as responding to dynamic mechanisms. Individual behaviors and identities were forged out of complex wholes, hence individual behaviors and subjects were viewed in ways that did not embody methodological individualism, enabling a reintroduction of individuals in relation to groups and social categories.

From Governmentality to the Hermeneutics of the Self as Education

The distinctiveness of Foucault's emerging problematic of governmentality, formulated in the years 1978–1979, also developed in a series of subsequent themes as "the government of the living," "subjectivity and truth," and "the government of self and others." These themes were also of relevance to education as a practice of struggle and engagement, and as such help us to understand the differences between a Foucauldian political economy of praxis, and that of Marxism. Of particular significance here are the themes on the problematic (and hermeneutics) of the subject and the relation between subjectivity and truth. These indeed signify a different relation between individual and collective in Foucault's work, compared to Marxism or Hegelianism. Although the individual subject is a product of social conditioning, they are always "free standing" in the sense they are never completely confined or defined by the determining structures from which they derive. Every convergence around culture or conditioning is also characterized by difference on the grounds that experience within space and time is itself individuating. Thus, while each individual is the product of class and culture each also has a situated and dated uniqueness. As Jane Bennett (1996: 660) states, citing Foucault (1989a: 313),

> There is no *escaping* a regime of power, but this does not mean that subjectivation is simply subjection, for there is always the possibility of "practices of liberation, of freedom, as in Antiquity, starting of course from a certain number of rules, styles and conventions that are found in the culture." (Italics in original)

This theorization of the individual within the whole as something that both reflects and yet exceeds it is distinctive of Foucault correction to Marxism, forcefully expressed in theoretical terms in works such as *The Archaeology of Knowledge* (1972).

Another difference from Marxism, springing from the rejection of Hegelian conceptions of a unilinear and progressive history, is Foucault's distain for utopian ideas that aim for the realization of a perfected or harmonious future society. To aim for a specific ordered ideal written "only on paper" as something that could exist harbors dangers associated with both "radical and global" forms of theorising. Foucault (1984b: 46) echoes liberal concerns with utopian engineering when he states that

> we know from experience that the claims to escape from the system of contemporary reality so as to produce the overall programs of another society, of another way of thinking, another culture, another vision of the world, has led only to the return of the most dangerous traditions.

Foucault puts more emphasis on local struggle and resistance on the basis that existing historical discursive systems (such as those concerning liberty, rights, democracy, etc.) that can be seen *already*—in the present horizon—to harbor a "repressed" or "sedimented" utopian vision. Such existing discourses that always operate in local sites can be seen to constitute the complex outcome of struggles in history. And it is in this sense that local struggles can be seen as the basis of revolutionary activity. To the question, "what replaces the system?" Foucault responds, "I think that to imagine another system is to extend our participation in the present system...I would rather oppose actual experiences, than the possibility of a utopia"

(1977c: 230–231). For Foucault, the revolution as Marxism conceived it as a single historical act of violence and transformation fails theoretically to be plausible in a global age because it can only be taken seriously within a unilinear and utopian frame of reference. What must be asked anew is how would such an idea of revolution can be conceived, planned for, and organized in a age that is both global and local. The notion of simultaneous and coordinated action across national contexts is no longer feasible in a global and "virtual" world. Indeed, to envisage a total sudden reconstruction or reordering of society given the dispersed digital character of financial and intelligence networks in advanced industrial nations boggles the mind. Hence, for Foucault, in a world that is both global and local the drive for change must take the form of resistance and struggle in specific sites, utilizing complex technologies and intellectual tools.

In addition to operating in specific sites, Foucault also emphasises the tasks of the "specific intellectual" as "sapping power" rather than "proposing alternative visions" or "telling people what to do." A critical interrogation of power is thus central as the basis of a critical education. As Foucault (1977e: 208) explains to Gilles Deleuze,

> [t]he intellectual's role is no longer to place himself "somewhat ahead and to the side" in order to express the stifled truth of the collectivity; rather, it is to struggle against the forms of power that transform him into its object and instrument in the sphere of "knowledge," "truth," "consciousness," and "discourse." In this sense theory does not express, translate, or serve to apply practice: it is practice. But it is local and regional...and not totalizing. This is a struggle against power, a struggle aimed at revealing and undermining power where it is most invisible and insidious. It is not to "awaken consciousness" that we struggle...but to sap power, to take power; it is an activity conducted alongside those who struggle for power, and not their illumination from a safe distance. A "theory" is the regional system of this struggle.

The impossibility of a discrete and total revolution also suggests the fact that on some issues and for some groups the revolution as Marxists conceived it might be considered in some senses as having already occurred. The theoretical task becomes in identifying those specific aspects or dimensions still not corrected such as climate change, the unjustified profits of many large multinationals, global inequality, the disadvantages of race, class, disability, and gender both within societies and globally. The normative standards constitutive of a good for mankind are already present, in some cases manifest, in some repressed, within the existing horizon based upon what is necessary for both collective and individual survival. This is to say that complex historically generated discourses on such things as rights, equality, democracy, and education for survival already exist and constitute the repositories of knowledge to become the curriculum for education for global citizenship in the future. Such a global polis is a thin, or decentered community, rather than a unified mutuality in Hegel's sense. It is motivated not by a model of the truth, but more pragmatically, by a normative conception of life and survival characterized by a common integrity and flourishing but that may take many different forms. Foucault's fellow Nietzschean thinker, and lifelong inspiration, Georges Bataille, theorized the importance of life as a force that guides ethics and education. For Bataille, like Nietzsche, the struggle for life represents a specifically nonmoral yearning once "God is dead," a goal that stretches out before one, as Bataille (2004: xviii) puts it, "independently of moral goals or of serving God," and yet paradoxically itself becomes interpreted as a moral obligation, imposing an object "that surpasses all others in value" (xvii) and translating as a "demand for definite acts" (xvii). As in a complex global world people are interdependent, individual and collective goals are intertwined: the well-being of one is inseparable from the well-being of all. John Dewey's model of problem solving for survival resonates a similar approach.

It is well known that Marx never theorized the nature of existence in the communist utopia in any detail, and he may well have lacked the tools for such an analysis,[16] but there is reason to believe that he held intuitively to a similar conception to Foucault. Foucault sees difference as manifesting itself, necessarily, within unity, the latter that is always precarious, always changing and never completely predictable. Just for instance, as "Britishness" defines a common attribute between a group of people, so within the group, and simultaneous or identical with it, there are a myriad of differences, pertaining to appearance, gender, age, or what have you. In this sense commonness and difference are co-present features of any phenomena. Within the existing horizon of survival, there are many legitimate yet different ways to live. It is in this sense also that in place of unilinear conceptions of change and causality, Foucault's model of historical materialism is consistent with twentieth-century conceptions of complexity theory. New realities, both physical and social, manifest themselves as emergent phenomena. While there is no necessary equilibrium that produces "happy endings" or "self-regulated markets," within limits we can understand the *affects* that particular combinations, alliances, and choices entail. Understanding possible *affects* of combinations, and alliances, is indeed the task of education and of ethics. To understand education as concerned with a *theory of affects* is of course to draw off Deleuze, as well as Spinoza and Nietzsche.[17] Yet it is an extension that we think (or hope) Foucault might approve.[18] Certainly it extends his thought in a way that brings out the important differences to Marxism, in a way related to how ethics and education would be conceived to have a role in the new era that is both global and national and that confronts a world that still awaits to be transformed.

The different way individual and collective are related in Foucault compared to Marxism indeed suggests a new order of ethics. In his later books, *The Use of Pleasure* and *The Care of the Self*, as Jane Bennett (1996: 655) explains, Foucault considers ethics as both a "code of morality" and in relation to "subjectivation." As a "code morality," ethics relates to justified moral precepts and rules. Christianity comprised one set of these, yet today, the new demands of survival, centering on issues such as climate change and the health of populations, presents a different set of precepts to guide action. In addition, however, such rules will not dictate or define conduct for each individual for every situation. While the tasks and requirements of survival will dictate some general, although historically specific, precepts and "rules," different contingent imperatives at different times and places will also operate, so the individuals' mode of ethical comportment will also reflect decisions, choices, and commitments that only he/she can decide in particular situations. For Foucault, as for poststructuralism in general, every action, like every statement, has a novel aspect, a situated and dated uniqueness, whereby ethical decisions and actions assume great importance for society and for the groups and individuals that comprise it. In this sense, as Bennett (655) explains, "Foucault finds 'code morality' insufficient." She cites Foucault (1985: 28),

> In short, for an action to be "moral," it must not be reducible to an act or a series of acts conforming to a rule, or a value. There is no moral conduct that does not [also] call for forming of oneself as an ethical subject; and no forming of the ethical subject without "modes of subjectivation" and an "ascetics" or "practices of the self" that support them.

One dimension of Foucault's writings on ethics that assumes importance for education today relates to global activism. This is related to the propensity or preparedness of groups and individuals to speak out, take a stand, or to join together to protest. Foucault, in his own life, manifested a constant preparedness and concern with many causes including the rights of prisoners, of lawyers who defended radical groups, and of the poor. Hence, he voiced concern for the rights of those on the high seas, against piracy, where he speaks of "human rights" to "confront governments" that are beyond the limits of nationality.[19] As he puts it (Foucault, 2001a: 474),

> There exists an international citizenship that has its rights and its duties, and that obliges one to speak out against every abuse of power, whoever its author, whoever its victims. After all, we are all members of the community of the governed, and thereby obliged to show mutual solidarity.... It is the duty of this international citizenship to always bring the testimony of people's suffering to the eyes and ears of governments.... The suffering of men must never be the silent residue of policy. It grounds an absolute right to stand up and speak to those who hold power.

Foucault's approach, as extended through concepts such as *parrhésia*, and contestation and resistance, also supports a model of global democracy and the role of grassroots critical social movements, which constitute a bottom-up theory of the democratization of world order and suggests a conception of education as instilling radical global citizenship. Such global protest movements constitute a check on nation states and put them under an obligation to respect human rights, as well as to refrain from exploiting or persecuting individuals within their borders. It creates, also, the conditions where all nations actions are monitored, and rendered accountable, at supranational levels, and where each is encouraged to adjust its own regime to accord with international standards and principles that have been deemed important at this time. Although the Iraq War has set back the cause of constructive international relations by decades, undermining the role of the United Nations, and having a hugely detrimental affect on producing a viable global approach to humanitarian intervention, notwithstanding such setbacks, it is toward a stable and just international order that Foucauldian political economy strives.

Notes

1. This introduction is based upon an editorial written for a special issue called "Marxist Futures" for the journal *Policy Futures in Education* (Peters, 2004).
2. References here and below except where otherwise indicated are from website: http://www.marxists.org/archive/marx/works/1859/critique-pol-economy/index/htm (accessed 12 September 2005), transcribed by Tim Delaney.
3. The full title of Darwin's masterpiece is *On the Origin of Species by Means of Natural Selection, or the Preservation of Favoured Races in the Struggle for Life*. Available at http://www.literature.org/authors/darwin-charles/the-origin-of-species/index.html (accessed 13 September 2005).
4. While Marx did indeed send Darwin a copy of his famous work in second edition to Darwin in 1873 it remained uncut on his shelves and there is no evidence that he read it. It also seems likely that the source of the myth is a confusion over a letter by Darwin published in a Soviet newspaper in 1931 that does not mention Marx but declines the offer of a dedication. The letter concerned Edward Aveling's (Marx's son-in-law) *The Students' Darwin* (Wheen, 1999); see also http://www.evowiki.org/index.php/Darwin_and_Marx (accessed 13 September 2005).
5. See also Bowles and Gintis (2001) "*Schooling in Capitalist America* Revisited" at http://www.umass.edu/preferen/gintis/soced.pdf#search=%22Herbert%20Gintis%2C%20Schooling%20in%20Capitalist%20America%3A%20Educational%20Reform%20and%22. Accessed 13 September 2005.
6. Political economy originally in the Greek had three related meanings: *oikonomia* meant the management of a household or family, *politike* meant pertaining to the state; and *ta oikonomika* or economics meant the art of household management. Thus *political economy* originally meant the management of the household of the state. It was used in this sense both by Jean-Jacques Rousseau and Adam Smith. Rousseau in "A Discourse on Political Economy" in 1755 defines political economy as "the government of the State for the common good." Rousseau says, "The word *economy*, or *oeconomy*, is derived from *oikos, a house*, and *nomos, law*, and meant originally only the wise and legitimate government of the house for the common good of the whole family. The meaning of the term was then extended to the government of that great family, the State. To distinguish these two senses of the word, the latter is called *general* or *political* economy, and the former domestic or particular economy" (http://www.constitution.org/jjr/polecon.htm accessed 14 August 2005).
7. New political economy seeks to combine "the breadth of vision of the classical political economy of the 19th century with the analytical advances of

twentieth-century social science"; to overcome old distinctions and divisions of the past (agency/structure; state/market) to provide an integrated analysis that draws on a range of concepts and methodologies without favoring adherence to one particular school, method, or theoretical approach. New political economy might draw on a range of theory: institutional-organizational approaches in economics; comparative theories of institutional and economic change in economics and economic history; structuration theory and strategic-relational theory in sociology; critical theories in international relations. This is drawn from the editorial by Andrew Gamble, Anthony Payne, Michael Dietrich, Ankie Hoogvelt, and Michael Kenny for the journal *New Political Economy* when it was established in 1996.

8. This paragraph is reformulated from the book by Mark Olssen (2006d), *Michel Foucault: Materialism and Education* (Boulder: Paradigm Publishers).

9. In his Résumé du cours for 1979 (in Foucault, 2004b: 323) Foucault indicates that the method he will adopt is based on Paul Veyne's nominalist history and in this respect he writes:

> Et reprenant un certain nombre de choix de méthode déjà faits, j'ai essayé d'analyser le <<libéralisme>>, non pas une théorie ni comme une idéologie, encore moins, bein entendu, comme une mannière pour la <<société>> de <<se\ représenter>>; mais comme une pratique, c'est-à-dire comme une <<manière de faire>> orientée vers objectifs et se régulant par une réflexion continue. Le libéralisme est à analyser alors comme principe et méthode de rationalisation de l'exercice de gouvernement—rationalisation qui obéit, et c'est là sa spécificité, à la règle interne de l'économie maximale.

10. The Foucault archives have been recently relocated from the IMEC (Institut Mémoires de l'Édition Contemporaine) Paris address (9, rue Bleue, F-75009 Paris) to Abbaye d'Ardenne (14280 Saint Germaine la Blanche-Herbe), e-mail: bibliotheque@imec-archives.com. The course Foucault delivered in 1975–1976 was translated by David Macey as *Society Must Be Defended* and was published in 2003 by Penguin (Foucault, 2003). While courses for 1977–1978, 1978–1979, 1981–1982 (*The Hermenuetics of the Subject*) have been recently published, courses for the years 1979–1980, 1980–1981, 1982–1983, 1983–1984 are still only available from the IMEC Foucault archives as recorded tapes.

11. The governmentality literature has grown up around the journal *Economy and Society*, and includes the work of Cruickshank, Hindess, Hunter, Larner, Minson, O'Malley, Owen, and others, as well as those referred to above, most of who have published in *Economy and Society* (for aims and scope, and table of contents, see http://www.tandf.co.uk/journals/titles/03085147.asp). Accessed 13 September 2005.

12. See Olssen, Codd, and O'Neill (2004: 167–171); and Olssen (2006: 29–30, 108, Chapter 10). Also see "Why Foucault?" (Peters, 2004) where Peters discusses Foucault studies in the English-speaking world by reference to the work of Marshall, Olssen, Ball, Popkewitz and Brennan, Besley, Baker, Middleton, and myself. My work on Foucault's governmentality dates from Peters (1994), with additional work in 1996 (with Marshall), Peters (1996), Peters (1997), and Peters (2001a, 2001b, 2001c). For additional work on Foucault see Peters (2003a and 2003b), Peters (2005a and 2005b). A special issue of *Educational Philosophy and Theory* published a special issue in 2006

entitled "The Learning Society and Governmentality" edited by Masschelein, Bröckling, Simons, and Pongratz.
13. As he writes in his Résumé du cours (in Foucault, 2004b: 323),

> Le thème retenu était doc la <<biopolitique>>: j'entendais par là la manière don't on a essayé, depuis le XVIII siècle, de rationaliser les problèmes posés à la pratique gouvenrement par les phénomènes propres à une ensemble de vivants constitutes en population: santé, hygiene, natalitié, longévité, races...

14. The remainder of this section is reformulated and drawn from Olssen, Codd, and O'Neill (2004).
15. This is the process that Ritzer (2000) describes as the "McDonaldization of Society."
16. The reason usually given, but which in any case would be consistent with our statement here, is that Marx considered theoretical speculation of this sort unscientific, indeed, utopian, because economic and social conditions would themselves change in ways that were unpredictable.
17. See Deleuze's (1988) book called *Spinoza: Practical Philosophy*, where he articulates the theory of affects.
18. Types of affects might include desire, sadness, or joy, as they did for Spinoza. However, the reader should entertain the possibility that Foucault would quite possibly object to this form of theorizing on the ground he eschewed normative theorizing of this sort. In extending Foucault in this way, it is thus in the spirit of the types of educational possibility and the types of normative theory that we are observing might "fit" with his Nietzschean approach. This seems to be a worthwhile way to extend Foucault if the possibilities of a Foucauldian radical political economy are to be developed. It should also be noted that while we are comparing Foucault here with Marxism, any affinities between the two systems are minimal. Foucault's system is not a form of Marxism, and indeed, he saw serious problems with Marxism. It is in many senses closerto Liberalism than to Marxism. For further writing on Foucault and Marxism, see Olssen 2006a, 2006b, 2006c, 2006d, 2006e and 2006f.
19. The occasion for the statement cited here, published in *Libération* in June 1984, was the announcement in Geneva of the creation of an international committee against piracy.

References

Barry, A., T. Osborne, and N. Rose. (eds.). (1996) *Foucault and Political Reason: Liberalism, Neo-liberalism and Rationalities of Government* (London: UCL Press).

Bataille, G. (2004) *On Nietzsche* (London: Continuum).

Becker, G. (1964) *Human Capital; A Theoretical and Empirical Analysis, with Special Reference to Education* (New York: National Bureau of Economic Research).

Bennett, J. (1996) "How Is It, Then, That We Still Remain Barbarians?": Foucault, Schiller, and the Aestheticization of Ethics. *Political Theory*, 24 (4): 653–672.

Bowles, S. and H. Gintis. (1976) *Schooling in Capitalist America: Educational Reform and the Contradictions of Economic Life* (New York: Basic Books).

———. (2001). Schooling in Capitalist America Revisited. Available at http://www.umass.edu/preferen/gintis/soced.pdf#search=%22Herbert%20Gintis%2C%20Schooling%20in%20Capitalist%20America%3A%20Educational%20Reform%20and%22.

Boyer, R. (1986). *La Théorie de la régulastion: une analyse critique* (Paris: La Découverte).
Brier, S. (1999) In: Roundtable on the Future of the Left. Transcript in Socialism and Democracy, 25 (Spring/Summer). Available at www.brechtforum.org/highlights/roundtable%20on%20%the%20future%20of%20the%20 left.htm.
Broyer, S. (1996) The Social Market Economy: Birth of An Economic Style. Discussion paper FS I 96–318, Social Science Research Center, Berlin.
Buchanan, J. (1991) *Constitutional Economics* (Oxford, UK; Cambridge, MA: Blackwell).
Burchell, G. (1991) Peculiar Interests: Civil Society and Governing "The System of Natural Liberty." In G. Burchell, C. Gordon, and P. Miller (eds.) *The Foucault Effect: Studies in Governmentality*, 119–150 (Chicago: University of Chicago Press).
———. (1996) Liberal Government and Techniques of the Self. In A. Barry, T. Osborne, and N. Rose (eds.) *Foucault and Political Reason*, 19–36 (Chicago: University of Chicago Press).
Burchell, G., C. Gordon, and P. Miller. (eds.). (1991) *The Foucault Effect: Studies in Governmentality* (Chicago: University of Chicago Press and Harvester Press).
Callinicos, Alex (2003) *An Anti-capitalist Manifesto* (Cambridge, UK: Polity Press).
Carver, T. (1998) *The Postmodern Marx* (Manchester: Manchester University Press).
Curtis, B. (2002) Foucault on Governmentality and Population: The Impossible Discovery. *Canadian Journal of Sociology*, 27 (4) (Fall): 505–535.
Day, R. B. (2002) History, Reason and Hope: A Comparative Study of Kant, Hayek and Habermas. *Humanitas*, 15 (2): 4–24.
Dean, M. (1999) *Governmentality: Power and Rule in Modern Society* (London: Sage).
Deleuze, G. (1988) *Spinoza: Practical Philosophy* (San Francisco: City Lights Books).
Derrida, J. (1994) *Specters of Marx: The State of the Debt, the Work of Mourning, and the New International*, trans. Peggy Kamuf, intro. Bernd Magnus and Stephen Cullenberg (New York; London: Routledge).
Desai, M. (2002) *Marx's Revenge: The Resurgence of Capitalism and the Death of Statist Socialism* (London: Verso).
Dosse, F. (1997) *History of Structuralism Volume 2: The Sign Sets, 1967—Present*, trans. Deborah Glassman (Minneapolis; London: University of Minnesota Press).
Du Gay, P., S. Hall, L. Janes, H. Mackay, and K. Negus (1997) *Doing Cultural Studies: The Story of the Sony Walkman* (London: Sage).
Featherstone, M. (1991) *Consumer Culture & Postmodernism* (London: Sage).
Ferguson, A. (1996) *An Essay on the History of Civil Society 1767*, ed. with intro. Duncan Forbes (Edinburgh: Edinburgh University Press).
Foucault, M. (1972) *The Archaeology of Knowledge*, trans. A. Sheridan (London: Tavistock).
———. (1977a) *Discipline and Punish: The Birth of the Prison*, trans. Alan Sheridan, 195–228 (London: Penguin).
———. (1977b) Nietzsche, Genealogy, History. In D. Buchard (ed.), D. Bouchard and S. Simon (trans.), *Language, Counter-memory, Practice: Selected Essays and Interviews*, 139–164 (Ithaca: Cornell University Press).
———. (1977c) Revolutionary Action: "Unitl Now." In D. Bouchard (ed.), D. Bouchard and S. Simon (trans.), *Language, Counter-memory, Practice: Selected Essays and Interviews*, 218–233 (Ithaca: Cornell University Press).

Foucault, M. (1977d) A Preface to Transgression. In D. Bouchard (ed.), D. Bouchard and S. Simon (trans.) *Language, Counter-memory, Practice: Selected Essays and Interviews*, 139–164 (Ithaca: Cornell University Press).

———. (1977e) Intellectuals and Power. In D. Bouchard (ed.), D. Bouchard and S. Simon (trans.) *Language, Counter-memory, Practice: Selected Essays and Interviews*, 139–164 (Ithaca: Cornell University Press).

———. (1978) Politics and the Study of Discourse, trans. C. Gordon. *Ideology and Consciousness*, 3 (Spring): 7–26.

———. (1980a) Prison Talk, trans. C. Gordon. In C. Gordon (ed.) *Power/Knowledge: Selected Interviews and Other Writings, 1972–1977*, 37–54 (Brighton: Harvester Press).

———. (1980b) Truth and Power. In C. Gordon (ed.) *Power/Knowledge: Selected Interviews and Other Writings 1972–1977*, 109–133 (Brighton: Harvester Press).

———. (1982) The Subject and Power. In L. D. Hubert and P. Rabinow (ed.) *Michel Foucault: Beyond Structuralism and Hermeneutics*, 208–226 (Chicago: University of Chicago Press).

———. (1984a) Politics and Ethics: An Interview, trans. C. Porter. In P. Rabinow (ed.) *The Foucault Reader*, 373–380 (New York: Pantheon).

———. (1984b) What Is Enlightenment? *The Foucault Reader*, ed. Paul Rabinow, 32–50 (New York: Pantheon).

———. (1985) *The Use of Pleasure: The History of Sexuality, Volume 2*, trans. Robert Hurley, 29–30 (New York: Pantheon).

———. (1986) Kant on Enlightenment and Revolution, trans. Colin Gordon. *Economy and Society*, 15 (1): 88–96.

———. (1989a) An Aesthetics of Existence. In *Foucault Live*, trans. John Johnston and ed. Sylverer Lotringer (New York: Semiotext[e]).

———. (1989b) *Résumé des cours 1970–1982* (Paris: Julliard).

———. (1997) The Ethics of the Concern for the Self as a Practice of Freedom. In Paul Rabinow (ed.) *and* Robert J. Hurley (trans.) *Ethics: Subjectivity and Truth. Essential Works of Michel Foucault, 1954–1984, Vol. 1*, 281–301 (London: Penguin).

———. (1998) Structuralism and Poststructuralism. İn J. D. Faubion (ed.) *Aesthetics, Method, Epistemology, Vol. 2, Power, The Essential Works of Foucault, 1954–1984*, 433–458 (New York: New Press).

———. (2001a) Confronting Governments: Human Rights. In J. D. Faubion (ed.), R. Hurley and others (trans.) *Power, The Essential Works 1954–1984, Vol. 3*, 474–475 (Allen Lane: Penguin Press).

———. (2001b) Interview with Michel Foucault. In J. D. Faubion (ed.), R. Hurley and others (trans.) *Michel Foucault: Power, Vol. 3*, 239–297 (Allen Lane: Penguin Press).

———. (2001c) Governmentality. In J. D. Faubion (ed.), R. Hurley and others (trans.) *Power: Michel Foucault, The Essential Works 1954–1984, Vol. 3*, 201–222 (Allen Lane: Penguin Press).

———. (2001d) The Political Technology of Individuals. In J. D. Faubion (ed.), R. Hurley and others (trans.) *Power, The Essential Works of Foucault, Vol. 3*, 403–407 (Allen Lane: Penguin Press).

———. (2004a) *Naissance de la biopolitique: Cours au collège de France (1978–1979)*, Édition établie sous la direction de Francois Ewald et Alessandro Fontana, par Michel Senellart (Paris: Éditions Gallimand et des Éditions du Seuill).

———. (2004b) *Sécurité, Territoire, Population: Cours au collège de France (1977–1978)*, Édition établie sous la direction de Francois Ewald et Alessandro Fontana, par Michel Senellart (Paris: Éditions Gallimand et des Éditions du Seuill).

Giddens, A. (1994) *Beyond Left and Right: The Future of Radical Politics* (Cambridge, UK: Polity Press).
Gordon, C. (2001) "Introduction." In J. D. Faubion (ed.), R. Hurley and others (trans.) *Power: Michel Foucault the Essential Works, Vol. 3*, xi–xli (Allen Lane: Penguin Press).
Gray, J. N. (1982) *F. A. Hayek and the Rebirth of Classical Liberalism, Literature of Liberty*, 5 (4) (Winter). Available at http://www.econlib.org/library/Essays/LtrLbrty/gryHRC1.html.
Harvey, D. (2006) *A Brief History of Neoliberalism* (Oxford: Oxford University Press).
Hayek, F. A. (1960) *The Constitution of Liberty* (Chicago: University of Chicago Press).
Jameson, F. (1991) *Postmodernism, or the Cultural Logic of Late Capitalism* (London: Verso).
Joerges, C. and F. Rödl (2004) "Social Market Economy" as Europe's Social Model? European University Institute (Florence) Working paper LAW No. 2004/8 at www.iut.it.
Lemke, T. (2001) "The Birth of Biopolitics": Michel Foucault's Lectures at the College de France on Neo-liberal Governmentality. *Economy and Society*, 30 (2): 190–207.
Lipietz, A. (1983) *Le monde enchanté: De la valeur à l'envol inflationniste* (Paris: La Découverte).
Marx, K. (1971) *A Contribution to the Critique of Political Economy*, intro. Maurice Dobb (London: Lawrence and Wishart).
Marx, K. and F. Engels (1965) *Selected Correspondence* (Moscow: Foreign Languages Publishing House).
Miller, D. (1995) *Acknowledging Consumption, a Review of New Studies* (London; New York: Routledge).
———. (1997) Consumption and Its Consequences. In H. Makay (ed.) *Consumption and Everyday Life*, 13–54 (London: Sage).
Miller, P. and N. Rose. (1990) Governing Economic Life. *Economy and Society*, 19 (1 and 3): 1–31 and 49–64.
Olssen, M. (2003) Structuralism, Post-structuralism, Neo-liberalism: Assessing Foucault's Legacy. *Journal of Education Policy*, 18 (2) (May): 189–202.
———. (2004a) *Education Policy: Globalisation, Citizenship, Democracy* (London: Sage).
———. (2004b) The School as the Microscope of Conduct: On Doing Foucauldian Research in Education. In J. Marshall (ed.) *Poststructuralism and Education*, 57–84 (Dordrecht: Kluwer Academic Publishers).
———. (2005) Foucault and Marx: Re-writing the History of Historical Materialism. *Policy Futures in Education*, 2 (3): 453–480.
———. (2006a) Foucault, Educational Research and the Issue of Autonomy. In P. Smeyers and M. A. Peters (eds.) *Postfoundationalist Themes in the Philosophy of Education* (Oxford: Blackwell).
———. (2006b) Foucault and the Imperatives of Education: Critique and Self-creation in a Non-foundational World. *Studies in Philosophy and Education*, 25 (3): 191–217.
———. (2006c) Invoking Democracy: Foucault's Conception (with Insights from Hobbbes). In M. Peters and T. Belsey (eds.) *Why Foucault*, 205–226 (New York: Peter Lang).
———. (2006d) *Michel Foucault: Materialism and Education* (Boulder: Paradigm Publishers).

Olssen, M. (2006e) Totalitarianism and the Repressed Utopia of the Present. In M. Peters and J. F. Moir (eds.) *Edutopias: New Utopian Thinking in Education*, 97–122 (Dordtrecht: Sense Publishers).

———. (2006f) Understanding the Mechanisms of Neoliberal Control: Lifelong Learning, Flexibility and Knowledge Capitalism. *International Journal of Lifelong Education*, 25 (3) (May–June): 213–230.

Osborne, D and T. Gaebler (1992) *Reiventing Government: How the Entrepreneurial Spirit Is Transforming the Public Sector, from Schoolhouse to Statehouse, City Hall to the Pentagon* (Reading, MA: Addison-Wesley).

Peters, M. (1994) Governmentalidade Neoliberal e Educacao. In T. Tadeu da Silva (ed.) *O Sujeito Educacao, Estudos Foucaulianos* (Rio de Janeiro: Editora Vozes).

———. (1996) *Poststructuralism, Politics and Education* (Westport, CT; London: Bergin and Garvey).

———. (1997) Neoliberalism, Welfare Dependency and the Moral Construction of Poverty in New Zealand. *New Zealand Journal of Sociology*, 12 (1): 1–34.

———. (2001a) *Poststructuralism, Marxism, and Neoliberalism: Between Theory and Politics* (Lanham; Oxford: Rowman & Littlefield).

———. (2001b) Education, Enterprise Culture and the Entrepreneurial Self: A Foucauldian Perspective. *Journal of Educational Enquiry*, 2 (2): 1–25.

———. (2001c) Foucault and Governmentality: Understanding the Neoliberal Paradigm of Education Policy. *The School Field*, 12 (5/6): 59–80.

———. (2005) Citizen-Consumers, Social Markets and the Reform of Public Services. *Policy Futures in Education*, 2 (3–4): 621–632.

Peters, M. and J. Marshall. (1996) *Individualism and Community: Education and Social Policy in the Postmodern Condition* (London: Falmer Press).

———. (2004) Editorial. Marxist Futures: Knowledge Socialism and the Academy, *PFIE*, 2 (3 and 4): 435–438.

Pignatelli, F. (1993) Dangers, Possibilities: Ethico-Political Choices in the Work of Michel Foucault. Available at http://www.ed.uiuc.edu/EPS/PES-Yearbook/93_docs/PIGNATEL.HTM.

Poster, M. (1984) *Foucault, Marxism, History: Mode of Production vs Mode of Information* (Cambridge, UK: Polity Press).

Ritzer, G. (2000) *The McDonaldization of Society* (Thousand Oaks, CA: Pine Forge Press).

Rose, N. (1999) *Powers of Liberty* (Cambridge: Cambridge University Press).

Schatzki, T., K. Knorr Cetiona, and E. Von Savigny. (eds.). (2001) *The Practice Turn in Contemporary Theory* (London and New York: Routledge).

Schrift, A. D. (1995) *Nietzsche's French Legacy: A Genealogy of Poststructuralism*. (New York, London: Routledge).

Thompson, K. (1986) *Beliefs and Ideologies* (Chichester: Ellis Harwood).

Vanberg, V. (2004) The Freiburg School: Walter Eucken and Ordoliberalism. Freiburg Discussion Papers on Constitutional Economics. Available at http://opus.zbw-kiel.de/volltexte/2004/2324/pdf/04_11bw.pdf.

Weber, M. (1921) *Economy and Society* (Totowa, NJ: Bedminster).

Wheen, F. (1999) *Karl Marx* (London: Forth Estate).

Williams, R. (1980) *Problems in Materialism and Culture* (London: Verso).
Willis, P. (1977) *Learning to Labour. How Working Class Kids Get Working Class Jobs* (Farnborough, Hants: Saxon House).
Witt, U. (2002) Germany's "Social Market Economy": Between Social Ethos and Rent Seeking. *The Independent Review*, 4 (3): 365–375.

Part IV

Politics, Divisions, and Struggle in Teaching, Learning, and Education Policy

Chapter 9

The Feminist Standpoint and the Trouble with "Informal Learning": A Way Forward for Marxist-Feminist Educational Research

Rachel Gorman

Introduction

The purpose of this chapter is (1) to argue for an educational research method that is both dialectical materialist and feminist; (2) to critique the way the current notion of informal learning both individualizes and depoliticizes the real lives and activities of learners; and (3) to argue that radical educators must integrate an understanding of *oppression*, *exploitation*, and *political consciousness* into our definitions of learning and education. I will make my arguments by presenting two case studies of informal learning. The first involves Kurdish women in diaspora. The second involves disability rights activists. Both case studies are located in the Greater Toronto Area (Canada).

Taking a Feminist Standpoint on Learning

Informal learning, conceptualized in contrast to formal (school-based) and nonformal (organized, noncredit) learning, has become an important topic in adult education literature. A central theme in the literature on informal learning is the rationale for, and the practice of, recording previously unrecognized learning experiences. Adult education theorists describe two overlapping purposes for researching informal learning: first, to identify more efficient ways for employees to learn at work and for the employer to realize the benefits of this learning (Watkins and Marsick, 1992), and second, to value the learning experiences of socially marginalized groups. This second project has gained momentum in the past decade, with the documentation of informal learning leading to educational credentials, as seen in the widespread development of *prior learning assessment* programs in further and higher education. Some theorists have also hoped this documentation would provide the rationale for increased workplace democracy and participation in decision making (Livingstone, 1997); and the recognition of the educational value of social movement activities (Foley, 1999). There are, however, serious problems with using the concept of informal learning to provide a way forward for marginalized groups. Theorizing "the learner" as an autonomous, competitive individual will not address the ways gender, race, and ability oppression play out in people's lives. It has also been noted that increasing the marketability of marginalized workers by emphasizing their learning achievements will not lead to a decrease the exploitation and oppression that these workers experience (Tobias, 1999).

The processes of educational research—the political economy of it, and our aims and intentions as educational researchers—are the key to understanding the emergence of new concepts and rubrics in adult education theory, such as "informal learning." As a feminist, prolabor, disability rights activist working in educational research from 1998 to 2003, I was in the position to participate in, and reflect on the pursuit of the goals of valuing and making explicit the educational goals and practices of marginalized communities. As I participated in two concurrent projects at the same large research institution, I experienced the tensions and difficulties, on the one hand, of being one of very few women researchers working on a project on "workers" (always conceptualized as white and male until proven otherwise), and on the other hand, participating in an informal group of feminist researchers who were trying to bring feminist, qualitative research methods to the task of intervening in this rapidly developing male theory of informal learning.

What follows is an attempt to document and reflect on the feminist research process and its potential for educational researchers interested in social justice. I argue that through conscious reflection, it is possible to significantly deepen our collective understanding of the purpose and method of feminist research. Through continually applying a dialectical materialist framework of analysis, we can expose contradictions in our work as educational researchers. A dialectical materialist approach to feminist research methodologies will allow us to uncover and reconsider some of our assumptions about the purpose of collecting interviews and testimony.

By reviewing two case studies from that time, I want to argue that as feminist researchers we need to become more critical in our process of deciding what counts as data—not by retreating further into faux-scientific practices of data coding and computer processing of our interview transcripts, but by analyzing what women are saying in a relational and highly contextualized way. I believe that close attention to our research processes, including the politics of the funding bodies, our own intellectual purposes and political desires, and the inseparable connection between research methodology and theoretical framework will help us to progress beyond our sometimes fuzzy attempts to make the feminist "process of knowing" or "process of knowledge creation" more explicit.

In order to facilitate a reflection on the feminist research process, I will attempt to document the process as I unfold the content of the study. The emerging understanding of informal learning as a concept and a process includes a critique of the ways that the literature on learning is classed, raced, gendered, and abled; and a critique of the ways that the literature assumes social exclusion and underemployment are a result of a lack of learning credentials. Many educational researchers focusing on marginalized groups are deeply committed to furthering equality and social justice. This makes the dearth of Marxist analysis in the literature all the more glaring.

Learning Survival, Resistance, and Struggle

In this section I will try to unfold a feminist critique of learning theory by describing the process of researching the learning practices of Kurdish women in diaspora.[1] Through the process of developing my critique, I hoped to gain deeper understanding of how women learn in social and political communities in a world where colonial, sexual, and racial violence are rampant. I was interested in developing a theory of how women both *learn to engage in*, and ultimately *learn through*, collective struggles against oppression.

This section draws on the results of a study of Kurdish women living in Canada, a community of women with a particular relationship to oppression and collective struggle. First, Kurdish women are from what has been referred to as a zone of genocide.[2] Second, once in exile, these women are subject to the experience of belonging to an immigrant community and the structural, ideological, and racial oppression that this entails. Third, as a result of their global/social location, Kurdish women are a highly politicized group, both in terms of their own participation in struggle, and in terms of the reification of "Kurdish women" as an ideological object by immigration policymakers, by U.S. foreign policymakers, by liberal feminists in the West, and by patriarchal Kurdish-nationalist groups.

In the process of developing this critique, I looked at a set of five interviews with Kurdish women living in the Toronto area. These interviews were completed in the summer of 1999.[3] The women had immigrated to Toronto in the past decade, and were recruited through the connections that the principal investigator, Dr. Shahrzad Mojab, had in the Kurdish community. The interviews were conducted in Kurdish (Sorani) or Farsi, and have been transcribed and simultaneously translated into English. The interviews are open-ended questionnaires, however, because Dr. Mojab has been actively engaged in Kurdish communities in diaspora and in the region for the past 25 years, and because she was interested in the women's analyses, the interviews are also to some extent dialogues. Another important feature of the interviews is that they also serve as testimony, both according to the women themselves, and in the sense that they recount stories of war and genocide that have been repressed and dismissed in the international media. The women were asked about the following topics in the interviews:

- experiences of education, formal and informal;
- family experiences, marriage;
- experiences in or of Kurdish political parties;
- experiences in or of armed struggle;
- the process of leaving Kurdistan and arriving in Toronto;
- adjusting to life in Toronto;
- experiences of being engaged in, or being in proximity to armed conflict;
- participation in community and political life in Kurdistan and Toronto;
- work situations in Kurdistan and Toronto

I chose Dorothy Smith's feminist standpoint methodology to proceed with this case study for three reasons. I originally adopted it because I was trying to uncover what was happening in women's lives that contradicted the literature on learning. Smith's standpoint method "begin[s] in people's experience" and "deals with what people do in the concrete settings of their lives, and how they talk about it" (1997: 128). Second, I found her notion that women's stories can be examined for relations of ruling to be extremely helpful. From this I was able to approach women's stories simultaneously as testimony, and as evidence of power relations. Third, her notion of a "trajectory of consciousness" provides a tool to help us deconstruct the swirl of ideologies surrounding Kurdish women from orientalism, to nationalism, to liberal feminism and postmodernism.

The region of Kurdistan is forcibly divided between four states, Iran, Iraq, Syria, and Turkey. Each of these states has engaged in various forms of ethnocide and genocide against their Kurdish populations. During the time of these interviews an autonomous Kurdish region existed within Iraq. It was governed by a power-sharing administration between two parties that had previously been engaged in an armed struggle with each other. To a large extent U.S. foreign policy interests have dictated the course of the struggle for Kurdish national independence. Prior to its invasion of Iraq in 2003, the U.S. supported the autonomous Kurdish region in Iraq because it was strategic for waging war against Iraq. Just over the border the Turkish government has been engaged in the brutal suppression, relocation, and murder of Kurds. However, based on its strategic relationship with Turkey, the United States condemns Kurdish organizations in Turkey as terrorist.[4]

One final contextualizing element I will mention is the language barrier among Kurds. First, there are two main dialects, Kurmançi (which has a related dialect called Zaza) and Sorani.[5] Second, due to a history of state-enforced programs of linguicide, there are Kurds who only speak the dominant language of the state that controls the part of Kurdistan they live in (that is Turkish, Persian, or Arabic, respectively). Third, Kurdish communities in diaspora must negotiate the dominant languages of their new locations, which would commonly include Dutch, English, French, German, Norweigian, and Swedish.[6]

These contexts create very different dynamics for both nationalist and feminist organizing in different parts of the region and in the diaspora. For example, Iran is ruled by a theocratic regime with institutionalized misogyny. In contrast women in Turkey are free to organize as Turkish women. However, anyone (Kurd or non-Kurd) who protests against the oppression of the Kurdish people, or who even discusses the existence of Kurds as a social group, risks imprisonment and torture. The situation is different again in the diaspora. Kurdish women have organized in feminist groups that specifically deal with the situation of women in Kurdistan.[7] When they try to link the feminist struggle to the struggle against national oppression, however, they are censured by the patriarchal hierarchies of Kurdish organizations.

I have raised all of this both to provide information about the political context in which Kurdish women have lived and struggled, and to stress that it is necessary to avoid treating the transcripts as *individual* stories of learning if the goal is to develop a critique of learning theory. If I consider each of the interview transcripts as isolated life stories of individual women, I am likely to find support for an individualized model of learning. I will take some excerpts from the transcripts[8] in order to exemplify my point. First, this quote in transcript #1 supports the claim that trauma impedes learning:

> I had depression as well. I was very sleepy all the time, I often fell asleep in the class.

Second, this quote in transcript #2 supports the claim that formal education leads to greater freedom and opportunity for women:

> Being a teacher was a good position for a woman in Iran. Whenever I used to see my aunt who was a teacher, I always dreamed of being a teacher.

Third, this quote in transcript #2 supports the idea that gaining more skills and finding a job is a way to "give back" to society:

> Everybody is so nice to you here. I was hospitalized here once and they were so nice to me. When I compare this with the way I was treated…I have no other choice but to be thankful to Canada. I am a burden now, I only worked five years since I have been in Canada…

and furthermore, that you cannot contribute without furthering your learning credentials:

> I only wished that I was younger so I could continue my education in order to be more helpful and help people. I have written poems about this. The reason I could not get my grade XII diploma was because of continuous harassment. My dream was to finish university and to be a teacher, and I am

working towards that goal. Wherever I went they required ECE [Early Childhood Education Certificate]. So I decided to get that after getting my degree.

These interpretations of the transcript data would be very much in keeping with qualitative research-based approach adult education literature that I was used to seeing in adult education conference proceedings and journals during the time I was engaged in this research.

In order to propose a different understanding of these transcripts, I would like to draw on the analysis developed in the feminist study group mentioned earlier. Through an analysis of the different cases we were working on, and through feminist dialogue, we were able to develop a framework for looking at women's learning in their communities. The study group's case studies included women of the Gozitan minority in Malta;[9] Spanish-speaking women in Toronto;[10] women union activists;[11] violence against women in the Azores;[12] Kurdish women in Toronto;[13] and Disability rights activists, which will be the subject or the next section. Our case studies involved the following issues: race and ethnicity; immigration and settlement; domestic violence and trauma; class and work experience; social exclusion; and rural versus urban contexts.

Through comparing our cases, we developed a hypothesis of learning. We posited three types of nonschool-based learning: survival, resistance, and struggle. *Survival* learning is how individuals develop strategies to cope in a world that has been constructed to exclude them. Individuals may figure out a coping strategy on their own, or may learn about it from other members of their community or social group. *Resistance* learning is how an individual or group develops strategies to resist the ways in which the world has been constructed to exclude them. *Struggle* learning is how a group develops an understanding of how their oppression has been constructed and reconstructed, and how that group develops counterarguments and strategies to dismantle the oppression. When we used these concepts in our discussions, resistance learning referred to the ways women meet and oppose violence as we encounter it in our daily lives, while "collective struggle" referred to the ways that women learn to identify and change the relations of oppression.[14]

Now I would like to once more present the quote that supported the claim that trauma impedes learning:

> I had depression as well. I was very sleepy all the time, I often fell asleep in the class. But I decided to continue. I even helped my sister in law and took her to the college to take the exam, I cheated and she was accepted, but at a lower level. This exam was in order to give us 26 credits towards our diploma. We only needed four more credits. We have now graduated.

Now we can see an individual story *and* a collective story. This woman is suffering from the effects of trauma, which is impeding her individual learning, and on a community level she has learned from her own experiences of trying to cope in a world that has been constructed to exclude her, and is using that knowledge to contribute to the survival of another. Now this quote is about both individual trauma and about survival learning.

These following excerpts are from transcript #1 and are spoken by a woman who has survived an extremely abusive marriage and was able to leave her husband. She recalls three different situations in which different relatives have asked her for help or advice about their situations. This women recommends different strategies for avoiding or surviving domestic violence, based on the options available to the family member she is advising.

> [My ex-husband] even hit the boys and they would complain to me and I would say that is OK, he is your father and you have to obey him...
> Last year my brother called and asked me to talk to his daughter who is in a bad marriage situation, and advise her to leave him. She said exactly what I used to say—I told her to leave him and learn from my lesson...
> Last year, my son's stepsister, who likes me very much, called me from Iran and complained that her father [my ex-husband] will not let her go to university. I told her don't listen to him. I talked to her mother as well and told her even if you sell your gold, do it and pay for the tuition, but let her attend university. That's what they've done. Why should she stay home and not be independent? When women are not literate, they have to obey their husband, but if they work, they will be more independent.

It is evident from these excerpts that this woman has taken what she learned from surviving a few years with an extremely abusive man, and she is using that experience to offer advice about surviving and/or resisting oppression (which is patriarchy in this case) at precisely the point where they experience it. That is, the advice is about how to get hit less, how to stop hitting that is happening, or how to avoid a situation in which you will get hit in the future. I would classify this as *survival* or *resistance* learning, but not *struggle* learning. Struggle learning around patriarchy might involve, for example, a dialogue about how to form a feminist collective; or how to lobby to change laws around violence against women; or how to educate the community about the issue.

Transcript #2 offers some hints about Kurdish resistance and nationalist struggle. The first is an incident that occurred when the speaker was a young woman in teachers' college.

> One of the girls reported to the headmaster that we were reading forbidden [political] books...The day after that a representative of the Ministry of Education, Savak, and the police came to the school...they deprived us from taking the final exam. I was one of them, but the rest of the girls in our class supported us by refusing to take the exam as well...When everybody refused to take the exam, they cancelled the exam and postponed it for a day. Then they let us take part in the exam the next day and we all passed that year.

The young women were not part of any political organization, and had not consciously chosen to engage in a political struggle, but when they experienced repression, they acted collectively to resist the specific repression (which was being forbidden to take the exam). The next excerpt is an incident that occurred a few years later. The speaker has been working as a teacher for a few years, and has been identified by the Ministry of Education as an activist and has been repeatedly harassed and threatened.

> In 1357 [1978], the prison experience was a collective experience. It was the burial ceremony of Aziz Yousefi. In that ceremony 200 were arrested in different places. After the burial, I was called in by the principal of the school. There was someone from the ministry and they blind folded me and took me in, I was in the prison for a month and a half. I was on hunger strike for eight days. I was the only woman, but the men did not support me. My crime was that during the ceremony I recited a poem called Dayke Nishteman ["The Motherland" by Ali Hassanyani].

This story is about struggle. Her act of reciting the poem was not a direct response to a specific repression she was experiencing, rather it was a political act that spoke about the collective oppression of Kurdish people. It also raises an interesting question about the role of gender in the nationalist struggle—questions that cannot be answered without much more contextualizing information. For example, why were the men in prison not supporting her hunger strike?

It will be difficult to move forward with this theory about learning for survival, resistance, and struggle without widening the scope. One possibility for doing this is to compare women's political demands in different locations and contexts. For example, at a conference in Paris, women from Turkey raised a vigoros critique of the masculinist politics of the political parties from the autonomous region. The feminist organizers living in Europe made their critique with much less force, probably because they want to have a working relationship with the male-run Kurdish Institute in Paris. Finally, the organized women representing the Kurdish regional government were apologists for the patriarchy in the Kurdish political parties.

It is very important to remember that the woman speaking in the excerpt above is not describing these incidents as learning, nor is she recounting this story in order to trace the development of her political consciousness.[15] I am not advocating that we should call the engagement in political struggle "learning." I am suggesting that when learning theorists talk about "learning from experience" and "tacit learning" they are not talking about experiences of collective struggle; and when they talk about the content of learning, they are not talking about political consciousness. I am suggesting that this is the direction our research must take in order to counter the idea that human learning is for the purpose of competing in the job market.

For educational researchers there is a useful application of this theoretical distinction between *resistance* that is about meeting and opposing oppression in specific places that it affects us, and *struggle* that involves identifying, grasping, and changing the conditions that produce the oppression we are experiencing, for developing a strategy to overcome a specific issue that affects us individually does not necessarily solve the problem. Yet the policy proposal outcomes of educational research are so often just that, individualized solutions that, if adopted in aggregate may *increase* the exploitation that members of the marginalized group are experiencing. Recall, for example, that the woman in transcript #1 believes that becoming a teacher leads to greater freedom and opportunity. Meanwhile, the speaker in transcript #2 testifies that the Ministry of Education is a central force in calling for and carrying out her harassment, imprisonment, and torture. It is clear that while being a teacher may be a useful survival strategy for preventing domestic violence, it does not help women avoid state violence.

Informal Learning Research and the Individualizing of Political Consciousness

In this section I will argue that the concept of informal learning does not provide an emancipatory way forward for adult education theory and practice. Rather, it individualizes collective political movements, bolsters the cult of credentials, and further excludes highly marginalized groups.[16] As in the section above, my critique emerges out of my own experiences as an educational researcher—in this case as one of the few women involved in a large-scale research project focused on documenting informal learning of workers.

In the process of developing this critique, I will draw on the transcripts of five interviews with artist/activists in the disability rights movement in Toronto.[17] I conducted these interviews in 1999 as a kind of pilot project to try and incorporate some disability-specific data into a multisite, multiyear study on the informal learning practices of unionized workers. Soon after conducting this pilot study, I left the informal learning project, and reframed my research with disability rights activists as an inquiry into class consciousness in the disability rights movement.[18]

I had been hired to work on the informal learning project as a research assistant to interview activists in a local (branch) of the Ontario Public Service Employees Union (OPSEU), my own former union. The majority of the local members were women, as was the case in my own local branch. However, from my own activist experience I knew that a majority of women did not guarantee a feminist analysis in the local's politics. As I struggled with the interview answers to the same set of questions the (majority) men were answering in the other sites of the study, I realized the questions being asked in the study presupposed an individualist, masculinist, and depoliticized approach to learning.

In the hopes of finding more politicized responses to the informal learning questionnaire, and of including disability issues in the study, I interviewed members of a group of cultural activists who were engaged in making art about disability oppression.[19] These participants were able to provide a basis for considering informal learning from the standpoint of social and economic exclusion. Also, because the participants are part of the same advocacy group, the interviews as a data set could provide some clues about the collective and political dimensions of learning. Due to the politicized nature of being a performer with a disability, the participants are all at various times activists and spokespeople.

There are several problematic assumptions built into theories of informal learning, the most obvious of which is the notion that informal learning is self-directed, and that home is a space where learners can take control of their own learning agenda. This notion contains a set of assumptions about gender, race, class, and ability. For example, Foley (1999) describes a family gathering in order to give two examples of informal learning. The first example is a group of male mine workers discussing and critiquing management practices. While this discussion is going on at the dinner table, two women share a recipe for cake (ibid.: 1–4). This account indicates that a highly gendered process is taking place, but Foley does not refer to gender. This story indicates a relationship between what things are learned, and the time and space (physical and intellectual) available to the learner. The male mine workers in the account have retreated to a safe place to reflect on their work experiences, while the women in the story are still "at work." They are not free for critical reflection on their own workday. Instead they are learning to make cake. Home is not a neutral space from the standpoint of disability either. For example, home may be an institution or group-home. Furthermore, people with disabilities who live in private homes may have support workers or family members organizing their time and space. At the other extreme, home may be a place of isolation and deprivation.

The second problematic assumption is that informal learning is an individual process. This implicit notion of the "individual learner" is part of a larger shift from education to learning (Ramdas, 1999), which depoliticizes learning. Applying theories of informal learning to social movements makes the theorization of political consciousness almost impossible. In his book on informal learning in social movements, Foley (1999) does not sufficiently deal with the power relations of race, class, and gender in the cases he provides. Church, Fontan, Ng, and Shragge (2000) have raised a critique of the individualized concept of informal learning in which they posit social learning as a way to reintroduce notions of diversity and community. However, the analysis lends itself more readily to a social work perspective on the benefits of socialization and social integration than to an analysis of political learning. Their account of workplace learning among psychiatric survivors depoliticizes a highly politicized group.

As with the case study in the previous section, I have chosen Dorothy Smith's (1997) standpoint method to analyze the interviews for three reasons: first, to uncover how the concrete reality of people's lives contradicts the literature on learning; and second, to approach participants' stories both as testimony and as evidence of power relations. The third reason is that her notion of a "trajectory of consciousness" provides a tool for acknowledging the role of ideologies in the reproduction of unequal social relations. Ideologies surrounding people with disabilities are particularly insidious and include ideologies of individual deficit, dependency, and differential social worth.

By using Smith's (1997) framework, participants' testimony can also be employed to develop a critique of the ways in which the literature on learning is classed, raced, gendered, and abled; and the ways in which the literature assumes social exclusion and underemployment are the results of a lack of learning credentials. The individualized concept of informal learning removes the learner from her social context, and her day-to-day life situation. Employing the standpoint method to understand the data can also move us beyond a critique of individual notions of learning by offering a theory of how people who experience disability oppression both *learn to engage in* and ultimately *learn through* collective struggles against oppression.

Participants' stories reveal experiences of segregation and discrimination within formal schooling, and of the ways in which disability intersects with notions of work and career. Participants described different ways that learning occurs: through individual survival, collaboration, and collective creation. Three of the participants had extremely limited access to informal learning opportunities at paid jobs. However, they were able to document learning experiences in nonprofit organizations. Since all five participants had postsecondary credentials, it would seem that structural ableism, rather than lack of individual credentials, may be the cause of their underemployment. This finding echoes in Robert Tobias' prediction that "[t]he vast majority are likely to become disillusioned with a search for qualifications within a shrinking global labour-market" (1999: 117). From this perspective the project of valuing the life experiences of marginalized groups becomes nothing more than a way to incorporate them into the global market as "skills" and "knowledges" that can be put to use in capitalist work relations.

The interview process itself proved to be problematic as it defined in advance that consciousness, or the process of "knowing," was to be counted as "learning." The Working Class Learning Practice (WCLP) interview questionnaire asked respondents to document informal learning that was identified by the respondent as learning *at the time of the interview*. This is an important criterion because it asks participants to retroactively consider their previous experience, in a world that is increasingly driven by what is called a "knowledge economy." Disability rights activist Spirit Synott (see quote below) told me she was accustomed to recounting her life experiences, including her advocacy experience, as a process of skills acquisition because she is trying to compete in an ableist labor market. At the OPSEU site the local president told me that the previous union leadership had been recruited into management because of their skills in contract negotiation and knowledge of labor legislation. These stories indicate how an individualized theory of learning encourages the fragmenting of knowledge created in collective social movements into individual skill sets. The process of reconsidering one's life experience and political commitments as ways to remain competitive in selling one's labor can be interpreted as *alienation*.

Marx theorizes that alienation results when a person is separated from her material world because she does not have control over the work she engages in, and more importantly, she has no control over what she has produced. She is also separated from her fellow human beings, through competition, and through the impossibility of most forms of cooperation (Ollman, 1971: 133–134). In order to see the role learning plays in alienation, it is necessary to consider how these new learnings are deployed within capitalist relations of production.

The documenting and credentializing of learning is underpinned by the logic of *human capital theory* that assumes that the more you have learned (or the more capacity you have for learning), the more of an asset you will be for your organization. In a human capital formulation the worker is compensated for the use of her critical thinking through higher wages and a higher position. Livingstone has argued that a human capital approach to learning disadvantages workers who have been marginalized from formal education. Further, if the life experiences and learning of marginalized workers were recognized, they could attain equality with the higher paid managerial employees (Livingstone, 1997: 11). However, this critique leaves the organization's ownership of workers' learning unchallenged.

In a Marxist understanding of relations of work, the worker cannot be confused with the idea of capital. To understand the relationship between the worker and capital, we must recognize that *labor power is a commodity* in the capitalist mode of production. As a commodity labor power is subject to the law of supply and demand, and workers are in direct competition with one another to sell their labor. In this configuration knowledge and skill acquisition can become part of the competition. The more the concept of learning becomes synonymous with market requirements, the more it becomes commodified, and alienated from the learner.

Taking "human capital" as the theoretical framework for understanding the role of learning in the political economy gives a contradictory analysis to the results gained from a "labor power as commodity" framework. If educational researchers take a 'human capital' approach, we assume that the greater one's learning capacity, the more wages one will be able to earn. If one subscribes to this understanding of learning, and one is concerned about equity, one might suggest several strategies: increase access to learning; document and give credit for informal learning; and increase workers' opportunities to use their learning at work so that their learning capacity and skills can be recognized. If we take a labor power as commodity approach, however, we recognize that workers are not paid according to their intrinsic "worth." As a commodity, labor power is subject to the requirements of capital. From this standpoint we can see that recognizing informal learning merely expands the sphere that capital can exploit, which on a grand scale, increases the alienation and exploitation of human beings.

These contradictory results are obtained by taking a dialectical approach to the issue, that is, by looking below the surface appearance to the underlying relations. On the surface it seems as though the documentation of informal learning should benefit the marginalized worker. However, without an analysis of workers' exploitation in the capitalist political economy, and an understanding of how disability oppression unfolds in relation to exploitation, it is difficult to explain why Spirit Synott's credentials are not getting her a full-time, well-paying job.

> Having such a difficult time finding work opened my eyes to how people view people with disabilities. I've got a *great* resume, with an excellent volunteer and paid work record, so getting job interviews was no problem—they'd be on the phone so fast it wasn't funny. But as soon as I arrived for the interview, they would come up with imaginary difficulties. I would give them solutions for those difficulties, but it was like pulling teeth to get work. (Synott, 1999, interview transcript, italics in original).

Using the standpoint method to examine stories about work and learning reveals that the notion of informal learning draws our attention away from recognizing social relations of oppression *and exploitation*.

Respondents' stories can also be examined in relation to the three types of nonschool-based learning outlined in the previous section: survival, resistance, and struggle. I would like to stress that it is important for people concerned with social justice to be able to make a distinction between *resistance* that is about meeting and opposing oppression at the specific sites where it is manifested, and *struggle* that involved identifying, grasping, and changing the conditions that produce the oppression that individuals and groups experience. Reflecting on his 30 years of experience as a disability rights activist, Kazumi Tsuruoka, notes that

> [t]he movement always focused on "the self"—talking about it as a group, but always the self. Like self expression—we write plays where disabled folks tell people how they feel. It's easy to talk about getting out of the institution, and other safe issues like "we want to get into buildings" and "we want to work." Since the 70's we've been talking about transportation and living on our own. What we need to focus on now is interpersonal relations, and political leadership, and mentoring our youth. That's how we will build the disabled movement. (Tsuruoka, 1999, interview transcript).

Again, I am not advocating that we should call engagement in political struggle learning. I propose instead a theory of learning that emerges from an understanding of consciousness, from the rich tradition of radical education theory and practice, and from participatory research methods that do not claim to be neutral. *Consciousness-raising* and *critical education* are the educational legacies of feminism, trade unionism, antiracism, and revolutionary struggle. Although feminist and Freirian educational theories have been staples of adult education theory in the past, they began to disappear from the literature in the 1990s (Ramdas, 1999). Feminist consciousness is rooted in collective debate, community problem solving, and solidarity. Consciousness-raising is a collective, grounded, group-defined process that is bound up in working toward a better existence for the group—*as a group*, not as the sum of its competing parts. Informal learning defined as skill acquisition and achievement is the antitheses of consciousness-raising.

Bertel Ollman urges us to recognize the vast qualitative difference between individual and group consciousness, and the differences between mechanisms of change in consciousness (or the processes of learning) for a group and for an individual. Group consciousness, whether it crystallizes around ideas of class, gender, race, nationality, ability, or sexuality, is always a dynamic, dialectic phenomenon. It is not a static or passive set of ideas, but a process or movement. Consciousness, then, has a direction. It can be progressive or regressive, and it is ever moving toward or away from goals of democracy, justice, and equality (Ollman, 1993: 147–179). By seeking to record examples of individual, politically neutral learning, the informal learning project becomes a *regressive* force in the struggle against social exclusion.

Recognition of informal learning, and its institutionalized process and prior learning assessment, is on the surface a way of leveling the privilege of credentials, and a way of helping workers get credit for what they know. It is important, however, to ask to what end are educational researchers interested in defining and quantifying informal learning. Livingstone (1997) argues that recording previously uncounted knowledge and skills reveals that employers under use their workers' skills, and that workers have many of the skills required to participate in the decision-making process at work. Another reason for recording informal learning, in the same vein, would be to combat the notion that there is a skills shortage in the labor market. Livingstone argues that the opposite is true, that unemployed and underemployed workers have many skills that are not being used. This is in fact true of many underemployed people with disabilities. However, I would argue that their underemployment is not simply a result of their learning not being recognized.

A second flaw in the rationale for credentializing informal learning is that "socially excluded" or marginalized people would presumably also have been excluded from the experiences through which people gain marketable skills. If, as educational researchers, we want to recognize informal learning for the purpose of decreasing social exclusion, we are quickly thrown into questions about what kinds of experiences to value, and whether we intend to privilege experiences of integration into mainstream society. If, as researchers, we shift the focus from *social exclusion* to *oppression*, then we would be talking about something that impacts life experience—something that harms the individuals and communities. Shifting the frame from exclusion to oppression would require us to go beyond valuing a different kind of experience to recognizing that oppression includes experiences of violence, isolation, and poverty. In other words, it has a material basis. When people have been excluded from workplaces, they may not have the knowledge that comes from being in a workplace. When people are socially, economically, and politically excluded, how can their life experiences be counted up and presented for credit? The literature on learning depoliticizes learners, and ignores ways that learning is organized by the social relations that the learners exist in.

Conclusion

Based on these two case studies, it is clear that informal learning as it is currently being conceptualized cannot provide an emancipatory way forward for adult education theory and practice. Current learning theories individualize collective political movements, bolster the cult of credentials, and further exclude highly marginalized groups. Further, the individualized concept of learning removes the learner from her social context, and her day-to-day life situation. Wilma Fraser points out that "[f]or those at the margins of political, economic and social power, the notion that they 'are what they have done' sounds more like a slap in the face than a term of encouragement..." (1995: 141). The literature on learning depoliticizes learners and ignores learning that is based in the social relations and collective agency of learners. I believe that it is necessary for feminist and social justice oriented educational researchers to find our way back to feminist and revolutionary theories of consciousness in order to raise a counterhegemony to the "knowledge economy" ideology that masks this particularly dangerous phase of capitalist conquest, militarization, and destruction.

Through the above example of a feminist research method for researching Kurdish women's learning, it is clear that working with interview transcripts is only part of the task at hand. In many ways the women's testimony presented above is both universal and particular. Some of the stories about family violence and community relations seem as though they could be located in a thousand different places and times over the past hundred years. Even the references to wars and repressive regimes sound similar to other wars and regimes in the past several decades. Yet the Kurdish women and the disability rights activists discussed above live, work, learn, and struggle in particular and specific contexts. As complicated as the context may be, I believe the social relations can be revealed, grasped, and acted on. If, as feminist and social justice oriented researchers, we can ground our analysis in the understanding that power relations unfold over time, and in relation to the global political economy, we can avoid the danger of dissolving into a sea of particularity and social relativity. We can also avoid the danger of describing women's social locations mechanically, as *additive* subject positions, thus obscuring and objectifying the women with a list of issue-based identities. Our purpose in feminist research is (to paraphrase Marx) to study the world in order to change it.

Notes

1. This case study is part of a larger research project called *War, Diaspora and Learning: Kurdish Women in Canada, Britain and Sweden*, which was led by principal investigator Shahrzad Mojab at the Ontario Institute for Studies in Education at the University of Toronto and funded by the Social Sciences and Humanities Research Council of the Government of Canada.
2. Mark Levene (1998) gives this name to the East Anatolia region (Turkey, Kurdistan, Armenia, and parts of northern Iraq). This region has seen a hundred years of genocide of Kurds, Armenians, and others, dating from the

late nineteenth century. Based on her analysis of the militarization of Afghanistan and Kurdistan, Shahrzad Mojab has put forth the argument that this region is simultaneously a region of gendercide.
3. In the summer of 2001, Dr. Mojab and I interviewed 15 women in the UK. In the fall of 2002, we interviewed 15 women in Sweden. We also conducted several group dialogues with women activists in both places.
4. Negotiations between the Turkish government and the European Union have shifted this relation slightly—in order for Turkey to join the EU it must have the appearance of granting formal rights to minority groups.
5. Kurmançi is the majority dialect. However, the regional government comprises mostly Sorani speakers. This has added to the issue of dialect a power dynamic attached to insinuations about who is "more Kurdish."
6. I met a man who described himself as being a political exile who had left Turkey, spent several years in Greece, then gone to Belgium, Germany, Holland, and then France, spending one or two years in each place. He did not feel he had had a chance to learn any of the languages adequately. We were able to converse because a woman who had settled in Belgium after fleeing Turkey was translating between Turkish and French for us.
7. For example, Kurdish Women against Honour Killing.
8. I will be referring to two specific transcripts. Both of the women's testimonies are difficult to read. In the first one, the narrator is describing extreme domestic violence, and in the second, the narrator refers to imprisonment and torture.
9. See Cutajar (2000).
10. See McDonald (2000).
11. Research I was undertaking at the time as part of a larger project called *Working Class Learning Practices*, which was led by principal investigator David Livingstone at the Ontario Institute for Studies in Education at the University of Toronto and funded by the Social Sciences and Humanities Research Council of the Government of Canada. This project will be discussed in greater detail in the next section.
12. Informal research I did as part of ongoing activism around violence against women.
13. See Mojab and McDonald (2001a, 2001b).
14. It seems the concept of resistance is much more prevalent than the concept of struggle in social movement learning theory. Much of this can be traced to Foucault's (1988) popular idea that multiple pathways of resistance exist on equal and opposite vectors as the multiple pathways of power. If that were true, would not oppression and exploitation be cancelled out in this equation? Clearly it has not been.
15. Although these are precisely the kinds of questions that we as feminist educational researchers should ask, and these are precisely the kinds of questions we did ask in our 2001, 2002, and subsequent interviews.
16. A version of this section has been presented at the *Canadian Association for Studies in Adult Education* (see Gorman, 2002).
17. All of the participants asked that they be credited when they are cited.
18. See Gorman (2005).
19. In conducting the interviews I used the same open-ended questionnaires on learning practices and time-use surveys as those employed in the *Working Class Learning Practices* (WCLP) project, a study of informal learning at six unionized work sites led by David Livingstone.

References

Church, K., J. Fontan, R. Ng, and E. Shragge. (2000) Social Learning among People Who Are Excluded from the Labour Market—Part One: Contexts and Case Studies. A working paper for the New Approaches to Lifelong Learning (NALL) Network Project (Toronto: Ontario Institute for Studies in Education).

Cutajar, J. (2000) *Widowhood in the Land Where Time Stood Still: Gender, Ethnicity and Citizenship in the Maltese Islands.* Doctoral thesis: University of Toronto.

Foley, G. (1999) *Learning in Social Action: A Contribution to Understanding Informal Education* (London: Zed Books).

Foucault, M. (1988) *Discipline and Punish: The Birth of the Prison* (New York: Vintage).

Fraser, W. (1995) Making Experience Count...towards What? In M. Mayo and J. Thompson (eds.) *Adult Learning, Critical Intelligence and Social Change*, 137–145 (Leicester: National Institute for Continuing Education).

Gorman, R. (2002) The Limits of "Informal Learning": Adult Education Research and the Individualizing of Political Consciousness. In S. Mojab and N. McQueen (eds.) *Adult Education and the Contested Terrain of Public Policy.* Canadian Association for Studies in Adult Education Conference Proceedings: Ontario Institute for Studies in Education of the University of Toronto.

———. (2005) *Class Consciousness, Disability and Social Exclusion: A Relational/Reflexive Analysis of Disability* Culture. Doctoral thesis, University of Toronto.

Levene, M. (1998) Creating a Modern "Zone of Genocide": The Impact of Nation-State Formation on Eastern Anatolia, 1878–1923. *Holocaust and Genocide Studies*, 12 (3): 393–433.

Livingstone, D. (1997) The Limits of Human Capital Theory: Expanding Knowledge, Informal Learning and Underemployment. *Policy Options* (July/August): 9–13.

McDonald, S. (2000) *The Right to Know: Women, Ethnicity, Violence, and Learning about the Law.* Doctoral thesis, University of Toronto.

Mojab, S. and S. McDonald. (2001a) Violence, Rights, and Law: Informal Learning Experiences of Immigrant Women. A working document for the New Approaches to Lifelong Learning (NALL) Project (Toronto: Ontario Institute for Studies in Education).

———. (2001b) Women, Violence and Informal Learning. Working document for the New Approaches to Lifelong Learning (NALL) Project (Toronto: Ontario Institute for Studies in Education).

Ollman, B. (1971) *Alienation: Marx's Conception of Man in Capitalist Society* (London; New York: Cambridge University Press).

———. (1993) *Dialectical Investigations* (London; New York: Routledge).

Ramdas, L. (1999) Climb Every Mountain, Dream the Impossible Dream: ICAE Past, Present and Future. *Convergence*, 32 (1–4): 5–16.

Smith, D. (1997) From the Margins: Women's Standpoint as a Method of Inquiry in the Social Sciences. *Gender, Technology and Development*, 1 (1): 113–135.

Tobias, R. (1999) Lifelong Learning under a Comprehensive National Qualifications Framework—Rhetoric and Reality. *International Journal of Lifelong Education*, 18 (2): 110–118.

Watkins, K. and V. Marsick. (1992) Towards a Theory of Informal Learning and Incidental Learning in Organizations. *International Journal of Lifelong Education*, 11 (4): 287–300.

Chapter 10

Myths of Mentoring: Developing a Marxist-Feminist Critique

Helen Colley

Introduction

The display of emotion is at the heart of the millennial zeitgeist. Presidents and prime ministers parade their passions and their frailties. World-famous footballers weep openly on the international pitch. Hundreds of thousands of ordinary people mourn a British princess they never knew. The fervor for "emotional intelligence" has fuelled a raft of initiatives, from circle time in infant classrooms to postcompulsory key skills curricula and business management theory. In team meetings at work, our managers are now as likely to ask us how we feel as they are to seek our opinions. Emotional openness has become a requirement of public citizenship for the twenty-first century.

Nowhere is this more evident than in the rise and rise of mentoring, in which a caring one-to-one relationship is portrayed as the instrument for transforming mentees, and the promise is that "love will win the day." Mentoring has become a key feature not only of training for the professions and in the academy, but also of policy initiatives directed at young people, particularly those categorized as "socially excluded." In the UK, this has culminated in the government's creation of a national youth support service, *Connexions*, which now employs thousands of "learning mentors" and "personal advisers," as well as in multimillion pound funding for myriad local programs coordinated through the National Mentoring Network. By 2000, 1 million mentors were already working in the largest two such programs in the United States alone, and President Bush had committed substantial funding to their further expansion. Similar developments have flourished internationally, particularly in the English-speaking world. I have termed this model "engagement mentoring," since its prescribed goals are focused on reengaging young people with the labor market and structured routes thereto, and in engaging their commitment to the needs and interests of employers (Colley, 2003a, 2003b).

Anyone who has encountered the mentoring movement will recognize its distinctive branding. Its logo is the figure of Athene, an Ancient Greek goddess. In introductions to the concept of mentoring and in explanations of the mentor's role, she is used to symbolize the ideal mentor through narratives linked to Homer's *Odyssey*. In that 5,000-year-old epic poem, Mentor is the elder whom Odysseus appoints as guardian to his infant son when he sets off to fight in the Trojan War; and 20 years later, Athene takes on Mentor's human form to prepare the boy for transition to adulthood, as his father finally returns home. I was struck by the ubiquity of this mythical image throughout the literature on mentoring—from leaflets aimed at prospective mentors, to academic research papers. It is not novel to observe that myths are commonly used to legitimate certain discourses and practices. A much more complex, interesting, and important task, however, is to explain exactly *what* is being legitimated, *how* that legitimation takes place, and *why*. This chapter begins with a brief account of my critique of the myth of mentoring, and offers some illustrations of how that myth influences practice, from my research in an engagement mentoring scheme. I then go on to discuss the Marxist-feminist analysis used to develop this critique, and to call for further research to advance a critical theory of emotion and learning.

A Mythical Role for Mentors

What stories, then, are told about mentoring through the myth of Athene? How is her image used to depict the mentor's role? In so many texts we hear how, knowing that Odysseus is approaching home, she is concerned that his son should be ready to meet him, to make him proud, and to help him overthrow the usurpers who have taken possession of his kingdom and his son's birthright. She is supposed to stir the young man to action, put mettle in his soul, lead him to find his long-lost father, help him face up to daunting challenges, and support him as his courage is tested in battle to save his nation. All this should a mentor do, in a compelling tale of rites of passage and heroic reconquest of power.

The most basic suggestion in the brand-name of Athene is, of course, that she is a goddess—all-seeing, all-knowing, and all-powerful—who can intervene into human affairs, with utter assurance in altering their outcomes. The mentor is thus portrayed as superhuman, a miracle worker able to transform the mentee and their lifecourse beyond all expectations.

This role is usually modified, however, in these narratives. They tend to focus on Athene's "specialness," and her inspirational character. Moreover, many use her story to evoke notions of selfless caring for the mentee, with a "readiness to go that extra mile 'beyond the call of duty'" (Ford, 1999: 13). Subtly, she is redrawn as a saintly rather than a godly figure: still capable of miracles, but further inscribed with devotional and self-sacrificing dedication to her task. Ford's account explicitly draws on the concept of *agapé*, which originated in Ancient Greek culture, but has crossed over into the terminology of Christianity to indicate selfless, platonic love. One implicit moral of this Christianised image is that (like that religion's saints) the mentor as well as the mentee may have to endure great trials in the course of mentoring, without ever flinching from their duty.

Even feminist critiques (e.g. Cochran-Smith and Paris, 1995; DeMarco, 1993; Standing, 1999) have bought into this myth, appealing to the icon of Athene as they advocated models of mentoring based on nurture—"women's ways" of mentoring—rather than patriarchal models based on hierarchy and control. All of these stories piqued my curiosity through an oddly serendipitous connection. In my youth, I had studied Homer in some depth, and these repeated tales jarred with my distant memories. One day, exasperated by this sense of incongruity, I reached for my copy of the *Odyssey*, blew the dust from its cover, and rediscovered its astonishing legend.

Present-ing a Past We Never Had

The original epic tells a very different and far more brutal story of mentoring. As it opens, the kingdom from which Odysseus has been absent for 20 years is now in utter disarray. His son is a feeble and depressed youth, and both the prince and his guardian Mentor are public laughing stocks before the unruly nobles who have occupied the palace. Athene does indeed step in as mentor to give advice to the young man, to act as an adult role model, to advocate on his behalf, and to encourage him to action. Yet, despite her female form, as the god of wisdom and of war, she is in all other respects a symbol of male rationality and aggression. Her relationship with the prince is entirely impersonal, with no sense of the caring love that is so strongly emphasized in modern re-tellings. Moreover, she undertakes this role not out of altruistic intent, but to serve her own ends among the gods.

It is, however, in the dénouement of the Homeric myth that I found the starkest contrast to today's happy accounts of successful transition to adulthood and a father-and-son reunion. The prince's attainment of adult status is actually achieved through the bloody slaughter of the usurpers, the sexualized torture and execution of household handmaids who had consorted with them, and the assertion of his and his father's rule in a final battle, in which Athene has to intervene to stop them short of total carnage. The outcome of her mentoring is political, military, and sexual domination by the rulers of a monarchic slave society.

The myth of kindly nurture and self-sacrificing devotion, then, is a creation of modern authors that bears little resemblance to the ancient myth. It stands as a simulacrum, a series of identical copies "for which no original has ever existed" (Jameson, 1984: 68). These modern re-writings teach us that mentoring was portrayed like this 5,000 years ago; that ancient myths supposedly tell us about universal human truths; so this is how we should act as mentors today. In this way, the conditions of the present are explained in terms of a past we never had. Rather than providing genuine antecedents, the "past" is reconstructed as a "prequel" that both depends on and reproduces taken-for-granted assumptions about contemporary society. This past(iche) then serves to legitimate current practice through a false appeal to antiquity, and to a notion of humanity undifferentiated by gender, race, or class. Such myths therefore serve to obscure and simultaneously reinforce unequal social relations. They deny the influence of context upon meaning, representing contingent *appearances* as an eternal and immutable *essence.*

The key to dominant discourses of Athene-as-Mentor lies in their emphasis not only on what the mentor *does*, but on the kind of person that she *is*, especially her emotional dispositions. This mirrors a shift in academic analyses of mentoring, which initially focused on the functions of a mentor, but have increasingly turned their attention (including in feminist critiques) to the nature of the mentoring relationship and the personal qualities and feelings demanded of a mentor. At the same time, partly because mentoring has proliferated in so many different contexts, there has been much debate about what constitutes its essence. Roberts (2000) reviews this literature to argue that specific forms of mentor activity are contingent on the particular location, but there is a broad consensus that its essential attribute is "a supportive relationship," regardless of the context in which it takes place. True to the zeitgeist, emotion has come to define mentoring practice. Let us look at how this happened in practice in the mentoring scheme I studied.

Love's Labor Lost?

My research was undertaken from 1999 to 2001 at a scheme anonymized as "New Beginnings," which was run by a local Training and Enterprise Council (TEC—these bodies have since been replaced by Learning and Skills Councils). It was aimed at 16–18-year-olds classed as "disaffected" since they were not in structured education, training, or employment. Funding had been obtained through the European Youthstart Initiative, and this required the scheme to focus on employment-related outcomes. It therefore combined mentoring with in-house basic and key skills training and work experience placements. The volunteer mentors were undergraduates at the local university, mainly from degrees in teaching and applied social science. New Beginnings was marketed to them partly as an opportunity to improve their own employability for the graduate labor market, and most of the volunteers aspired to careers in the caring professions. The data were generated primarily through in-depth semistructured individual interviews with participants in nine established mentor relationships, and the stories of the female mentors were remarkably similar in their accounts of feeling under pressure to transform their mentees, feeling frustrated and guilty when this did not happen, and feeling the tension of having to conceal their emotions within the relationship.

Jane was a mature Social Sciences student, and the mother of a small child. She had previously studied counselling, and was very concerned to demonstrate empathy and exercise a nonjudgmental attitude in her mentoring practice. Her mentee, Annette, had been in care after losing her mother at an early age. Annette was doing well in her training at New Beginnings, but did not plan to continue into employment. Instead, she was preparing to have her first child and become a full-time mother. Jane found it difficult to reconcile her feelings about this resistance to the scheme's targets with her desire to be nonjudgmental. When asked about her views on the long-term prospects for Annette, her reply was angry and frustrated:

> It seems to be, the more I talk to them, I didn't realise how much, but it's "Oh, I'll decorate the baby's bedroom when this cheque comes"... "My boyfriend's gone on the sick, so when he gets his big cheque, we'll do this"... The boyfriend is off work...a 22 and a 17-year-old, fully healthy people, but they've no intention of doing anything, and not an education to get them where they want to go... It's this cheque, that cheque, social, income support, but it's the only thing they know, and to me...like you said, what's the future? Who can make such a difference to make them change? Who can make such a big impact to say "That is not the way you're going to go for the rest of your life?" Who can do that? I don't know.

She felt hurt as Annette began to loosen their relationship once she had left New Beginnings and had her baby. She saw her mentee integrating instead into a network of friends, family, and other young mothers on the estate where she lived.

> Now with Annette it's different, because it's not a dead-set arrangement, you know, I go one day and she's not in cos baby's ill or what-have-you, so I try again the next week and she's not there, but with other people. I leave a note telling her to phone me. I always put a 20-pence piece in it, and I say "Give me a ring and we can arrange to meet up," but she never does. She's a naughty girl [annoyed tone, shaking her head].

Jane threw herself into mentoring another two young people through the local Probation Service. With far fewer restrictions than New Beginnings, this program allowed Jane to spend long hours with her new mentees, doing tasks for them and giving them "a bit of nudging" to ensure they met the requirements of their probation orders. This too exacted a price at times:

> Like we'd go to court, she [the mentee] was in court, and if I've got to be somewhere, I'm there on the dot, or 10 minutes early if it was court or something, and she arrived 20 minutes late, and I thought, "I can't do this any more, you know, if you keep making a fool of me," cos I'm making excuses for her all the time, but I constantly...I kept thinking, you know, "Bite your tongue," and maybe...just by building a relationship in that way...[...] I thought that day in court, I think it was the 22nd of December, so obviously, you know, I was handing in chapters of my dissertation, I was really hectic busy, plus obviously, you know, my daughter was off school, so I'd arranged child-care to get to court at ten in the morning, and then when she didn't come, I thought, "I can't...you know, I can't keep doing this"...

She explicitly describes here how she "builds a relationship" through the forcible repression of her own feelings, and through her patient self-sacrifice that at times seems unbearable.

Yvonne, whose mentee Lisa kept failing in a series of work placements, had similar feelings. They had been meeting together for 18 months at our second interview, and part of the difficulty for Yvonne was absorbing the frustrations and disappointments of her mentee:

> What is a mentor? Sometimes I think I'm just a verbal punchbag, and that's what I'm there for. She can come in and say, "The whole world's shite and I don't want to do it," and just get it off her chest.

She talked about her disillusionment with the promises of benefits for the mentors that the university had promoted to recruit them:

> It [mentoring] has brought me a lot of stress. [...] I can't remember half the promises that the university made, and I just sit there and I think, "Why did I do this?" I put it on my CV, and then I dread anybody asking me about it in an interview. I really dread it, because I think, well, what do I say, you know?

Yvonne judged herself, as well as her mentee, by the expected employment outcomes of engagement mentoring, and felt others would judge her by this criterion too. As mentoring failed to transform Lisa, their relationship seemed to be grinding to difficult halt, but Yvonne felt trapped, and afraid of moving to end it:

> At the end of the day, I've just sort of had to cope with it myself...I just have to switch off, otherwise I'd just crack up, you know. [...] I don't want to be the one that says to Lisa, you know, "You're doing my head in, you're not getting anywhere, go away." I think in some ways I'm scared of bringing it up in case she thinks I'm pushing her away.

These violent metaphors were strongly typical of most mentors' accounts of struggling with their feelings, and they portray vividly the painful consequences of internalising the impossible myth of mentoring. But how can we go beyond this myth, when even feminists have embraced it? I turn now to an explanation of how my own Marxist-feminist critique of mentoring developed.

Developing a Dialectical Materialist Analysis

Feminist educational research does not itself comprise a universal approach to understanding experiences of learning, but contains at least four different strands of thinking. Liberal feminism accepts the status quo in general, but seeks a more equal place for women within it. Radical feminism avoids issues of class, seeing the world in terms of male dominance and the undifferentiated oppression of women. Feminist models of mentoring to date have tended to draw on these two approaches. Postmodern feminism has resisted the universal category of "woman" to insist on the distinct lived experiences of different groups, but at the same time construct difference as a universal in ways that make both analysis and resistance difficult (Haber, 1994).

Marxist feminists, by contrast, reject unitary notions of "women's ways" of knowing or acting, but argue that class, race, and gender all interact to shape our social world in complex, relational ways. Unfashionable though Marxist theory has been in educational research in recent years, it is experiencing a significant resurgence already discussed in the introduction to this book. In particular, its dialectical materialist philosophy promotes a reflexive analysis of social phenomena. Such reflexivity is far from the personal (even, at times, narcissistic) confessions of ultrarelativist postmodern theorists, but turns upon the logic of the moral and political arguments used to justify explanations and strengthen their academic value. It delves beneath overt meanings, and "instead enquire[s] into who is served and who is not" by the particular ways in which a practice is constructed (Gaskell, 1992: 33). This entails a "radical doubt" (Bourdieu, 1992) that seeks to break with common sense and official representations enshrined in social structures, practices, and institutions, as well as in people's minds. Such (mis)representations often conceal the perpetuation of social inequalities beneath a veneer of democracy.

Marxist philosophy suggests that any social phenomenon has both an essence and an appearance. It is interested in the dialectical relationship and interplay between social relations, the material world, and the evolution of thought (reflected in cultural elements, human consciousness, and agency for transformation). However, Marxist interpretations of essence and appearance are very different from those of bourgeois logic. (This in turn shows that essence and appearance are themselves socially constructed categories, which convey different meanings in different contexts.) The notion unique to dialectical philosophy is that *essences are neither eternal nor immutable*, expressed in Hegel's dictum that "[i]n essence, all things are relative" (Novack, 1986: 121). Marx took up this philosophical revolution, while rejecting Hegel's idealism, and created a radically different form of materialism, in which essences are neither absolute nor foundational, and in which anything can be transformed, under given conditions, into its opposite (Lenin, 1964: 203, cited in Novack, 1986: 47). Essence maintains a complex relationship with *appearances*, which are themselves immediate and absolute when considered in abstraction from essences:

> The essence of a thing never comes into existence by itself and as itself alone. It always manifests itself along with and by means of its own opposite. This opposite is what we designate by the logical term *appearance*. It is through a series of relatively accidental appearances that essence unfolds its inner content and acquires more and more reality until it exhibits itself as fully and perfectly as it can under the given material conditions. (Novack, 1986: 113, italics in original)

This is, however, a purely logical construction of opposition. As Novack goes on to argue, the complexity of the relationship between essence and appearance raises two problems.

The first is to avoid the superficiality of assuming that the essence of a thing is one and the same as its particular appearance at any time. To avoid misrecognition, we need to *distinguish* essence from appearance. The second is more difficult, since appearances will change and even contradict each other as the relative essence of a thing develops. In doing so, appearances may coincide, interplay, or overlap with essence. There is therefore an "equally urgent need to see their unity, their interconnections, and their conversion—under certain conditions—into one another" (ibid.: 114). This opposition *and* identification between essence and appearance throughout the development of a phenomenon is described as a process moving from an initial point of unity, at which the appearance subordinates the essence, through a phase of divergence, to the apogee of development at which essence and appearance are reunited, but in which the essential nature of the thing becomes transparent and dominates all of its particular appearances.

Novack (1986: 116–121) gives the example of money to illustrate these difficult philosophical concepts, and it is worth briefly recounting some of his argument—not least because it will lead us back to the concept of mentoring. Money has not always existed. The need for it only arose with the beginnings of trade and the need for a common equivalent in the exchange of diverse products between different communities. In the early stage of its existence, money took on many appearances, as any individual commodity (various forms of food, raw materials, tools, or artifacts) could be used as a measure of value. But the nature of money changed as trade widened and markets grew. Precious metals replaced this diversity of commodities. With the introduction of coins, money became a universal form of exchange, and its essence as coinage (or ingots or, later still, notes) was established. A further development was the gold standard, which asserted itself as the most essential form of money. This still allowed other relatively less essential forms of money to circulate, with which gold maintained definable relations, in particular determining their specific value. In one sense, gold represents the height of essence coinciding with appearance in the case of money. Yet this proposition is itself superseded if we consider that, in capitalist society, money in fact "represents specific economic relations between people. *These relations constitute the essence of money*" (Novack, 1986: 121, italics added).

Money has been used here as a philosophical example—but this consideration of its various transformations reveals an important aspect of the current social and economic context in which mentoring takes place. The role of money indicates that we live in a society where exchange-value has replaced use-value (Marx, 1975). This applies to labor power itself, which has become exchangeable for money, and as result of becoming a commodity, has become alienated and dehumanized, so that there is "no other nexus between man and man than...callous 'cash payment'" (Marx and Engels, 1977: 44). Money, through exchange relations, has thus become the essential social bond, as well as determining social power relations (Bottomore and Rubel, 1956). In this way, social bonds have been reified. They appear as independent things, and as the direct personal relationships implied by the concept of "community" are ruptured, "society" has come to represent impersonal economic relationships.

Harvey's (1997) insightful interpretation of Wim Wender's film *Wings of Desire* underlines this point. The story is of an angel, Damiel, who chooses to come to earth. His transformation into human status is symbolized repeatedly in the film by the need for and exchange of money. He has to borrow money from a stranger and sell the ancient piece of armor that fell to earth with him, in order to buy food, drink, clothes. He seeks out another former angel, played by Peter Falk, for support and advice on how to function in this new, human form. Falk's response is not to befriend him but to offer him money: "Damiel's entry into this human world is now firmly located within the co-ordinates of social space, social time, and the social power of money" (Harvey, 1997: 319). In an important sense, Falk's act is one of mentorship—the exchange of money as social bond and as initiation into the human condition under capitalism—bringing my argument full circle.

A Marxist-Feminist Critique of Mentoring

How can this dialectical materialist analysis be applied to the academic constructs and social practices of mentoring today? In particular, how can we understand their special emphasis on the deployment of emotion and nurture by mentors? The work of earlier Marxist-feminists has been of particular help in addressing these questions, particularly through the theorizing of emotional labor.

In her book *The Managed Heart: The Commercialization of Feeling* (1983), Arlie Russell Hochschild uses Marxist theory to present a very different analysis from the trite proclamations of "emotional intelligence" that currently abound. She argues that much of the work allocated to women within the patriarchal capitalist division of labor, particularly in caring services, involves not only physical and intellectual labor but also emotional labor. (We can note here that, as at New Beginnings, around 80 percent of mentors working with "socially excluded" young people are women.) Such labor enables capitalists to make profit, because producing a certain affective state of mind in the client is a necessary part of the service. However, the worker also has to manage her *own* feelings in order to manage those of the client. When the production and management of feelings has to be sold on the labour market in this way, where it clearly has an exchange-value rather than a use-value, it then becomes susceptible to direct prescription and control and is transformed into a source of alienation rather than human connection.

As I have already argued, the appearance of mentoring is weak and fragmented in terms of its functions across a wide range of contexts. Its appearance is strong, however, in terms of the *emotional disposition* it demands of mentors, not least through the rhetoric of myths. Engagement mentoring, as I have shown in detail elsewhere (Colley, 2003b), seeks to reform young mentees' dispositions in line with employers' demands for "employability." But it also seeks to engender devoted and self-sacrificing dispositions in mentors through its discourse of feminine nurture. My research revealed many stories of mentors who struggled to manage their own feelings while working with young people from very different social backgrounds—including their guilt at failing to live up to the image of the ideal mentor.

A Marxist-feminist analysis suggests a radically new definition of mentoring: its distinctive, historical essence constitutes *a form of emotional labor*, since its purpose is to transform the dispositions of those on both sides of the relationship, in ways that are determined by the needs of dominant groupings rather than by the needs or desires of mentors or mentees themselves. Our very selves are treated as the raw material of this emotional labor process, and as a result, are dehumanized as capital. In mentoring, the greatest contradiction is that this brutal commodification of the self is cloaked in the guise of human relationships based on warmth and compassion.

Of course, some men too are engaged in mentoring and other caring occupations, and the triumphalist discourse of globalization in the workplace has, as Rikowski (2002) demonstrates so well, made the personhood of all workers more deeply exploited than ever by capitalist modes of production. But Hochschild (1983) argues that emotional labor and its costs have a differential impact and create further inequalities according to gender and class, since women, and working-class women in particular, are socially conditioned within the home and the family to care for the needs of others at the expense of themselves. They have to rely on the exchange-value of their emotional capacity, since they have limited access to economic and material resources. Women's subordinate status as a gender renders individual women more vulnerable to the displacement of feelings by others. Furthermore, patriarchal power relations mean that such labor, characterized as "women's work," is more likely to remain unrecognized.

Toward a Critical Theory of Emotion in Education

Hughes (2005) and Martin et al. (2000) argue that current interest in the management of emotion in the workplace—including the invitation to express emotion more openly and authentically—represent a contradictory form of control that both obscures alienation while exposing its emotional costs. Cotter (2002) and Thompson (1998) both question, in different ways, how ethics of care have been constructed to imply the neutrality of care as a concept, and the potential for a benign, caring capitalism. Tronto (1989: 172) similarly cautions against assumptions that any attribute of women is "automatically a virtue" and that "whatever women do is fine because women do it." Cotter (2003) in particular points to the fact that revived interest in the material, through interest in embodied emotions, does not guarantee a helpful perspective on class and gender. She notes the growth of what she terms "delectable materialism"—in which the material is treated through the prism of idealist, dehistoricized subjectivity—in opposition to dialectical materialism. All of these perspectives suggest that we need to unpick the assumptions and the aporia underlying celebrations of emotion.

These are not just theoretical or ideological contradictions, however. Despite the celebrations, there are some harsh realities about the use of emotion in education, training, and employment, which stand in stark contrast to the rhetoric. Caring occupations continue to be extremely heavily gender-stereotyped, overwhelmingly employing women, and they are accordingly low-status and low-paid in our patriarchal capitalist labor market. Women continue to do the bulk of unpaid caring work in the home and family too. Much of the emotional work done by women in either context remains invisible:

> the problems of specifying caring work, and particularly the emotional labour which is part of caring work, [can be attributed] to the minimal attention which has been paid to it. The low profile and low status historically attributed to such work contribute to this, for it is a form of labour which is recognised not when the outcome is right, but on those occasions when it goes wrong. The product itself is invisible...the value of the labour is as hidden as the value of the routine management of emotion. (James, 1989: 28)

Despite alternate celebrations and outcries in the media that feminine skills and qualities are now key to the success of top managers—and despite proclamations that mentoring can overcome barriers in employment—women continue to face a glass ceiling in many institutions, and few employers actually use emotional intelligence or related categories as criteria for hiring or promotion (Alimo-Metcalfe, 1995).

Marx contended that the dualist separation of emotion and rationality is a relatively recent historical development; and that their counterposition in bourgeois thought, which treats rationality as superior and emotion as inferior, is a false one that is key to the modern fragmentation of organic and holistic understandings of human existence. As Agnes Heller (1979) argues in her seminal application of Marxist thinking to elaborate a theory of feelings, emotions are *not* a set of undifferentiated resources to which different social classes or genders have differentiated access and differentiated affordances to deploy. Despite the tendency to characterize emotions as both essentialized and unpredictable, they are in fact highly predictable, because they are prescribed by normative frameworks dictated by bourgeois culture and morality. These prescriptions are expressed through dominant "configurations" or "repertoires" of feelings and emotional display that are regulated relationally by social structures—they vary according to class, strata within classes, and gender (I might also add here, following Thompson, 1998, according to culture and race).

These repertoires of feeling are fundamentally related to the mode of production in any given society, to divisions of labor within it, and to different relations to the means of production and to the reproduction of labor power.

Because emotion is assumed to belong to the private rather than public domain, liberal feminist authors have tended to treat emotion only as a sub-category of caring or mothering, rather than as a key factor in the social regulation of work. In predominantly female occupations like childcare and care of the elderly, the management of one's own and others' feelings is not a private adjunct to work, nor simply an aspect of caring. It is a key feature of the workplace, a form of paid labor, or to be more accurate, of labor *power*. Guiding "ideologies of practice" attached to particular spheres of work help to constitute this regulation (James, 1989), and the image of Athene in mentoring can be seen as a symbol of just such an ideology.

This is not to suggest, however, that such emotional configurations are deterministic in an absolute way. Resistance and revolutionary movements exist across the globe, in part at least because we have the capacity also to feel differently: to desire social justice, to direct anger at the ruling classes and their inhuman system, to feel solidarity with others. However, as Cotter points out, emotion is all too often understood as abstract, universalized *affect*, rather than as concrete *effect* of capitalist social relations:

> Contemporary feminism has all but abandoned the question of mode of production—especially the relation of labor-capital and its impact on gender and sexuality... [It] transforms the laws of motion of capital into sentimental codes of affect, caring and civility, and, therefore, advocates primarily for changes in *behavior* as a means for social transformation... In doing so, [it] puts forward the understanding that the social relations of reproduction are not only autonomous from the relations of production but also the root social relations that need to be transformed in order to emancipate women. (2002: 1–2)

What difference does it make to see emotion as an effect of the capitalist mode of production? Most importantly, I would argue, it helps us to see that the shift in some capitalist workplaces from the repression and denial of emotion to the open expression of emotion, including in the widespread practice of mentoring, represents simply a flip from one to another side of the same coin of intensified productivity. Rikowski describes this as our "predicament" in the post-Fordist workplace, which increasingly demands the display of feelings and dispositions, and thus "incorporates a social drive to recast the 'human' as human capital" (2002: 196). A Marxist-feminist analysis suggests that this commodification of the most intimate parts of our personhood may always have been an aspect of women's labor, but also refocuses our attention on feelings as part of the labor process. Given that the mode of production has intensified these processes in the 20 or 30 years since these ideas were first aired, much more research needs to be done on the part that education and training plays in preparing people for the work of emotional labor. In short, we need a critical theory of emotion in education. Rare but important examples of such research can be found in the work of Bates (1994) and Skeggs (1997) on the training of care workers, in that of Smith (1992) and James (1989) on nursing, and I have developed this analysis further in my own research on childcare education (Colley, 2006). At the same time, researchers in this field need not only to promote better understandings of the problem by theorizing it more adequately. We also need to find whatever opportunities we can to support struggles by those who labor with their feelings, since only these can change the social and economic conditions that keep us captive to emotive myths.

Note

The research on which this chapter is based was funded by a PhD bursary from Manchester Metropolitan University, and I am grateful for this support. Discussions with many colleagues and students have helped to develop these ideas, but I particularly wish to thank my supervisors Jane Artess, Phil Hodkinson, and Mary Issitt, as well as Ian Stronach, for their challenge and insights.

References

Alimo-Metcalfe, B. (1995) An Investigation of Female and Male Constructs of Leadership and Empowerment. *Women in Management Review*, 10 (2): 3–8.

Bates, I. (1994) A Job Which Is "Right for Me"? Social Class, Gender and Individualization. In I. Bates and G. Riseborough (eds.) *Youth and Inequality*, 20–34 (Buckingham: Open University Press).

Bottomore, T. B. and M. Rubel. (eds.) (1956) *Karl Marx: Selected Writings in Sociology and Social Philosophy* (London: Watts).

Bourdieu, P. (1992) A Radical Doubt. In P. Bourdieu and L. J. D. Wacquant (eds.) *An Invitation to Reflexive Sociology*, 119–128 (Cambridge, UK: Polity Press).

Cochran-Smith, M. and C. L. Paris. (1995) Mentor and Mentoring: Did Homer Have It Right? In J. Smyth (ed.) *Critical Discourses on Teacher Development*, 142–151 (London: Cassell).

Colley, H. (2003a) Engagement Mentoring for "Disaffected" Youth: A New Model of Mentoring for Social Inclusion. *British Educational Research Journal*, 29 (4): 505–526.

———. (2003b) *Mentoring for Social Inclusion: A Critical Approach to Nurturing Mentor Relationships* (London: RoutledgeFalmer).

———. (2006) Learning to Labour with Feeling: Class, Gender and Emotion in Childcare Education and Training. *Contemporary Issues in Early Childhood*, 7 (1): 15–29.

Cotter, J. (2002) Feminism Now. *The Red Critique*. Available at www.geocities.com/redtheory/redcritique/MarchApril02/feminismnow.htm (accessed 14 July 2003).

———. (2003) The Class Regimen of Contemporary Feminism. *The Red Critique*. Available at www.geocities.com/redtheory/redcritique/Spring2003/theclassregimenofcontemporaryfeminism.htm (accessed 14 July 2003).

DeMarco, R. (1993) Mentorship: A Feminist Critique of Current Research. *Journal of Advanced Nursing*, 18 (8): 1242–1250.

Ford, G. (1999) *Youthstart Mentoring Action Project: Project Evaluation and Report Part I* (Stourbridge: Institute of Careers Guidance).

Gaskell, J. (1992) *Gender Matters from School to Work* (Milton Keynes: Open University Press).

Haber, H. F. (1994) *Beyond Postmodern Politics: Lyotard, Rorty, Foucault* (New York: Routledge).
Harvey, D. (1997) *The Condition of Post-modernity* (Cambridge, MA: Blackwell).
Heller, A. (1979) *A Theory of Feelings* (Assen: Van Gorcum).
Hochschild, A. R. (1983) *The Managed Heart: Commercialization of Human Feeling* (Berkeley; Los Angeles: University of California Press).
Hughes, J. (2005) Bringing Emotion to Work: Emotional Intelligence, Resistance, and the Reinvention of Character. *Work Employment and Society*, 19 (2): 603–635.
James, N. (1989) Emotional Labour: Skill and Work in the Social Regulation of Feelings. *Sociological Review*, 37 (1): 15–42.
Jameson, F. (1984) Post-modernism or the Cultural Logic of Late Capitalism. *New Left Review*, 146: 85–106.
Lenin, V. I. (1964) *Collected Works*, Vol. 19 (London: Wishart and Lawrence).
Martin, J., K. Knopoff, and C. Beckman. (2000) Bounded Emotionality at the Body Shop. In S. Fineman (ed.) *Emotion in Organizations*, 72–81 (London: Sage).
Marx, K. (1975) *Wages, Price and Profit* (Moscow: Progress Publishers).
Marx, K. and F. Engels (1977) *The Communist Manifesto* (Moscow: Progress Publishers).
Novack, G. (1986) *An Introduction to the Logic of Marxism*, 5th ed. (New York: Pathfinder Press).
Rikowski, G. (2002) Fuel for the Living Fire: Labour-Power! In A. C. Dinerstein and M. Neary (eds.) *The Labour Debate: An Investigation into the Theory and Reality of Capitalist Work* (Burlington: Ashgate).
Roberts, A. (2000) Mentoring Revisited: A Phenomenological Reading of the Literature. *Mentoring and Tutoring*, 8 (2): 145–170.
Skeggs, B. (1997) *Formations of Class and Gender* (London: Sage).
Smith, P. (1992) *The Emotional Labour of Nursing: How Nurses Care* (Basingstoke: Macmillan).
Standing, M. (1999) Developing a Supportive/Challenging and Reflective/Competency Education (SCARCE) Mentoring Model and Discussing Its Relevance to Nurse Education. *Mentoring and Tutoring*, 6 (3): 3–17.
Thompson, A. (1998) Not the Color Purple: Black Feminist Lessons for Educational Caring. *Harvard Educational Review*, 68 (4): 522–554.
Tronto, J. (1989) Women and Caring: What Can Feminists Learn about Morality from Caring? In A. M. Jaggar and S. R. Bordo (eds.) *Gender/Body/Knowledge: Feminist Reconstructions of Being and Knowing*, 32–45 (New Brunswick; London: Rutgers University Press).

Chapter 11

Popular Press, Visible Value: How Debates on Exams and Student Debt Have Unmasked the Commodity Relations of the "Learning Age"

Paul Warmington

Introduction

Why should we, as Marxist educators, expend energy examining the news media's construction of education issues? In the UK vast quantities of education coverage emerge each year. Stories include hardy annuals, such as the predictable August uproar over exam results, perennials, such as the relative performance of boys and girls, and sudden sproutings of concern over, say, the fall in the numbers of undergraduates studying sciences or modern languages. Despite this, the body of academic literature exploring the relationship between the UK's education sector and news media processes is fairly modest. This is unsurprising. As with other public services, relationships between the education sector and largely right-wing news media remain fractious; the quality of education news coverage is generally held in low regard by educationalists. It is important, though, that our wariness of the media should not lead us to ignore the role of news coverage within contemporary capitalism's education settlement.

This chapter examines the ways in which reporting of education news expresses and reinforces the contradictory social drives that characterize education under capitalism. Its particular concern is the news media's depiction of the "value" of proliferating educational qualifications. Discussion focuses on the years 2002–2003 when media panic in the UK over post-16 qualifications reached unprecedented levels of intensity, impacting upon public debate and government policy. These panics were expressed in a set of related news narratives painfully familiar to educationalists in the UK: rising exam pass rates, the unstable value of A-Levels and degrees and the cost of widening participation in higher education. These particular news issues exist in close proximity. First, they invariably provide headline stories in the weeks immediately following the publication of A-Level exam results each August. Second, over recent years they have been increasingly bound together by a metanarrative of "falling standards," in which A-Level exams are derided as becoming progressively easier and the worth of postcompulsory qualifications is questioned. The critique offered here is predicated upon the view that the news media are a significant shaping factor within the political economy of education and that the discourse the media produce and inhabit is a material force within the social universe (as opposed to existing in the parallel universe of "discourse" routinely proposed by postmodernists). Print and broadcast media have, in the current decade, played a key part in intensifying debates around exams, credentialisation, and mass access to postcompulsory education. However, while these populist education debates are often decked out in the language of educational "standards" or "quality," it is anxiety over fluctuations in the *use-value* and *exchange-value* of qualifications that remains the news media's unspoken concern: the poltergeist in the basement.

In short, what has marked news coverage of exam pass rates and massification in the early 2000s is an increasing, albeit obliquely expressed, concern about the commodity value of educational qualifications and a simultaneous clouding of actual social relationships by the frequent depiction of students as "investors" in their own education. News media coverage has simultaneously exposed and veiled the commodification of educational qualifications; in turn, as always, the commodity form exposes and veils the social relations of capitalism. It is in this context that the bitter debate on "inflated" exam pass rates, "meaningless" qualifications, and proliferating degree courses must be understood. What often appears as scapegoating of the teaching profession, exam boards, and students is, in fact, largely an expression of a drive to inscribe new hierarchies of qualification value at a moment of social reconfiguration: that reconfiguration being the massification of postcompulsory education and training (hereafter PCET). Thus the sterility of education news coverage does not derive merely from those journalistic tendencies usually held responsible: misunderstanding of data, crude summarizing, lazy populism, or party political grudges. The discursive contest that pervades media coverage of education issues is truncated because it is in thrall to the commodification of educational qualifications but incapable of properly defining that commodity.

Qualifications Anxiety

Since Collins (1979) christened "the credential society" and famously speculated about the consequences of "credential inflation" it has become commonplace in academic critiques of PCET for educational qualifications to be described as "cultural capital" (Ainley, 1994), as "currency" (Warmington,

2003) or as putative "passports" to economic and social opportunity (Riseborough, 1993). These are terms loaded with notions of exchange, acquisition, value, and markets. They also suggest a concern to understand the shifting social significance of educational qualifications and the implications for social justice of differentiation between the credential haves and have-nots. The New Labour government's target for participation in higher education (50 percent of 18–30-year-olds by 2010) has underlined the proliferation and consumption of qualifications as central to its socioeducational vision of a "learning age" (DfEE, 1998; Labour Party, 2001; National Audit Office, 2002). The term learning age as a variant on the notion of a learning society was adopted as the self-celebratory title of one of New Labour's earliest Green Papers (DfEE, 1998); the document signalled a clear continuity with previous Conservative Party policy in relation to promoting the upskilling and certification of the labor force.

However, as a result of this semiplanned expansion of PCET, one of the contradictions of the, so called, learning age has been exposed. To stimulate participation in higher education and in other areas of PCET, the desirability of qualifications must be *driven up* but, in extending credentialisation, the value of qualifications as a commodity risks being *driven down* by qualification inflation. The massification debate is one point at which academic concerns over the commodification of qualifications have bled into popular discourse. This is apparent in the news media's recurring depiction of the supposed devaluation of A-Level qualifications. First introduced in the 1950s (and revised in 2000 into a unitized, two-part qualification), A-Levels remain the key post-16 academic qualification in England, Wales, and Northern Ireland.[1] Notwithstanding recent reforms aimed at diversifying post-16 qualifications and widening participation in PCET, A-Levels still provide the "standard entry" route into undergraduate higher education. The high stakes nature of this academic "gold standard" is underlined in the final fortnight of August, which sees the publication of A-Level exam results. This has become a major diary item in the UK news media's calendar (Warmington and Murphy, 2003, 2004). Each summer news items about "exams that no-one can fail," "soft" subject options, and "Mickey Mouse degrees" cross the boundaries of tabloid and broadsheet, traverse the party political allegiances of particular newspapers and provide headlines in both print and broadcast media.

The often incoherent nature of the exam results news coverage, wherein stories celebrating students' A-Level triumphs run side by side with columns condemning "exams you can't fail" (*Daily Mirror*, 15 August 2002), reflects a political moment characterized in the UK by what Cole and Hill (1999: 39) have described as New Labour's carrot and stick policy. Here "inclusive" social policy, as exemplified by the work of the Social Exclusion Unit or the *Every Child Matters* Green Paper (DfES, 2003) or the target of 50 percent participation in higher education, is twinned with divisive economic policies that represent only a slight softening of neoliberal imperatives (as in New Labour's championing of deeply hierarchical consumer choice in education). To adopt Jessop's (1990) terms, the hegemonic component of the current

UK state project juggles "one nation" strategies, which aim to secure popular support by more equitable distribution of material rewards and concessions with "two nations" strategies that are concerned to mobilize "the support of strategically significant sectors and pass the costs of the project to other sectors" (211).

In this good cop-bad cop hegemonic setting it is little wonder that the news media, regardless of the microloyalties of party politics, are caught between conflicting drives and audience appeals. On the one hand, the aim of increasing participation in PCET and extending the ownership of credentials plays well, representing as it does an inclusive one nation social policy aspiration, albeit one that also adheres to the market-led belief that a more highly skilled and qualified workforce is better placed to compete in global markets. On the other hand, the prospect of wider distribution of credentials raises questions about how the costs of the widening participation project will be spread: not only in economic terms (with regard to fees, taxation, and central funding) but also in relation to the potentially diminished cultural capital of qualifications. Hoisted by these contradictions, the media response to rising exam pass rates has assumed unprecedented levels of sound and fury in recent years.

Given its necessarily limited space, this chapter's discussion inevitably tends to homogenize its description of news media. Clearly, there are important variations between broadcast and print forms; legislation requires broadcast media to adhere to stricter divisions between news and editorializing. This is in obvious contrast with print journalism. Print and broadcast media, tabloids and broadsheets, and specialist and general journalism all exhibit presentational and discursive peculiarities. Yet there is enough in common between these subsets in their treatment of rising exam pass rates and questions about supposedly declining standards to make encompassing analysis viable. Cottle (1995), for example, offers useful insights into the increasing similarities between broadcast and print media's preferred discursive forms. Moreover, Wallace (1993) and Jeffs (1999) analyses of news media coverage of other aspects of the education sector suggest that the journalistic discourse of falling standards is widely pervasive and structures much of the UK's educational debate.

Rising Stakes, Media Panic

The key vision statements of New Labour's learning age have customarily conflated the language of social justice and the drive for economic expansion into a nonconflictual image of one nation social inclusion. Yet the outcry that has met successive rises in A-Level pass rates and increases in the numbers entering higher education provides stark illustration of Hatcher's (1998: 207) dictum that "in education there is no classless universal interest around which national consensus can be constructed." Typically, exam news narratives have taken the rise in A-Level pass rates (to 96 percent in 2004) as indicators of falling standards in curriculum, examination, and awarding practices. Since the start of the 1990s, the stakes of the debate have been raised annually by a series of factors. These include the increasing importance of degree-level education as a career precondition, the growth in numbers entering higher education and the proliferation of alternative routes into higher education. In addition, McCaig (2003) points out that, before coming to power in 1997, New Labour began to promote a higher standards agenda as a political signal to aspirational parents that the state system was a safe place to school their children. The standards agenda was coupled with support for the retention of A-Levels. This policy context heightened expectations from parents and students alike, feeding into the feverish media reportage of A-Level results each August, which reached a peak in 2002, with marked consequences for both the government and the boards responsible for administering and monitoring the A-Level exam system.

It should be noted that, in the two years prior to the media designated "exam fiasco" of 2002, the boards responsible for administering exams had begun to attract negative media attention (Warmington and Murphy, 2004). In 2000 the Scottish Qualification Authority (SQA) had faced criticism over its management of the Higher Still, its equivalent of A-Level. In 2001 the Edexcel exam board had been the subject of news stories about errors in exam papers and other administrative failures. From 2002 onward media anxieties over exam pass rates begat concerns over grading mechanisms and the role of awarding bodies. Claims that the gold standard A-Level system was being undermined and that school leavers were racking up devalued qualifications, in turn, provoked bitter debate over the numbers and nature of students gaining access to higher education and consternation over who should pay for the privilege. However, reporting of the putative decline in standards represented something more than a concern over quality; these news narratives connoted an anxiety about the unstable *value* of educational qualifications.

Initially the news media's response to the publication of A-Level results in August 2002 focused on concerns about the record 94.3 percent pass rate (Warmington and Murphy, 2004). In the following month complaints were made in the press by a number of schools (supported by the Secondary Heads Association, a key professional body representing independent schools) about A-Level grading practices. At this point a media-fuelled debate was set in motion, culminating in the resignations of the secretary of state for education and the chairman of the Qualifications and Curriculum Authority (QCA), the quasi-independent body responsible for developing and monitoring the national qualifications framework. The major policy consequence was the initiation of the Tomlinson Review of secondary level curriculum and assessment. In 2003, with A-Level pass rates reaching 95.4 percent, populist celebration of student success almost immediately gave way to the business of moral panic. There was a rush to account for the continuing rise in A-Level passes. Students were accused of cynically opting for soft subjects, such as Business Studies and Psychology (which now superseded Media Studies as favored caricatures); the competence of markers was questioned; the impartiality of awarding bodies was impugned. In the press (and not only in its more right-wing quarters) there were flights of paranoid whimsy, in which exam boards and markers were fingered as part of a New Labour plot to push higher education entry toward its 50 percent target.

Little attention was given to the fact that the actual increase in numbers gaining A-Levels had been fairly modest over the past three years.[2] The 95.4 percent pass rate was enough to fix the image of a wildly inflated credential system in which swarms of 18-year-olds armed with certificates that they had racked up like scouting badges were descending on a higher education system that, itself, was bloated on a diet of golf course management degrees. Five months later, in January 2004, the lingering question of where the onus of "investment" in mass higher education should lie, ultimately a question about the nationally aggregated value of degree qualifications, provoked New Labour's largest ever rebellion over home policy. This was quickly followed by the first leaks of the Tomlinson Report, which suggested the phasing out of A-Levels. Details aside, the banner headline was clear: qualification inflation was rampant, the postcompulsory system was devalued and more A-Level passes, undergraduates, and degree courses could only mean worse.

Newsprint and Ideology

In addressing the specific forms via which ideological debates on qualifications are mediated, it is important to acknowledge that the annual exam furore is, in part, due to journalistic custom and culture. This is not to deny the extent to which news coverage is shaped by explicit political agenda but, rather, to recognize that

> ...even explicit ideological statements must be organised and presented through media formats and...such formats contribute to the shape and texture and emphasis of...coverage. (Altheide, 1987: 164–165)

The falling standards debate is a template populated by iconic language: gold standards, traditional subjects, soft subjects, dumbing down. Williams (1997: 27) makes a number of important points about the way in which iconic terms (or condensation symbols) of this kind operate within education debates: first, as formats that structure and shorthand complex arguments (organizing them around "polarised and simplistic dichotomies") and, second, in relation to how they "link with much wider educational and political discourses upon which they draw and into which they feed" (ibid.). One structural cum ideological feature of such terms is that they interlink discourses of derision and idealization, as a means of legitimizing and normalizing preferred viewpoints and practices (Kenway, 1990; Wallace, 1993). Another is that icon terms "can allow for contradictory policies and practices to be reconciled" (Williams, 1997: 27).

Moreover, it is important to acknowledge that news media may indeed serve dominant social interests, while also recognizing that these dominant ideologies remain contested in specific, localized contexts, examples of which are the processes and formats of news item production. In short, we should heed Miller and Philo's (1999) dictum that, while the news media do not necessarily tell people what to think, they certainly impact on what people think *about*. Insights into the narrative and presentational processes via which falling standards debates are structured in the popular media can be gained by analysis of three particular elements of print and broadcast forms (Warmington and Murphy, 2004). The first of these is the distribution of different headline categories and narrative themes. The second is the role of *news templates* in shaping the coverage of A-Level exam results, in which news templates are defined as the structural, narrative, and technical formats that exist *prior* to the emergence of specific news events and that are drawn upon by the news media in order to present news issues and debates in readily consumable form. Third, through headline and content analysis, it is possible to identify the discursive features that populate what has become the dominant template for A-Level coverage: the claim that exam standards are falling and are dragging down standards in higher education along with them.

Headline News: Falling Standards

Warmington and Murphy's (2003, 2004) analyses of headline and narrative themes in print and broadcast coverage of A-Level results in 2002 and 2003 reveal that in both years falling standards headlines comprised the largest category of headline types. This encompassed headlines suggesting that it was now easier to obtain A-Level pass grades, items referring to declining curriculum standards, headlines calling for reform of the A-Level system and (one of 2003's key themes) the claim that pass rates were rising because students gravitated toward supposedly less demanding subject options. Within this headline category, there were gradations of distance between the falling standards thesis and editorial lines. Nevertheless, in each case an editorial decision had been made to prioritize falling standards as the key news issue.

In 2002 falling standards stories accounted for 31.5 percent of print and broadcast news headlines about A-Level results (Warmington and Murphy, 2003). However, the types of comparison upon which claims about declining standards were based were, to say the least, flexible. By turns, blame was laid on incomparability between the newly designated AS and A2 [1] stages ("The root of the debacle is the ridiculous fiction...that an AS-Level is equivalent to half an A-Level," *Daily Telegraph*, 18 September 2002); on lack of comparability over time ("I got 6 A Grades but I wouldn't have passed them all 10 years ago," *Daily Mirror*, 16 August 2002) and unreliability in the grading of different student groups. It was the latter that generated the most explicit anxieties over the relationship between widening participation and qualification value. The *London Evening Standard*, for instance, made specific accusations about the practices of the Oxford, Cambridge, and Royal Society of Arts Exam Board (OCR) exam board.

> OCR felt that it was losing credibility among ministers for recklessly awarding higher marks to brighter pupils. It therefore engaged in some "reverse discrimination." (*London Evening Standard*, 18 August 2002)

There were other ideologically motivated claims that suggested the UK's exam boards were involved directly in liberal social engineering:

> ...the state system was made to look as if it was narrowing the gap (with private schools)...it stinks of class prejudice of the worst kind. (*Sun*, 19 September 2002)

Ultimately, the *Daily Telegraph* suggested that the devaluation of A-Levels might be gauged by incomparability between systems, as "good schools," opted out of the A-Level system, in favor of the International Baccalaureate:

> What a statement it would be about our education system if its "gold standard" was judged to be so fatally compromised that the brightest and best wanted nothing more to do with it. (18 September 2002)

In 2003 the falling standards template was by far the largest category of print headline (39.4 percent) and was almost as prominent in broadcast news headlines (36 percent). Moreover, 2003's reporting added a third, widely aired variant: the claim that the rise in pass rates (from 94.3 percent to 95.4 percent) was fuelled by a trend for students to opt for "easier," "soft" subjects. The hard/soft dichotomy appeared in a quarter of 2003's print narratives and was clearly a variant of falling standards types. First, in depicting the A-Level popularity of subjects such as Psychology as a new trend, a decline in standards over time was implied (these 'easier' subjects were often described as new or nontraditional subjects). Second, the purported lack of comparability between, say, Maths A-Level (hard) and Business Studies (soft) suggested a lower standard of syllabus content in the latter, and consequently also indicated the erosion of a single A-Level gold standard ("We do still have to address the issue of equal standards for all A-Level subjects," *Guardian*, 26 August 2003). Where comparability is undermined, so is the capacity for A-Level qualifications to function as a nationally agreed form of exchange and performance prediction.

Reading Between the Lines: Use and Exchange-Value

Media panic about "declining" exam standards is sinewed with anxieties about qualifications as highly unstable commodities: that is, as bearers of shifting use and exchange-value. As previously suggested, there are two broad senses in which the news media depict the rise of exam pass rates as indicating diminishing value. First, the rise in pass rates upward is often depicted as an indication that the exams are easier than they were in the past ("Heads hit out at easy courses as A-Level passes rise again." *Guardian*, 14 August 2003). This narrative has several variations: that curriculum content is now less demanding, that exam formats are less testing (an accusation made frequently following the reconfiguration of A-Levels into AS and A2 components), that grading practices are over generous, that schools and colleges increasingly "teach to the exam," concentrating on exam technique as opposed to the development of a broad knowledge base.

These are commonly presented by the media as issues of curriculum and assessment "quality." However, all of these criticisms also refer to what might be broadly described as the *validity* of qualifications: that is, the extent to which possession of a qualification accurately reflects the level of knowledge and skills (or, at least, the performance under assessment conditions) of the holder. Clearly, one use-value of educational qualifications is their function as valid predictors of the holder's performance in higher education or employment. A common criticism is that pass rates are such that even top "A" grades no longer distinguish between the good and the excellent ("A-Levels will be a success when more pupils fail," *Sunday Times*, 17 August 2003). In addition, each August, there is no shortage of higher education tutors ready to opine that the quality of undergraduate entrants has declined over the years. Ultimately, as a predictor, the use-value of qualifications lies in its function as an indicator of the holder's labor power potential. By labor power potential, we refer to the nexus of knowledge and skills that configure the holder's capacity to labor (Rikowski, 2002), whether in further educational endeavor (wherein the student's potential is usually characterized as the capacity to benefit from higher education) or in the labor market (wherein the emphasis is upon the employer's capacity to benefit from the holder's labor).

However, as with all commodities, qualifications exist in doubled form; they are both exchange-values and use-values (Marx, 1976). There is a second cluster of news narratives that are organized around a qualitatively different set of criticisms, which assert that, as the currency of A-Levels and degrees becomes more widely held, so their value is undermined. It is an argument that makes scant reference to the fact that, despite the headlining of 95 percent plus A-Level pass rates, only around a third of 16–18-year-olds in the UK study A-Levels. Moreover, this line of argument refers to something more than a concern with the diminished *use-value* of A-Levels and degrees as predictors; here media concerns also tend to focus on qualifications that are "meaningless" or "devalued" in terms of *exchange-value*. The system has, it is held, betrayed the 18-year-old Left holding a 'worthless' qualification in Media Studies, because it counts for little with employers or higher education institutions. The exam system has also betrayed the bright school-leaver who finds her string of A grades is still not sufficient capital to allow her entry to Oxford or Cambridge.

> More than 5,000 students achieving three straight A's at A-Level were rejected by Cambridge University last year, fuelling concerns that the education

system's so-called gold standard has been devalued...new figures reveal how universities are facing even more difficulty in choosing between large numbers of well qualified candidates for a limited number of places. (*Guardian*, 22 February 2005)

Here the rising pass rates/falling standards thesis characterizes qualifications as currency: as a socially accepted general equivalent, but one being undermined as a form of exchange by credential inflation.

Visible Value

As previously noted, 2002 was a pivotal moment in the ongoing debate over falling exam standards. However, as the year drew to a close, higher education funding emerged as the dominant education story, bound up with speculation about the effect that the imposition of tuition fees or graduate taxes might have on students' decisions about undertaking higher study. This was often expressed via attempts to calculate average levels of student debt against the increased earning power of graduates over their working lifetimes. Cohen (2003: 27) was one of a very few journalists to reflect upon the iconic terms of the fees debate:

> Throughout the debate on the funding of universities, one suspiciously precise figure has been cited by nearly every politician and pundit. A university education guarantees the "average" graduate an extra £400,000 over his or her lifetime. Not £300,000 or £500,000, but £400,000 on the button. (*Observer*, 26 January 2003).

As Cohen explained, such averages "inevitably hide wild variations." Yet the promise that each student's initial outlay in fees and living expenses will be amply repaid in the graduate labor market is more than a piece of statistical crudity. Whether graduates can be expected to earn an extra £400,000 or £300,000 is less significant than the news media's explicit expression of degree qualifications as an exchange-value. Moreover, the averaging of the exchange-value of a degree conceals the social relationships of education under capitalism behind the smoke and mirrors of consumer fetishism. The exchange-value repeatedly posited during the fees debate constructs students as "consumers" or "investors" within a crude human capital model, in which initial investment in one's own education is repaid by the handsome accumulation of a graduate salary. The fact that the earning potential of graduates remains deeply furrowed by inequalities of social class, age, gender, race, and of course job status is rhetorically dissolved by the averaging, one nation fictions of the consumer model.

> ...as citizens we can balance the rights we can expect from the state, with the responsibilities of individuals for their own future, sharing the gains and the investment needed. (DfEE, 1998: 8)

The consumer-investment model inevitably raises the stakes in relation to qualification value. Qualification commodities bear exchange-value (for instance, a supposed lifetime graduate earnings advantage) and use-value (as a measure of past competence and a predictor of future performance). In addition, of course, educational qualifications configure a nexus of other use-values: "cultural capital" derived from participation in a prized form of social activity, a kind of "self-evident" social inclusion, self affirmation. It is in this intensified context that the heated debate around qualifications must be understood: as a discourse that both reveals and conceals the social relations that underlie qualifications as commodities.

Conclusion: Standards and Scapegoats

The intense news media debate over falling exam standards is pervaded by dualities. Even where its language is swathed in apparent nostalgia for a time when an elite minority entered higher education ("Turning the clock back may not be the worst option," *Independent*, 19 September 2002), capitalism's real thrust is always forward. Therefore, what appears to be a harking back to the standards of yesteryear is, in fact, a drive to inscribe new hierarchies of exchange-value and cultural capital. All A-Levels, all A grades and all degrees are not, say the headlines, created equal. The media stimulated anxiety over the diminished value of A-Levels as a selector of "the brightest and best" was a driver behind the Tomlinson Report (early leaks of which suggested, for example, that current A and B grades might be replaced by a detailed set of scores organized around four or five levels, thus enabling fine distinction between bright HE (highes education) applicants with identical top scores (*Guardian*, 12 February 2004).[3]

Also it is tempting to read falling standards claims simply as yet more scapegoating of the public sector by a hostile right-wing press. Yet this shows an insufficient grasp of the complexity of what is taking place. First of all, there is a real power attributed by the news media in the falling standards debate to a range of actors: the exam boards, curriculum designers, widening participation advocates, teachers who teach cynically to the exam, students who craftily manipulate the A-Level system. Their power is depicted as *actual* power, rather than the *potential* power of the scapegoat (a distinction borrowed from Postone, 1986). Allegedly, the power of these destabilizing forces is so great that "[t]he system must be liberated from the terrible incubus of political control" (*Daily Mail*, 19 September 2002).

In A-Level news coverage the undermining of the gold standard is bound up with diminishing exchange-value but it is not identical with it. This ambiguity is revealed in the shifting expressions of idealization and derision present in the falling standards template, wherein the putative diminishing of the gold standard is also identified with social restructuring and dislocation. It is hardly surprising, then, that news coverage of A-Level exam results and related issues has become highly charged and often incoherent, as it navigates the contradictory drives of postcompulsory education in the learning age. While it is understandable that many educationalists regard the annual A-Level debate, in particular, as ritualistic and often trivialized, this chapter suggests that there are genuine reasons why the debate remains ritualistic and trivial, in that each year's coverage represents another veiled navigation of the fundamentals of value and commodification, concealed behind the euphemisms of standards and quality. This chapter does not offer a naive hope that leftist educators can, at the touch of a keyboard, "influence" the news media. It is, however, concerned that scepticism of news coverage should not breed denial of its place within the contemporary educational settlement. Instead, it suggests that the value of analyzing the construction of education news lies in potential insights about the complex negotiation of the value of qualifications within capitalism's education settlement. There is no doubt that the news media are often guilty of caricaturing their subjects but that is no reason for Marxists' analysis of the media to fall into caricature also.

Notes

This chapter draws, in part, from the two *News Media Depiction of A-Level and GCSE Examination Results* studies, conducted at the University of Nottingham in 2002 and 2003, the second of which was funded by the Economic and Social Research Council. This chapter is indebted to Professor Roger Murphy, who directed both projects. Acknowledgment is also due to Colin McCaig who provided insightful analysis of New Labour's manifesto contents.

1. The reform of A-Levels in 2000 reorganized the qualification into two parts: the AS (Advanced Subsidiary) and the A2. The AS is a stand-alone qualification, valued as half a full A-Level. The more demanding A2 is the second half of the qualification; those achieving A2 are awarded the full A-Level.
2. In 2001 the 89.8 percent A-Level pass rate represented 672,500/n 748,886 (subject entries). By contrast, 2002's 94.3 percent pass rate for A2 represented 661,401/n 701,380. In 2003 the A2 pass rate was 95.4 percent (716,012/n 750,537). The A2 pass rate in 2004 was 96 percent (735,597/n 766,247).
3. In early 2005, as this chapter was being prepared, the first news of the current education secretary's response to the final Tomlinson report began to emerge. Despite Tomlinson's recommendation to develop a new overarching baccalaureate qualification, initial government statements strongly emphasized that A-Levels are to be retained as the standard academic qualification for 16–18-year-olds. It would be difficult to furnish stronger evidence of the continued attachment of UK governments to the A-Level gold standard.

References

Ainley, P. (1994) *Degrees of Difference: Higher Education in the 1990s* (London: Lawrence and Wishart).

Altheide, D. (1987) Format and Symbols in TV Coverage of Terrorism. *International Studies Quarterly*, 31: 161–176.
Cohen, N. (2003) The Miserly Generation. *The* Observer, 26 January, 27.
Cole, M. and Hill, D. (1999) Into the Hands of Capital: The Deluge of Postmodernism and the Delusions of Resistance Postmodernism. In D. Hill, P. McLaren, M. Cole, and G. Rikowski (eds.) *Postmodernism in Educational Theory*, 31–49 (London: Tufnell Press).
Collins, R. (1979) *The Credential Society: An Historical Sociology of Education and Stratification* (New York: Academic Press).
Cottle, S. (1995) The Production of News Formats: Determinants of Mediated Public Contestation. *Media, Culture and Society*, 17: 275–91.
Department for Education and Employment (DfEE). (1998) *The Learning Age: A Renaissance for a New Britain* (London: DFEE).
Department for Education and Skills (DfES). (2003) *Every Child Matters*. Green Paper (London: DFES).
Hatcher, R. (1998) The Limitations of the New Social Democratic Agendas. In S. Ranson (ed.) *Inside the Learning Society*, 205–213 (London: Cassell).
Jeffs, T. (1999) Are You Paying Attention? Education and the Media. In B. Franklin (ed.) *Social Policy, the Media and Misrepresentation*, 157–173 (London: Routledge).
Jessop, B. (1990) *State Theory: Putting Capitalist States in Their Place* (Cambridge, UK: Polity Press).
Kenway, J. (1990) Education and the Right's Discursive Politics: Private versus State Schooling. In S. Ball (ed.) *Foucault and Education: Disciplines and Knowledge*, 167–206 (London: Routledge).
Labour Party. (2001) *Ambitions for Britain: Labour's Manifesto 2001* (London: Labour Party).
Marx, K. (1976) *Capital*, Vol. 1 (London: Penguin).
McCaig, C. (2003) School Exams: Leavers in Panic. *Parliamentary Affairs*, 56: 471–489.
Miller, D. and G. Philo. (1999) The Effective Media. In G. Philo (ed.) *Message Received: Glasgow Media Group Research 1993–1998*, 21–32 (Harlow: Longman).
National Audit Office. (2002) *Widening Participation in Higher Education in England* (London, NAO).
Postone, M. (1986) Anti-Semitism and National Socialism. In A. Rabinbach and J. Zipes (eds.) *Germans and Jews since the Holocaust: The Changing Situation in West Germany*, 302–316 (London: Holmes and Meir).
Rikowski, G. (2002) Methods for Researching the Social Production of Labour Power in Capitalism. Research Seminar, University College Northampton, 7 March.
Riseborough, G. (1993) Learning a Living or Living a Learning? An Ethnography of BTEC National Diploma Students. In I. Bates and G. Riseborough (eds.) *Youth and Inequality*, 32–69 (Buckingham: Open University Press).
Wallace, M. (1993) Discourse of Derision: The Role of Mass Media within the Education Policy Process. *Journal of Education Policy*, 8: 321–337.
Warmington, P. (2003) "You Need a Qualification for Everything These Days." The Impact of Work, Welfare and Disaffection upon the Aspirations of Access to HE Students. *British Journal of Sociology of Education*, 24: 95–108.

Warmington, P. and R. Murphy. (2003) News Media Depiction of A-Level Results in 2002. Paper presented at the British Educational Research Association Conference, Heriot-Watt University, Edinburgh, 11–13 September.

———. (2004) Could Do Better? Media Depictions of UK Educational Assessment Results. *Journal of Education Policy*, 19: 285–299.

Williams, J. (1997) *Negotiating Access to Higher Education: The Discourse of Selectivity and Equity* (Buckingham: Open University Press/Society for Research into Higher Education).

Part V

Labor and Commodification in Education: Theory, Practice, and Critique

Chapter 12

Academic Labor: Producing Value and Producing Struggles

David Harvie

Introduction

Do teachers labor within that "hidden abode of production" in which value and surplus-value are produced, in which "capital is itself produced"? (Marx, 1976a: 279–280.) Classical Marxism, following Marx's apparently explicit categorization of reproductive labor as unproductive (e.g., Marx, 1969: 161 and 172), has tended to answer in the negative. For Kevin Harris (the vast majority of) teachers are unproductive since they "are employed by the State and they are paid out of revenue" (Harris, 1982: 57). And Simon Mohun has suggested that "labour-power is not a produced commodity in the same sense [as other commodities]. It is a capacity or potentiality of people, and people are not (re)produced under capitalist relations of production.... Labour-power...is a commodified aspect of human beings, and human beings are not produced in any valorisation process" (Mohun, 1994: 398 and 401). Activities such as the "daily and generational reproduction of labor-power,"

> do not produce value, because there is no social mechanism for commensurating different labor activities, and so there is no way in which the time taken in such activities can be regarded as "socially necessary."...Such labor is non-productive; indeed, in value theory terms it does not count quantitatively at all. (Mohun, 1996: 38)

I suggest here that it may in fact be more useful to understand teachers' labor as productive of value and surplus-value for capital. *First*, teachers' activities are increasingly taking the form of alienated and abstract labor, where abstract labor is the substance of value. That is, teaching labor is, in itself, becoming directly productive of value. *Second*, teachers do *produce* the commodity labor power and, through this production (or rather coproduction), they also produce surplus-value, though this surplus-value is only realized through the exploitation of new labor power. To the extent that teachers produce value and surplus-value, they exist within capital. But teachers are also human beings with their own needs and desires. Teachers struggle. They struggle against this tendency of their labor to become more alienated and abstract. (And it is primarily for this reason—struggle—that this is a tendency only, that teachers' labor, like all labor, is contradictory.) They struggle against their roles as mere producers of labor power. They may also struggle to posit alternative ways of being, alternatives that go beyond the capital relation. Thus, I suggest that teachers are *also* unproductive workers. Or rather, teachers are also productive of struggles and, to this extent, they exist against-and-beyond capital.

In the chapter I attempt to analyze and illustrate both the ways in which teachers (re)produce the capital relation and value, that is, exist within capital, and the ways in which their already existing practices rupture this (re)production and thus produce struggle. I treat these two antagonistic moments—value production and struggle—separately, but this is for ease of exposition only; in reality they cannot be disentangled. Finally, I suggest that this understanding of value and its production is useful since recognizing these modes of existence, and the forms of activity that constitute them, can be an important tool in our struggles against capital and to expand the spaces and possibilities for alternatives ways of being.

Academic Labor: Becoming-Alienated, Becoming-Abstract

Productive labor is that which produces value and surplus value for capital. But what is value? Value is embodied labor that is also abstract labor. The substance of value, then, is abstract labor. For Marx, abstract labor is "human labour-power expended without regard to the form of its expenditure." Like labor, like result, "All its sensuous characteristics are extinguished" (1976a: 128). Massimo De Angelis suggests that abstract labor is labor that is "alienated, imposed, and boundless in character" (1995: 111; see also 2004). Labor is alienated because the work activity appears to the worker as an external power, outside their direct control: it is not "the satisfaction of a need but a mere means to satisfy needs outside itself" (Marx, 1975: 326). Since alienated labor appears as an external power, such labor is "not voluntary but forced, it is forced labour," that is, it is imposed (ibid.: 326). And since abstract labor, by definition abstracts from concrete labor and from the useful character of concrete labor, it cannot be limited by a set of needs. It is thus boundless, "production for production's sake" (De Angelis, 1995: 111–113).

Our understanding of abstract labor—and thus value—is of a tangible reality, a lived experience, the "sensuous-less" of alienated, imposed, and boundless activity (see De Angelis, 1996, 2004). And this tangible reality is as applicable to the labor performed in capital's reproductive circuit, $LP-M-C\{MS\}\ldots P\ldots LP'$ (see Cleaver, 2000: 123), as it is to that performed in its industrial circuit, $M-C\{LP, MP\}\ldots P\ldots C'-M'$. In his *Economic and Philosophical Manuscripts of 1844* (Marx, 1975), Marx distinguishes four aspects of alienated labor. Under the capitalist mode of production, workers become alienated from (1) the act of production, that is, from their activity; (2) the product of their labor; (3) their own species being; and (4) their fellow workers. Each of these aspects of alienated labor is increasingly applicable to the labor of teaching (and also to that of studying), which, moreover, frequently appears as imposed and boundless, as I discuss in more detail below. We can thus characterize teaching labor as *becoming-alienated*.

First, as curriculum content and teaching methods are increasingly dictated by external powers, teaching labor becomes an activity that is alien, an activity that does not *belong* to the teacher. On the face of it, academics retain considerable control over the content of their teaching. But university curricula in the UK are now required to conform to so-called Subject Benchmarks, produced and policed by the Quality Assurance Agency for Higher Education (QAA)—which is itself part of an International Network of Quality Assurance Agencies in Higher Education. Besides setting-out guidelines for "subject-specific learning outcomes," Subject Benchmark Statements catalogue so-called generic learning outcomes, which all university degree programs must produce. These include written and oral communication skills, numeracy, the ability to work in groups, IT skills, critical thinking: in short, "desirable labor power attributes" (Rikowski, 2002/2003). Further, many academic disciplines are themselves under attack and have been forced to reinvent themselves as a response to market pressures. In many universities arts, humanities, pure and social science degree programs have contracted or even been completely closed down; others have been forced to refashion themselves as branches of business or management studies. In both cases, course content is imposed on academics by an alien power, whether state bureaucrats or (global) "market forces" (though the extent of this imposition is tempered by negotiation, mediation, and other struggle).

And of course, it is not sufficient that teachers teach: in addition, university lecturers (along with schoolteachers) are required to perform the alien and imposed labor of categorizing students and ordering them into hierarchies. This activity of hierachizing and categorizing—principally writing references and grading—new labour power is a central function of educators within the capitalist mode of production. Harry Cleaver summarizes it nicely:

> In the language of George Caffentzis's [1980/1994] essay on "The work/ energy crisis and the apocalypse" [teachers] are expected to play the role of "Maxwell's Daemon": sorting low from high entropy students—giving high grades to the former because they have demonstrated their ability and willingness to make their energy available for the work they are assigned and giving low grades to the latter who either won't or can't. (2004: 6)

Thus teachers are under increased pressure to subsume their own interests as human beings, whatever these may be, to those of capital and, in such circumstances, their labor becomes forced, an alien activity.[1]

Besides external forces acting on work *content*, both school and academic work*loads* have expanded sharply as student numbers have grown and as the quantity of both tests and "quality assurance"–related administration work has proliferated.² No one who has completed the paperwork associated with QAA "audits" (or Ofsted inspections in schools) can doubt the alienated, imposed, and boundless nature of this activity! And the number of tests and other forms of assessment proliferates: the phrase *testing for testing's sake* expresses well the boundless nature of this system of imposing work.

The *second* aspect of alienation concerns the alienation of workers from the product of their labor. Now the "product" of a university lecturer is the graduate who, as a result of their education is now supposedly the bearer of a range of knowledge, skills, and attributes. But as control over curricula is increasingly determined by the needs of capital (whether mediated by the state or the market), this knowledge and these skills and attributes will increasingly correspond to desirable labor power attributes. Teachers help produce new labor power. But this new labor power—new workers—will in turn be employed to produce value and surplus-value, that is, to produce and reproduce capital, the very social relation that exploits human beings, including teachers. Thus, the product of teachers' labor is turned against them and they thus become alienated from this product.

Third, the *becoming-alienated* labor of teaching includes the alienation of teachers from their own species being. In all human societies there have been mechanisms, customs, and social codes to regulate the transmission of knowledge, skills, behavioral norms, and so on. It is the existence of culture, understood in this sense, that distinguishes humans from (most) other animals.³ Clearly education, in its broadest sense, is part of the cultural transmission mechanism. However, when education becomes subordinated to capital's need for particular subjects to be taught, in a particular context, then teachers—those employed to teach these subjects—become estranged or alienated from their human species being, their natural propensity to impart knowledge and nurture thinking in others. This species being then becomes "a being *alien* to him and a *means* of his *individual existence*" (Marx, 1975: 329, italics in original).

Fourth, as a result of teachers' alienation from their activity, from the product of their labor and from their own species being, they also become alienated from their fellow workers. The workers with whom teachers most closely associate are of course their students. But the relationship between teacher and student is mediated by their mutual imposition of work. On the one hand, the teacher imposes work on the student—through assignments and tests, requiring attendance in lectures, and so on. The hierarchical nature of the relationship here is clear. On the other hand, the student may impose work on the teacher—for instance, seeking extra help, using various tactics to apply pressure on teachers to do additional work in order to make tedious material more interesting. And, for some teachers, positive student feedback is essential for continued employment. This situation arises because this work of teaching and of studying is externally imposed: teachers must teach particular material, whilst students face various economic pressures to obtain academic qualifications in particular subjects, which may themselves hold little interest. It is no surprise that teachers and students often have an antagonistic relationship as they become increasingly alienated from one another.[4]

Teachers also become alienated from their fellow teachers as their mutual relationships become more and more mediated by external pressures (powers). Such pressures or powers can take the form both of bureaucratic structures and of "market forces." Here is an example of the former: a teacher may be forced to justify to their peers (organized into some committee) how a new teaching arrangement conforms to a particular set of guidelines. An example of the latter is the competition, within and between institutions, to attract students to particular courses or universities. And though not the main focus of this chapter, it is also the case that student becomes alienated from fellow student, as competition for good grades intensifies. For the student seeking "employability," the goal is less to become good at something; rather it is to become better than one's peers. A first-class degree is meaningless if everyone gets one! Antagonism may also arise between students who simply want to be taught only whatever is necessary in order to obtain a good grade and those who wish to spend class time exploring issues of interest to them that lie outside the examined curriculum.

Thus the work of teachers (along with that of students) is increasingly taking on the characteristics of alienated labor; teachers' labor is *becoming-alienated*: teachers are becoming estranged from their activity, the product of their activity, their fellow humans, and their own species being. Their labor is increasingly imposed and, moreover, is boundless, without limit. As Glen Rikowski (2000) suggests, "school improvement" is a concrete expression of capital's *"social drive to enhance the quality of human labour power* [which] like all of capital's social drives is *infinite"* as a process of commodification (italics in original). Teaching labor, then, increasingly takes on a twofold nature, becoming a contradictory unity of concrete labor *and* abstract labor, the substance of value.

Commensurability and Measure

Many of the tendencies forcing teaching labor to assume the qualities of abstract labor also facilitate the establishment of *socially necessary labor-time* (SNLT) for various teaching activities. This, in turn, allows, *first*, the *commensuration* of teaching labor, vis-à-vis other teaching labor and completely different concrete labor, and *second*, the driving down of these socially necessary labor-times. Other important tactics include the construction of league tables and fostering cultures of "best practice" and "efficiency." League tables facilitate quantitative comparisons of schools, universities, and/or departments and, coupled with their use in determining funding, become a key tool in capital's strategy of marketization. SNLT is driven down as individual "units" strive to become "more efficient," or to catch up with current "best practice."

In terms of teachers" teaching and students" studying, structures employed to impose content are also used to enforce quantities: to define, for example, *how much* a student should know in order to be awarded a particular qualification. Across the university sector, the role of external examiner, now augmented by the QAA, is to ensure comparability across institutions. Within universities, it is the role of "quality managers" (or similar) to ensure this commensurability. Such individuals may monitor course content, modes of assessment, and the location and dispersion of grades awarded.[5] The acquisition of knowledge and development of critical faculties are broken up into discrete steps as "learning outcomes" are codified across levels of a degree program. For instance, first-year undergraduate students are usually only required to "explain" concepts or theories, whilst finalists will be expected to "critically evaluate." There is a clear tendency here toward rigid definition of the quantity of work (number of hours) required of a student of "average ability" in order to achieve a certain qualification and grade, that is, to define socially necessary labor-time for the labor of studying.

There exist parallel tendencies toward the definition of SNLT for the labor of teaching. If class contact hours and assessment methods are standardized across courses or modules for students (say, x lecture hours and y seminar hours per week for 20 weeks, assessed by 2500-word essay and 3-hour examination) then this standardization frames workload calculations for teachers too—the other key variable is the number of students taught. Managers can (and do) construct workload models on this basis, from which emerges a norm for the average number of hours required to teach a course unit or module of so many credits to a certain number of students. Such norms are, of course, ridiculous, but they are also *real*, as "inefficient" teachers—those unable to meet the norm—come under pressure to become "more efficient" or else work in their own time. Thus SNLT emerges for teaching labor.

"Research-active" academics have to "meet or beat" another set of norms—SNLT—those emerging from the machinations of research selectivity and the various funding councils (see Harvie, 2000). Spanning academics' activities as a whole is the series of Transparency Reviews, imposed on English and Welsh universities by the UK Treasury and implemented by the Higher Education Funding Council (HEFC). The aim here is to discover how much time academic staff devote to various aspects of their work—teaching, teaching-related, administration, research, and so on. This is yet another *metric*, which can be used to make commensurable these various activities. And the metrics developed within the education sector do not stand apart from wider capitalist society. A fast-growing economics literature is concerned with estimating "returns to schooling," both private (to the individual) and social (to the economy as a whole).[6] Using such studies, and experimenting with alternative funding models, such as granting universities the freedom to charge variable top-up fees, capital can attempt to link education labor to the wider economy. For capital, the ideal is to make commensurable the concrete labor of any individual academic or schoolteacher with that of any other social subject. Of course, in such a model teacher remuneration is tied to productivity and performance. We can already see examples of this strengthened link between teaching work and income in the proposals for advanced skills or super teachers and teachers' discretionary payments (see, e.g., Boxley, 2003), and in leading universities' demands for the freedom to tear up national bargaining agreements in order to set market rates for academics' pay.[7]

Finally, the culture of best practice embodies many of the tendencies described above. *First*, the requirement to conform to best practice imposes work. *Second*, this labor is alien since best practice is defined by an external power. *Third*, since best practice is usually defined in generic terms, it also facilitates comparison of the performance of teachers across subject areas and across institutions. We can also note the frequency with which best practice is actually the creativity of teachers themselves, appropriated and turned against them. Given a specific task that needs to be accomplished, an individual teacher or group of teachers may invent a new, better, or more efficient way of performing it, a way that is simply quicker or preferable for them perhaps. However, in time, managers may define this new method as best practice, read *minimum acceptable* practice. This has the effect of imposing more work on teachers in general and/or of driving down socially necessary labor-times. "The best practice always hints at a better one, even as it winks at the question of 'Better for what?'" (Martin, 2003: xi)

The Reproduction of Labor Power and the Production of Surplus-Value

In *The Arcane of Reproduction*, Leopoldina Fortunati (1995) makes a forceful argument that housework is productive of value and surplus-value. The reproductive labor of the (usually female) houseworker, which reproduces the (male) wage-worker, increases both the value of the latter's labor power and its use-value. The value of this labor power is increased since this reflects its costs of reproduction, which include the houseworker's labor, whilst it has enhanced value in use since a well-fed worker, one whose emotional needs have been met and so on, is likely to be more productive. But the houseworker's labor has some special characteristics: it is (1) unwaged, (2) largely not recognized as labor, being considered rather as a "natural" activity and perhaps performed "for love," and (3) does not appear to be organized capitalistically. Finally (4) her product is inseparable from a commodity, the male's labor power, which must be owned by someone else, that is, the male, since a defining feature of the capitalist mode of production is that one's own labor power is exactly that, one's own (until sold, for a particular and limited time period, to an employer). Because of these characteristics, which largely concern the hidden nature of the houseworker's labor, capital is able to systematically pay the male worker a wage (exchange-value of his labor power) below the value of his labor power. That is, his wage reflects the costs of reproduction of the wage-worker and his family: this wage is sufficient only to pay the houseworker for her necessary labor; however she also performs surplus labor, which increases the value of the male's labor power. Thus, according to Fortunati, capital exploits two workers with one wage; both workers produce value and surplus-value and hence valorize capital.

We can easily extend Fortunati's argument to apply it to teachers' labor, which, although waged, shares with housework the other special characteristics. At least historically, teaching has appeared less like labor than other forms of labor. Many teachers have seen themselves, and been seen, as engaged in a vocation, their activity a natural part of themselves. This is particularly true for academics and certainly, until the current restructuring, universities have not appeared to have been organized in a capitalistic manner. Finally, like housework, the product of teachers' labor cannot be separated from a commodity, labor power, owned by someone else. Given these characteristics of teaching labor, the exchange-value of, say, a university graduate's labor power (the wage they can command), may reflect the cost of their education, that is, the *value* of their lecturers' labor power, rather than the value *produced* by their lecturers and embodied in the graduate. Yet, the graduate is, at least potentially, more productive by virtue of their education: by employing the graduate, capital is thereby able to appropriate not only the surplus-value produced by the graduate, but also that produced by those who taught them.

But this argument hinges on the *produced* nature of labor power.[8] Although it is undoubtedly true that, as Simon Mohun (1994) suggests, labor power is a commodified aspect of human beings, this aspect is not *natural*:[9] this capacity or potentiality of people must be shaped. In general, labor power—the capacity to labor—does not simply mean the *ability* to perform physical or mental work. It means, in addition, the willingness to do so under another's control, regardless of whether this control is direct or indirect and whether exercised by a private capital or by social capital. It means the willingness, even if reluctant and "unwilling," to comply with capital's discipline. Nor are the different abilities, which distinguish one form of concrete labor from another, natural either. These will tend to vary across historical periods, geographical locations, and strata of workers. For most labor powers in advanced capitalist economies, a certain degree of literacy and numeracy are desirable aspects, but this has not always been the case and it is still not everywhere the case. Some labor powers are expected to possess the ability to think creatively and to solve problems; others are expected to be able to follow instructions without question. In some labor powers the ability to kill other human beings efficiently is a desirable characteristic, while others must be tender and caring. Although all of these characteristics—the ability to read and write, to be creative, to kill, to be tender, and so on—have existed in human beings prior to the capitalist mode of production, one cannot say that capital has simply appropriated certain of them, without changing and developing them. We can think of this process of capital's development and shaping of labor power, as opposed to mere appropriation of it, as part of the *real*, as opposed to merely formal, subsumption of labor under capital (Marx, 1976b). Really, it is impossible to disentangle any natural, human "essence" of people from their existence as labor power within capital. As John Holloway (2002) has argued, our humanity is "damaged"; or in Cyril Smith's (1996: 69) formulation, humanity is "encased within...inhuman forms of life."

At the time when Marx was writing, capital made far more use of overt force to coerce people to submit to its discipline and *Capital*'s first volume contains a mass of historical information detailing the pressures on reluctant labor power. More recently and particularly since the social struggles of the 1960s and 1970s, at least in the developed capitalist countries of the North, capital has tended, and has certainly preferred, to use far more subtle methods of ensuring compliance. But this compliance must be produced, and a key figure in this process of production is the teacher, along with new labor power itself—a teacher can only teach: the student must also be willing to learn. What the teacher produces, or rather coproduces, then are this compliance and other desirable labor power attributes, whether general or particular, such as those discussed above. These produced attributes of new labor power enhance its value in use, but are the (coproduced) product of teachers' labor and surplus labor. Capital realizes the value and surplus-value produced by teachers—and is thus valorized—only through the exploitation of the new labor power.

Academic Labor and the Production of Struggle

I have argued that teaching labor has, or is tending toward, having, the following characteristics: (1) it is (becoming-)abstract labor, where abstract labor is the substance of value; (2) it is becoming commensurable with other teaching labors and other labors in general, facilitated through processes associated with the marketization of education; (3) it produces a commodity, namely the special commodity labor power; and (4) it produces surplus-value and valorizes capital, but that capital appropriates this surplus-value only through exploiting the labor power of school or university graduates. In as much as teaching labor has these characteristics, we can say that teachers are productive laborers—are productive of value for capital—and labor within the *school-as-factory*. To the extent that their labor has these characteristics, teachers exist *within* capital.

But, teachers are productive laborers only *inasmuch as* their labor has these characteristics. Teachers exist within capital only *to the extent that* their labor has these characteristics. In part, educators' labor *does not* have these characteristics because the processes and tendencies discussed in the above sections are continually contested: for this and other reasons, they are by no means complete and nor could they ever be complete. Given the *"affective,"* or *"immaterial,"* nature of education labor and the degree to which education and socialization is a social process, involving the cooperation of numerous subjects, how could the contribution of an individual teacher ever be measured? In this way then, Hardt and Negri's suggestion that the production of value now takes place "outside measure," where this "refers to the impossibility of power's calculating and ordering production at a global level," can be applied to educators' labor (2000: 357). But, far more important than this are the actions of educators themselves to resist measure, to refuse to be productive of value for capital, that is, the struggles of educators to be productive of struggle, to exist against-and-beyond capital.

Academics' and schoolteachers' struggles against the imposition of work (and hence against the production of value) take myriad forms. Many such forms are individualized. Most—simple work-reduction tactics or attempts to develop *human* relationships with students—are probably not consciously anticapitalist; rather they may be "mere" expressions of species being. These are nevertheless *against* capital. Indeed, some such activities may result in the greater imposition of work on students or colleagues. Other forms of struggle may be more collective. And some struggles may be conscious attempts to undermine capital or to transcend capitalist education, that is, they may go *beyond* capital.

The most visible struggles to oppose capital within education are those collective ones organized through various teachers' (and students') unions, though these unions' opposition is sometimes rather ambivalent. For instance, in the UK, National Union of Teachers members recently voted on a possible boycott of Sats (national standard assessment tests) for 7- and 11-year-old children, and teachers have been joined by parents in their opposition to testing with the launch of a "Stop the Sats" campaign (Escobales, 2003; Coughlan, 2003). The University and College Union campaigns on pay and conditions within universities, but appears less concerned about testing and performance indicators in themselves. In the United States, a large proportion of university and college teaching is done by casual labor—graduate students and "adjuncts"—and these workers are in the forefront of a growing academic labor movement (see Johnson et al., 2003). These teachers are leading unionization drives, challenging the "apprenticeship" model, used by administrators to continue graduate-student exploitation, and demanding collective bargaining. Besides such basic issues as pay and job security, activists are also taking on such issues as freedom of expression and the whole concept of "academic citizenship." Meanwhile in Africa, students and teachers have engaged in countless struggles—student strikes, teacher strikes, exam boycotts, demonstrations, road blockades, occupations of school and university buildings—against Structural Adjustment Programs adopted by their governments and education institutions. As Silvia Federici and George Caffentzis recount,

> In country after country, demonstration after demonstration, in its slogans, flyers, and position papers the African student movement has shown a remarkable homogeneity of demands. "NO to starving and studying," "NO to tuition fees," "NO to cuts in books and stationary," "NO to Structural Adjustment, to corrupt leaders, and to the recolonization of Africa" are slogans which have

unified African students in the SAP era to a degree unprecedented since the anti-colonial struggle. (2000: 115)

These collective, and sometimes very militant, struggles are clearly immensely important. For the "corporate university" chasing "global competitiveness," low labor costs and "market responsiveness" are key requirements. Hence its need for a casualized, flexible labor force. For global capital more generally, reducing social spending, including spending on education—that is, shifting the costs of reproduction back onto labor power itself—is essential to the success of its neoliberal project to restore capitalist profitability following the crises of the seventies. Capital must also strengthen the link between money and work. In other words, it must make more labor productive. Thus, we can understand all of these organized struggles—against structural adjustments, against funding cuts, against limitations on academic freedom, casualization; in Europe, North America, Africa, and elsewhere—as directly anti-capitalist and as struggles against the becoming-productive of the labor of teaching and studying.

But also important are the very varied, less obvious, and frequently hidden ways in which education workers, acting outside of any formal organizations, refuse capitalist work and seek to satisfy and develop their own needs and desires. In some senses, these forms of struggle are more interesting because of their near invisibility.

"Good" teachers are enthusiastic about their subject. They deliver stimulating, original lectures and they tailor their classes to meet students' interests. They set interesting exercises, welcome questions and discussion, and are always available to provide additional help. But all this takes considerable time and energy, and is particularly difficult when teaching activity must be subordinated to the external requirements of curriculum and assessment. One obvious work-reduction tactic, adopted by many teachers and academics, is to simply teach from the textbook and to constantly reuse old material. But while this may make life easier, though probably duller, for the teacher, it will almost certainly make classes far more tedious for students, and hence increase the amount of labor they will have to perform in order to achieve a certain grade (since tedious work is more laborious than stimulating work). Teachers may also make themselves as unavailable to answer student queries as possible, and be purposively unhelpful—which may include being condescending, arrogant, or aggressive—if and when they are tracked down. Again, this individual work-refusal tactic shifts labor from teacher onto student.

Lecturers or school teachers may reduce their own burden of marking/grading work by setting multiple choice question (MCQ) test and examination papers, which can be computer-marked. Besides reducing teacher work, this can also make life easier for students, many of whom prefer MCQ-type tests. As such, this tactic can be adopted as a means to free-up class time to pursue teachers' and/or students' own interests. Of course, such work-reduction tactics can also be perceived (by managers) as "efficiency gains," and thus a means to drive down (socially necessary) labor-times. Consequently there is a strong incentive for individuals who employ such work-reduction tactics to conceal their full effects from management. Teachers may also refuse the labor of imposing labor by providing their students with "hints" as to examination questions, making courses easier or systematically "marking-up." Another individual work-refusal tactic is that of refusing the labor of writing comprehensive and reliable student references. Here teachers are refusing their hierachizing and categorizing function. Many teachers and academics simply have too many students to know any very well. Rather than performing the laborious task of consulting files and colleagues, and then attempting to compose an accurate portrait that meets the employer's needs, it is far easier to simply continually recycle a few stock—and frequently glowing—reference letters.

Students' own varied collective and individual work-reduction tactics—cooperation to share work; plagiarism, made easier by the proliferation of internet sites; selective studying of topics; smuggling notes into examinations and other forms of cheating—are contributing to and exacerbating teachers' refusal and/or inability to order students into hierarchies and categories adequate to capital's needs. We can gauge the scale of this problem for capital from the latter's response. In addition to general laments over "falling standards" and grade inflation, and the growth in resources devoted to detecting plagiarism, many employers have resorted to increasingly sophisticated procedures to recruit new workers. These include two-, three-, or more-stage recruitment processes, interview-cum-assessments that extend over several days and psychometric testing. Clearly, these protracted recruitment procedures impose considerable labor on students/applicants, but we should understand them as part of capital's response to educators' refusal and/or inability to provide reliable information on job candidates' intellectual and character attributes. In effect, teachers' and university lecturers' (more or less individual) struggles are pushing the costs of screening, ranking, and categorizing new labor power back onto capital.

Besides individual and collective actions *against* capital, teachers may struggle to go *beyond* capital. Again this activity may be consciously anti capitalist, but more likely it is not. Many teachers and academics seek time and space to teach material that lies outside or beyond externally imposed curricula. Indeed some of this material may be implicitly or explicitly critical of capitalist social relations. In universities, lecturers may even be able to offer whole courses in Marxist or other critical theory. Teachers who are often all too aware of the power relationships that separate teacher from taught may explicitly explain these relationships in the classroom as a first step toward dissolving them, at least partially. It may also be possible for teachers adopt various tactics in order to weaken the link between labor performed in the classroom and the grade awarded.[10]

Conclusion

Within the capitalist mode of production, the education system performs a key function in the (re)production of that special commodity labor power.

I have suggested here that we can understand teachers' labor as productive of value and surplus-value for capital and, to this extent, teachers exist within capital. But the school or university is also a site of struggle. To the extent that teachers struggle—*against* the imposition of work in the school-as-factory and *for* activities and relationships that transcend capitalist social relations—their labor is unproductive, or rather it is productive of struggle, and they exist against-and-beyond capital.

I have treated the two antagonistic moments—on the one hand, that of producing value and, on the other, that of producing struggle—separately, but in reality it is difficult to disentangle them. Capital develops in response to struggle and, in turn, new forms of struggle emerge in response to capital. So, for example, capital's response to the problem of plagiarism is to invent detection software and best practice procedures for dealing with it. Capital's response to teachers' refusal (or inability) to perform categorizing and hierarchizing functions is to develop alternative processes to recruit new labor power. On their part, teachers have employed various forms of subterfuge in response to the imposition of workload models and bureaucratic structures. It is also the case that many teaching practices may be ambiguous, simultaneously combining *within*-capital elements with *against-and-beyond* elements. One such example is provided by radical academics who offer courses in Marxism or other critical theory, yet take pride in their strictness as graders. In this way their Marxist theory is used as a means of imposing additional work! It is for this reason that understanding teachers as productive of *both* value *and* struggle is useful. For the Marxist who believes that the function of education is purely "ideological," being strict may be necessary to ensure that students learn the "correct" ideology, even the "lazy" ones!

We should also recognize that education is a collective activity: many social subjects—teachers, students, administrators—must cooperate.[11] Just as no education system can guarantee the production of human beings with desirable labor power attributes, nor can radical teachers produce revolutionaries. It is quite possible to draw on Marx's insights, say, and skills in critical, imaginative thinking, and harness them in the interests of capital. Indeed, capital *needs* such thinkers. Clarke and Mearman, for instance, in arguing that "Marxist economics should be taught," suggest that "students [might] become more creative, better problem-solvers, which can raise their productivity"! (2003: 69).

I have attempted to analyze and illustrate here, *first*, the ways in which teachers exist within capital, that is, the ways in which they (re)produce (the) capital (relation) and thus produce value, and *second*, the already-existing ways in which they undermine and rupture this reproduction, and thus produce struggle. Most examples were drawn from the higher education sector in the UK, but the arguments apply to the labor of schoolteachers and education workers in other countries. Recognizing these modes of existence, and the forms of activity that constitute them, can be an important tool in our struggles against capital and to expand the spaces and possibilities for alternative ways of being.

Notes

A version of this chapter was presented at Marxism and Education: Renewing Dialogues II, Institute of Education, University of London, 1 May 2003, and I am grateful to other participants for their useful remarks. Thanks also to Kevin Harris and, especially, Glenn Rikowski for their comments and encouragement. I am responsible for the remaining errors, of course.

1. Academics' research activity is also becoming increasingly alienated, at least in the UK, through the process of so-called research selectivity. See Harvie (2000) for an exploration of these issues.
2. See De Angelis and Harvie (2006) for a discussion of the increasing burden of academic labor, including greater quantification, standardization, and surveillance.
3. Relatively recent work suggests that other primate societies, in particular those of chimpanzees, possess culture. Whiten et al. (1999) offer an overview and synthesis of a number of field studies of chimp societies.
4. Teachers' unions, for example, are increasingly concerned about violent incidents in the classroom. See, for example, Curtis (2003).
5. The mean mark measures the average quantity of work imposed by the lecturer. The standard deviation of marks measures the lecturer's relative contribution to the hierarchizing of students. Lecturers may be required to provide an explanation if the mean mark for their module or the standard deviation of marks lie outside certain parameters.
6. See, for example, Card (2001); Harmon et al. (2003); or Chevalier et al. (2004).
7. This understanding of commensurability differs from that of classical Marxism, which suggests that commensuration of diverse labors can only take place through the market (see, e.g., Mohun, 1996). In contrast, I argue that the market is simply one tool amongst several by which capital can attempt to make diverse labors commensurable.
8. It should be noted, though, that labor-power, unlike most commodities within the "general class," is never "finally" produced. Rather, it is constantly reproduced (or not) within each human being. Labor power must be maintained over time and it may be "improved" or it may "deteriorate." This special feature arises because this commodity labor power "is an aspect of the person; it is *internal* to personhood, in a special sense. It is a unified *force* flowing throughout the person" (Rikowski, 2000, italics in original). This force must constantly be replenished. Also important is the fact that new labor power itself must *always* be coproducer of itself.
9. Mohun does not himself use this language.
10. See Harvie (2004) and Harvie and Philp (2006) for a more extensive discussion of the communities that may exist in universities, including classroom communities, and the ways in which they seek to transcend capitalist education. Cleaver (2004) also explores individual and collective struggles against work of university students and, to a lesser extent, "professors," whilst Ovetz (1996) discusses organized resistance, by students, staff, and local communities, to the "entrepreneurialization" of universities. Many of the questions raised are not new: see, for example, Caffentzis (1975) and Thompson (1970) for earlier treatments of the struggle between capital and labor as it is acted out within universities.

11. Marx (1976a: 474) notes "that the specialized worker produces no commodities. It is only the common product of all the specialized workers that becomes a commodity," and here quotes with approval Thomas Hodgskin: "There is no longer anything which we can call the natural reward of individual labour. Each labourer produces only some part of a whole, and each part, having no value or utility in itself, there is nothing on which the labourer can seize, and say: It is my product, this I will keep to myself."

References

Boxley, S. (2003) Performativity and Capital in Schools. *Journal for Critical Education Policy Studies*, 1 (1). Available at http://www.jceps.com (accessed 8 August 2007).
Caffentzis, G. (1975) Throwing Away the Ladder: The Universities in the Crisis. *Zerowork*, 1: 128–142.
——. (1980/1994) The Work/Energy Crisis and the Apocalypse. *Midnight Notes*, 3. Rpt. Midnight Notes Collective (eds.) *Midnight Oil: Work, Energy, War 1973–1992*, 215–271 (Brooklyn, NY: Autonomedia).
Card, D. (2001) Estimating the Return to Schooling: Progress on Some Persistent Econometric Problems. *Econometrica*, 65 (9): 1127–1160.
Chevalier, A., C. Harmon, I. Walker, and Y. Zhu. (2004) Does Education Raise Productivity, or Just Reflect It? *Economic Journal*, 114: 499–517.
Clarke, P. and A. Mearman. (2003) Why Marxist Economics Should Be Taught but Probably Won't Be! *Capital and Class*, 79: 55–77.
Cleaver, H. (2000) *Reading Capital Politically* (Leeds: Anti/Theses and Edinburgh: AK Press).
——. (2004) *On Schoolwork and the Struggle Against It* (Canberra: Treason Press).
Coughlan, S. (2003) "Stop the Sats" Campaign Launched. *BBCi News*. 28 June. Available at http://news.bbc.co.uk/1/hi/education/3028766.stm (accessed 8 August 2007).
Curtis, P. (2003) Let Battle Begin. *Guardian Unlimited*. 21 October. Available at http://education.guardian.co.uk/classroomviolence/story/0,,1067169,00.html.
De Angelis, M. (1995) Beyond the Technological and Social Paradigms: A Political Reading of Abstract Labour as the Substance of Value. *Capital and Class*, 57: 107–134.
——. (1996) Social Relations, Commodity-Fetishism and Marx's Critique of Political Economy. *Review of Radical Political Economics*, 28 (4): 1–29.
——. (2004) Defining the Concreteness of the Abstract and Its Measure: Notes on the Relation between Key Concepts in Marx's Theory of Capitalism. In A. Freeman, A. Kliman, and J. Wells (eds.) *The New Value Controversy and the Foundations of Economics*, 167–180 (Cheltenham: Edward Elgar).
De Angelis, M. and D. Harvie. (2006) Cognitive Capitalism and the Rat Race: How Capital Measures Ideas and Affects in UK Higher Education. Paper presented at Conference on Immaterial Labour, Multitudes and New Social Subjects: Class Composition in Cognitive Capitalism, Cambridge, 29/30 April. Available at http://www.geocities.com/immateriallabour/angelisharviepaper2006.html (accessed 8 August 2007).
Escobales, R. (2003) NUT Test Boycott Fails to Win Enough Votes. *Guardian Unlimited*. 16 December. Available at http://education.guardian.co.uk/sats/story/0,13294,1108117,00.html.

Federici, S. and G. Caffentzis. (2000) Chronology of African Students' Struggles: 1985–1998. In S. Federici, G. Caffentzis, and O. Alidou (eds.) *A Thousand Flowers: Social Struggles against Structural Adjustment in African Universities*, 115–150 (Trenton, NJ; Asmara, Eritrea: Africa World Press).

Fortunati, L. (1995) *The Arcane of Reproduction: Housework, Prostitution, Labor and Capital*, trans. Hilary Creek (New York: Autonomedia).

Hardt, M. and A. Negri. (2000) *Empire* (Cambridge, MA: Harvard University Press).

Harmon, C., H. Oosterbeek, and I. Walker. (2003) The Returns to Education: Microeconomics. *Journal of Economic Surveys*, 17 (2): 115–156.

Harvie, D. (2000) Alienation, Class and Enclosure in UK Universities. *Capital and Class*, 71: 103–132.

———. (2004) Commons and Communities in the University: Some Notes and Some Examples. *The Commoner*, 8. Available at http://www.commoner.org.uk/.

Harvie, D. and B. Philip. (2006) Learning and Assessment in a Reading Group Format. *International Review of Economics Education*, 10 (2): 98–110. Available at http:www.economicsnetwork.ac.uk/iree/v5n2/harvie.pdf.

Harris, K. (1982) *Teachers and Classes: A Marxist Analysis* (London: Routledge and Kegan Paul).

Holloway, J. (2002) *Change the World without Taking Power: The Meaning of Revolution Today* (London; Sterling, VA: Pluto Press).

Johnson, B., P. Kavanagh, and K. Mattson. (eds.). (2003) *Steal This University: The Rise of the Corporate University and the Academic Labor Movement* (London; New York: Routledge).

Martin, R. (2003) Foreword. In M. Bousquet, T. Scott, and L. Parasondola (eds.) *Tenured Bosses and Disposable Teachers: Writing Instruction in the Managed University*, ix–xi (Carbondale, IL: Southern Illinois University Press).

Marx, K. (1969) *Theories of Surplus Value*, Part 1 (London: Lawrence and Wishart).

———. (1975) Economic and Philosophical Manuscripts. In *Early Writings*, 279–400 (Harmondsworth: Penguin).

———. (1976a) *Capital: A Critique of Political Economy*, Vol. 1 (Harmondsworth: Penguin).

———. (1976b) Results of the Immediate Process of Production. Appendix to *Capital*, Vol. 1 (Harmondsworth: Penguin).

Mohun, S. (1994) A Re(in)statement of the Labour Theory of Value. *Cambridge Journal of Economics*, 18: 391–412.

———. (1996) Productive and Unproductive Labor in the Labor Theory of Value. *Review of Radical Political Economics*, 24 (4): 30–54.

Ovetz, R. (1996) Turning Resistance into Rebellion: Student Struggles and the Entrepreneurialization of the Universities. *Capital and Class*, 58: 113–152.

Rikowski, G. (2000) Messing with the Explosive Commodity: School Improvement, Educational Research and Labour-Power in the Era of Global Capitalism. Paper presented at the British Educational Research Association Conference, Cardiff University, 7–10 September. Available at http://www.leeds.ac.uk/educol/documents/00001610.htm.

———. (2002/2003) That Other Great Class of Commodities: Labour-Power as Spark for Marxist Educational Theory. Paper presented at the Education Research Centre, Research Seminar Series, University of Brighton, 6 November; and at Marxism and Education: Renewing Dialogues II, University of London, Institute of Education, 1 May. Available at http://www/leeds.ac.uk/educol/documents/00001624.htm (accessed 8 August 2007).

Smith. C. (1996) *Marx at the Millennium* (London: Pluto Press).

Thompson, E. P. (ed.). (1970) *Warwick University Ltd: Industry, Management and the Universities* (Penguin: Harmondsworth).

Whiten, A., J. Goodall, W. C. McGrew, T. Nishida, V. Reynolds, Y. Sugiyama et al. (1999) Cultures in Chimpanzees. *Nature*, 399: 682–685.

Chapter 13

Marxist Political Praxis: Class Notes on Academic Activism in the Corporate University

Gregory Martin

As a student of critical social theory who works within the radical pedagogical tradition influenced by Freire and Marxist currents, I am fundamentally concerned with revealing structural explanations for inequality in power relationships in society and doing so in ways that enable oppressed and exploited groups and classes to change the relational basis that underpins and constitutes this historical set of social fetishes and structures (the objectified effects of concrete and abstract human labor). I chose my career in education based upon the personal example of people who I admire such as Freire, Fanon, Lumumba, Luxemburg, and Che who committed their individual energies and capacities for thought and action to collective endeavors that might serve to transform this totalizing system of oppression that structures our everyday lives. In the late 1990s, student and social activism were cresting in the United States with the "Battle of Seattle" and the antisweatshop movement (Rikowski, 2001a). In Los Angeles, Justice for Janitors and the Bus Riders Union/Sindicato de Pasajeros (BRU) had won significant victories. It was during this period of acute struggle that I began my doctoral studies at the University of California, Los Angeles (UCLA). Faced with the urgency of change required, I wanted to pursue a scholarly career with both a social purpose and a decidedly activist edge. Against the backdrop of today's conservative anti-intellectualism, I do not wish to devalue the dignity and importance of theoretical work. On the contrary, as Marx emphasized, the struggle over theory is socially practical, especially as ideas are altered, modified, or perhaps discarded (if they are wrong) in the course of struggling to put them into practice. In directing our outrage against the crimes of capitalism, the invariable conflict between theory and social activism in the bourgeois university is based upon a false distinction, which sets the limits of the "knowing" subject by devaluing forms of intellectual engagement that are activist (Maxey, 2004: 159).

In this chapter, I discuss the contradictory positions and complex situations I experienced, interpreted, and acted upon as I sought to develop my identity as an academic/activist in and against the value system embodied within the labor-capital relation of the modern corporate university. Chouinard (cited in Fuller and Kitchin, 2004) writes that being an activist/academic

> means putting ourselves "on the line" as academics who will not go along with the latest "fashion" simply because it sells, and who takes seriously the notion that "knowledge is power." It means as well personal decisions to put one's abilities at the disposal of groups at the margins of and outside academia. This is not taking the "moral high ground" but simply saying that if you want to help in struggles against opposition you have to "connect" with the trenches. (5)

With a focus on how the critical and self-reflexive subject can begin the process of building a new set of social relations through this intimate form of intersubjective and collective struggle (Freire, 1993; Rikowski, 2001b), this narrative "snapshot" explores how I embraced the principles, aims, and commitments of a revolutionary critical pedagogy to reach out to the BRU, an experimental political working-class organization built to fight against the government for civil rights in the form of an environmentally sustainable first-class, clean fuel, mass public transportation system (Allman, 2001; Mann, 1999, 2002). The BRU was established in 1993 as an experimental project of the Labor/Community Strategy Center, a "think tank-act tank" also based in Los Angeles that has a history of initiating and building left organizational forms (Mann, 2002: 4).

From the time it burst onto the local scene in Los Angeles, struggling to improve the daily conditions for low-income working-class people through its militant, visible, and influential campaigns such as "Billions for Buses, Fight Transit Racism," the BRU has distinguished its character and basis of appeal from most mainstream trade unions by taking a progressive stance "on almost every major issue facing the working class as a whole (for example, immigration, affirmative action, crime, police, prisoner rights, and tax policy)" (Mann, 1998: 3). By offering such a comprehensive and competitive political alternative, the BRU claims to have successfully built one of the largest social movements in the United States. It claims to represent an estimated 3,000 dues-paying and 50,000 self-identified members, representing 400,000 Los Angeles bus riders (Mann, 1998). Although those numbers are based upon a relaxed definition of membership, at a time when most Americans are soundly asleep at the wheel about the threat of capitalism, the BRU is involved in a collective experiment to build new forms of social cooperation and organization across space around issues such as public transportation, which serves as a link to work, school, services, and to community

(Hurst Mann, 1998; Mann, 1996, 1998; Pastor, 2001). In fact, although popular accounts of the BRU give it a larger than life portrayal, its homegrown success has spun off variants in other parts of the United States as well as in Canada.

Which Side Are You On?

Within the university, the core of the problem at this historical time of transition is that scholarly research involves political choices made manifest in the clash between professional/entrepreneurial and activist discourses. In the contemporary political climate, intellectuals on the academic Left have increasingly found themselves on the "bad-edge" of academic entrepreneurialism as research agendas, now forced to compete for limited resources, are increasingly freighted toward greater market compliance. This, of course, has left progressive scholars with even fewer funded opportunities to support community-based organizations working to secure social and economic justice.

The commodification of education stems in part from its subordination to both the ideology and practice of neoliberal economic policy. Given this situation, it is a sobering fact that the decision to devote one's energy to developing models of intellectual engagement that are personally and politically committed to changing the material conditions the current social order forces upon us in life and struggle does not occur in a social vacuum (Martin, 2005). As our cash-starved public schools and universities are subjected to the discipline of market forces, many functions and activities are becoming commodified as consumer goods (objects of trading) and individual forms of investment (human capital) related to the "buy now-pay later" credit system (Lapavitsas, 2003: 65). In a nutshell, the social character of education in this pragmatic, anti-intellectual paradigm is being reduced to the delivery of marketable, prepackaged knowledge (e.g., online courses, videos, and CDs) to be added onto "useful" skills and practices such as "problem solving" and "critical thinking" deemed to be of relevance and economic benefit to all the differing national capitals (Ebert, 1997: 47; see also Hill, 2004b; Rikowski, 1999).

During the course of this process, more than a few careerist academics with fading socialist credentials have rushed to "adjust" and "update" their course offerings and research with a liberal "equality" agenda (empowerment rather than power), which has facilitated the flow of profit into their pockets (often as a side benefit), whether by way of tenure, a good review, and/or external sources of funding (e.g., industry-community partnerships, consultancy and advisory services) (Martin, 2005). What should not be forgotten here is the multiple and ritualistic ways in which the reification of our teaching (the fetishization of performance appraisals) and research (the study of social life restricted to the "bottom line" drive of business) is transposed into a finished commodity (as the abstract universal equivalent of money) within the "ideas factory" of the university. The control of the labor process through the use of internal and external performance indicators (e.g., "benchmarking" to measure production costs, market share, workloads, employment rate of graduates) imposes a value system upon which we base our social relations.

These fetishised relations reflect the structural subordination of labor to capital in the academy, and its vehicle of expression, absolute surplus-value (Ebert, 1997). This system of value production conceals the social character of private labor and the underlying structures of exploitation associated with it. On the ground, changes in the management techniques of the academic workplace to do with the technical division of labor (a core cabal of managers, fewer full-time academic wage earners and a peripheral, casualized labor force) and the objective material conditions of work (speed up, overwork, burn out) function to reproduce the actual, local, concrete conditions of capital accumulation (Meyerson, 2004). More fundamentally, the seductive ideological impulse of competitive capitalism that underpins the governing discourses of merit, efficiency, and accountability within these bureaucratic command structures has pulverized and atomized "an instinctive sense of solidarity and antagonism" against management (Callinicos, 1999: 212).

With regard to Marx's (1988) notion of alienation, the constant reshaping of internal conflicts within our personhood whenever we directly confront the agonizing contradiction between the source of our material privilege and our political convictions is symptomatic of how our lives, work, and consciousness are being "remade" under the negative influence of these capricious forces inside the prison structure of the corporate university. Under this dehumanizing value system, education and pedagogy have been reduced to commodities that are being put forward as instruments to suppress critical thought and creativity (Hill, 2004a; Martin, 2005; McLaren, 2005; Rikowski, 2001b). This negative trend forces academics on the progressive Left to walk the political tightrope between living our own struggle as individuals who are being robbed and deprived of our lives and trying to search out ways to connect this internal struggle, with the collective struggle of the most exploited (Landsteicher, 2004–2005). Within this violent energy field of value, what matters is that if education is to become a stronger intellectual framework for individuals and groups struggling for a truly free existence, then graduate students and faculty will need to decide on the type of research they will practice and whose interests they will serve (Feagin and Vera, 2001). This is a daunting choice for anyone working within the university, especially given the pressure placed on graduate students and new faculty alike during this phase of purgatory to do conventional teaching and research (Feagin and Vera, 2001; Martin, 2005).

Despite this difficult political terrain, Rikowski (2002) asserts that the creation of possible worlds "beyond capital" cannot be reduced to a purely remote form of speculative activity but must be struggled and fought for by the creative activities of human minds and hands (8). Given, as Rikowski (2002) puts it, that the underlying contradictions of our social existence in the university "screw us up, individually and collectively," he argues that we need to overcome our alienation within the labor process itself by resisting our self-reduction to the "peculiar" form labor power takes as human capital under the alien and hostile powers of money and the state (Marx, 1967: 167). Without overlooking how the interests of academics are counterpoised to those of management, Rikowski (2002) identifies labor power as "capital's weakest link." Expanding upon Marx, Rikowski notes that as a distinctly human force, labor power—which he defines as our "capacities" to labor in the form of epistemological paradigms, language codes, technical skills, attitudes, dispositions and behaviors—has reality only within the human subject. Without falling prey to the illusion of "free will," what is important to remember here is that although capital depends upon the subject who actively participates in his or her subjection, this social relation is ultimately "precarious" because the subjective potential of living labor power is "fundamentally indeterminate: It can always work for or against capitalist accumulation" (Read, 2003: 136). But what does this mean in terms of beginning the process of deterritorializing the fetishised structures that hold our "congealed identities" and alienated subjectivities in place within the hellish nightmare of hierarchical and competitive capitalism? (Ghosh, 2004: 4). Working furiously behind the scenes for a new generation of politicized academics and radical activists, Rikowski (McLaren and Rikowski, 2001) writes,

> We require a *politics of human resistance*. This is a politics aimed at resisting the reduction of our personhoods to labor power (human-capital), thus resisting the capitalization of humanity. This politics also has a truly negative side: the slaying of the contradictions that screw-up, bamboozle and depress us. However, only collectively can these contradictions constituting personhood (and society: there is no individual/society duality) be abolished. Their termination rests on the annihilation of the social relations that generate them (capitalist social relations), the social force that conditions their development within social phenomena, including the "human" (capital) and the dissolution of the substance of capital's social universe (value). A collective, political project of human resistance is necessary, and this goes hand-in-hand with *communist politics*, a positive politics of social and human re-constitution. (Italics in orginal)

Without romanticizing a naive reflex of activism at the expense of theory and the exchange of ideas, I believe that in terms of a "politics of location" (McLaren, 1999: lxiv), the understandings and social implications of Marxism as a mode of political discourse and an axis of working-class political organization ought to be enlarged and fleshed out by linking theory, politics, and practice through the development of praxis or what Marx defined as "revolutionary, critical-practical activity" (cited in Dunayevskaya, 2001). Unfortunately, all too often, academics on the progressive Left (let alone Marxists) treat activism as an object of study and their self-authored calls for resistance only grow to be materialized in scholarly journals as "high knowledge," which contributes to the demonization of academic discourses and practices on the "street." In acknowledging this failure to unite theory with practice at the everyday level of *habitus* and embodied subjectivities (Hunter, 2004), Raduntz (1998) rightly points out,

> The effect of this isolation on Marxist theory has been to undermine it within academic life and to marginalise its impact beyond it. In terms of the dialectic principle Marxist scholarship has become one-sided and as a consequence abstract.

While I do not wish to dismiss the impressive efforts to develop and preserve ideas and hope in the knowledge of history, as understood by Marx, this dilemma raises sharply the question of how do you begin to transform the totality of social relations (property, family, and work) from what Giroux (2003) calls "a language of resistance and possibility"? (5). I believe that the answer to this question can only be worked out dialectically through participation at the most intimate and "sensuous" level of class struggle (Marx, 1988), as complexly registered in the emotions, imaginations, and practices of "real" social actors. At the daily level of self-activity this requires academics to begin "a process of renewal" by uniting the individual with the social at the local scale of community conflict (Lefebvre, 1984: 43). Place-based struggles are important because they constitute embodied spaces of experience, knowledge, and activity that can "jump scale" to energize and transform internationalist organization (DeFilippis, 1998). On this note, rather than engage in a superficial show of principles not actually implemented in practical activities, I believe that academics committed to struggles for social change (capitalism to socialism) must do more than engage in an abstract form of theorizing about questions that spring out of the everyday life and death struggles of the world's gendered and raced proletariat.

Decapitalizing Methodologies

Where a politics of class struggle is concerned, a realistic form of socialism must grow out from experiments in building actually existing praxis-communities, where workers are actively developing skills to participate in running society. As Moss (2004) notes, this requires academic/activists to find "positions for intellectuals in radical social movements" (104). This is a departure from the normative strategies of political propagandistic and information-based practices produced within the university. Conscious of the ambiguous personal commitment in the *habitus* and field of radical theorists to social practices produced in the day-to-day practices of militant labor struggles as well as the leftward moving layer of social movement organizations, this narrative "close up" provides a concrete account of academic/activism as I sought to resist the reifying impulse of the corporate university through forms of collective engagement based in the material interests and class conflict shaping political forces in Los Angeles. Since there is no division of space between the individual and social, the focus of my one-year engagement at the BRU was my personal commitment to critical/revolutionary praxis organized around the need for individual and social transformation, which is the aim of revolutionary critical pedagogy.

Over the past decade, radical educators such as Paula Allman, Mike Cole, Ramin Farahmandpur, Dave Hill, Peter McLaren, Peter Mayo, Helen Raduntz, Glenn Rikowski, and Valerie Scatamburlo-D'Annibale have grappled with and responded to the demands of social movements for political and educational action by bringing Marxist theory back into conversation with the field of critical pedagogy. This has helped to reverse critical pedagogy's degeneration by providing an alternative conceptual framework in the favor of working-class struggle. With a keen eye to the parlous state of critical pedagogy, Peter McLaren (1997a, 1997b, 1998, 2000) has recognized that after suffering the corrosive effects of liberalism and the sugarcoated bullets of postmodernism, it was left with barely a pulse in the 1990s. As a reaction to this primarily institutional form of co-option and vulgarization, revolutionary critical pedagogy (a term first coined by Paula Allman, 1999, 2001), is a relatively new field of materialist intervention in the field of regular and adult education.

Despite the provocative claims of critics on the Left-liberal educational Left, who have abandoned Freire like a burnt match, revolutionary critical pedagogy does not stem from utopian optimism but rather from radically democratic praxis (Martin, 2005). Roughly speaking, it seeks to create the pedagogical conditions for the production of a "proletarian" subjectivity through a dialectical process of genuine dialogue, critical analysis, and problem solving based upon peoples' everyday experience and knowledge of capitalism. Regrettably, few radical pedagogues who work within this tradition have declared how such theory can be fused with embodied practice in order to build new forms of noncapitalized human sociability, especially through close contact with students (potential laborers) and workers. As Ghosh (2003) writes, "Revolutionary practice, according to Lenin, is impossible without a revolutionary theory. The question that one needs to ask today is: what will the revolutionary theory serve if there is no revolutionary practice"? (16).

With the benefit of a whole tradition of Marxist educational theory developed and debated in books, conference papers and leading academic journals, I believe that a major task of intellectuals who engage in Left politics today is to be accountable for the "real world" implications of theory by working out ways to connect education with struggles on the street for social and economic justice. Without overlooking how research/social relations are shot through with power, the emphasis here is not on the "proper" degree of participation or immersion but on renewing dialogue between academics and activists toward "a new relation of theory and practice" (Hudis, 2003).

As a relatively privileged academic, I have decided to make a conscious choice in my day-to-day practices to avoid commodified activities (organized to yield a profit) so that, whenever possible, I am able to devote my time and energy to historically concrete individuals and groups condemned to the economic margins of society. As a site for the "enfleshment" of a politics of human resistance (Cruz and McLaren, 2002: 191), at the intersection of where individuals construct themselves subjectively within capitalist institutions, this social commitment has evolved dialectically out of my own direct exploitation and estrangement in the labor process, as well as through my engagement with the genealogy of critical pedagogy associated with Freire and Marxist currents. Given that a greater degree of face-to-face interaction between academic and activist elements is required in order to bridge the gap between theory and practice, my social/research practices embody the activist and goal-orientated epistemology of Marxism, which following Freire (1993) is not only focused on studying power relations but also dismantling them through collective praxis. A distinctive feature of a Freirian methodology is its ontological commitment to shared participation in learning and problem-solving processes as well as capacity building (political organization). Without fleshing this out further, it asserts that the social power embedded in the experience and knowledge distributed amongst exploited and aggrieved individuals and groups can guide revolutionary struggle. Since this methodology has no relation to capital but is rather driven by the exploited class through the course of shared struggle, concepts such as "cooperation," "dialogue," and "problem solving" are not objectified and capitalized activities but rather forms of counterpraxis harnessed to promote social democratic action (ibid.).

Breaking Ranks

Living and studying in Los Angeles, I considered myself to be a member of the local community. Against the backdrop of anti-globalization protest and renewed student activism on college campuses in the United States during the 1990s, I was interested in expanding my involvement in some of the struggles and enterprises of this community, which "exists only in and through the division of labor" (Marx, 1972/1978: 189). After two years studying in a small college town in the Midwest, I recognized that living in a large and diverse city would provide fertile soil for developing both a theoretical and practical understanding of the different class forces operating in the field including their motion and direction. Although I was engaged in a strong critique of the United States and imperialism, the undemocratic nature of capitalism, its global hegemony, and the social machinery of state enterprises such as schooling in my scholarly writings, I fell into a deep despair over the differing worlds I negotiated in my daily life in Los Angeles, a sprawling megalopolis that exhibits all the contradictions of modern capitalism. Without overlooking how the "street" erupts violently into life within the carefully manicured lawns of the "Ivory Tower," I asked myself the following: do I stand at the side criticizing and gesturing? Or do I engage in the protracted and often gut-wrenching process of throwing myself into the midst of the struggle? At the same time, given the lack of social and political space for a politics of human resistance I was confronted with the question of how to shift my identity, strategies, and values at the structural level of social relations within the university to the most grassroots level of dialogue, analysis, collective decision making, and direct action.

Despite my lack of certainty on where to begin, as I did not have any "organic roots" in the Los Angeles community, through my prior participation in social justice movements (antiracist, environmental, antinuclear, labor) in both Australia and the United States, I understood that it does not take much money or imagination to refocus the goals of human activity on meeting human and nonhuman nature needs. Anyone who wants to overcome what Rikowski (1999) refers to as "the lived tensions arising from ourselves as capitalised life-forms" can find an outlet, right in themselves and in their backyard through an embodied commitment to any number of social and ecological forms of political engagement (52–53). Rather than adopt a pessimistic outlook regarding the decay of the revolutionary Left and "old" workers' organizations in the United States, as an index of action I began to look around to find out what the most militant and emergent social forces were doing to resolve this historic crisis of leadership. In this effort, I spent long hours conducting research on the Internet and talking to graduate students and academics who had a prior background or continuing involvement in community organizing in the Los Angeles region.

What instantly captured my interest in the BRU was that the organization had energized the working class as an agent of social change, thereby altering the structure of the field of knowledge, which enabled it to leverage power over the state system (Mann, 1996, 1999). Contrary to what one might suppose, the primary actors in this historic struggle were not the trained organizers who helped to build this apparatus from the ground up but rather the hundreds of ordinary, spirited women and men, who were the ultimate resources of power at the BRU. Indeed, recognizing that the organic knowledges, strengths, and capacities of its membership constituted the integral component (moving) parts of its day-to-day organizing strategies, the BRU was able to achieve a string of concrete objectives and demands (Fight Transit Racism, 2002). To take just one example, in 1994, the BRU filed a civil rights lawsuit against one of the most powerful county agencies, the Metropolitan Transportation Authority (MTA), winning a precedent setting Consent Decree, which forced it to improve and expand the dismal public transportation system in Los Angeles. Part of the provisions contained in this legal action required the MTA to reduce monthly bus fares, resulting in a 10 percent increase in bus ridership since 1996. Along the way, the BRU has also forced the MTA to replace 1158 old diesel fuel buses with new Compressed Natural Gas (CNG) buses. Significantly, it was anticipated that these victories would result in the hiring of 1200 newly unionized bus drivers, mechanics, and maintenance people during each of the next three years (Fight Transit Racism, 2002).

Across the Lines

Without entering into the rich tangle of detail, I want to begin by emphasizing the "huge credibility problem" academics face in bridging the academic/activist divide (Smith, 1999: 118). Historically, researchers operating within all spheres of institutional and bureaucratic class society including the university have treated "community" as a data plantation in the aid of imperial and colonial designs under the universal guise of the "social good." (Smith, 1999). Owing to this institutional history, as Kim England (1994) wryly observes, academics cannot just expect to "parachute into the field" with notebooks, pencils, and a desire to "make a difference" (84). Given the legacy of distrust between oppressed and exploited individuals and groups and state apparatuses such as the university, bridging the academic/activist divide means bridging one's locality, history, and politics. For example, seeing the personal as political, Marxist academics who wish to avoid the divisions that often emerge at the grassroots out of empty slogans such as "working class unity" ought not reduce the category of class to an abstract intellectual construct but rather challenge themselves in terms of their inner politics, activities, and social relationships through class struggle based in genuine dialogue and collective action.

As an English-only-speaking white male in a place/locality with a deeply embedded history of racism, my experience at the BRU confirmed the importance of embracing and internalizing a commitment to honesty, humility, and openness as I developed cooperative interpersonal relations based in mutual respect and solidarity with staff and members through a number of collective actions (Allman, 1999, 2001). Although I shared a basis of political unity with BRU staff and members around issues to do with capitalism, I came to understand the practical importance of recognizing each person's individual strengths in building a mass movement from the ground up. For example, when organizing on the buses and street corners, I learned about the importance of language, as I relied on Spanish and Korean speaking staff and members to dialogue with riders about their experiences, concerns, and perspectives on a whole host of issues such as the bus system, the war in Iraq, and the U.S. economy. Here, as always, BRU staff paid close attention to issues to do with composition and representation in order to reflect a gender, race, and language balance in all of its mass and legal work. Needless to say, this applied to the implementation of procedures and practices designed to develop the knowledge, self-confidence, and leadership potential of BRU members.

Clearly, the BRU did not emerge fully formed from the sea and inevitably reflected the standards of society around it. Yet, as an organization it consistently acted in principled opposition to the values, beliefs, and norms embedded in these standards. Adopting an "inclusive" stance toward class, the BRU has built up a shared system of meaning by striving to implement a culture inside the organization that is multilingual (all meetings are conducted in English, Spanish, and Korean, and all flyers are considered public educations and are printed in these languages), antiracist, and supportive of women's liberation in its day-to-day work (Mann, 2001). One of the strengths of learning organizations such as the BRU is that it does not view education as a tool to teach people how to sit still and "get along" but rather as a generative process that encourages participants to make a difference by examining their lives critically and taking action to change social conditions (Lenin cited in Kelsh, 1998).

The BRU rejects what Freire (1993) termed the "banking" method, which reduces learners to empty containers for readymade "deposits" (53). With a focus on individual and collective agency, BRU organizers, who are exposed to radical theory and pedagogy during their training, direct their creative capacities and energies toward the social production of a dialogical space that encourages not simply the exchange of peoples' experiences but also critical discussion, problem solving, and direct action on a range of issues (e.g., the war in Iraq, Valley and Hollywood Secession and the occupation of Palestine) (Bus Riders Union Asks, 2002). To provide just one example, at monthly meetings, full-time organizers regularly hold legal workshops and "teach-ins" to explain the litigation process. Instead of a sole focus on the transmission of "facts" related to the legal case, the BRU recognizes who drives the entire process and also holds open microphone sessions to allow for new knowledge and insights to emerge. Recognizing this space as a terrain of struggle over issues to do with truth and power, "Open Mic" sessions are not entirely improvisional as the chair and monitors (one woman and one man for gender balance) routinely intervene to discipline the flow of interaction and talk in order to prioritize the voices of working-class women, undocumented immigrants, and oppressed nationalities, for example, Spanish and Korean speakers. This practice is necessary, for example, because of the culturally enacted tendency of English-speaking-only members toward domination, which renders invisible the languages and perspectives of the "Other." Given the material ways in which these sorts of linguistic practices define the content and boundaries of action, I self-consciously distanced myself as a speaking voice in such participatory activities as I did not want to interrupt or shut down emerging narratives or new frames of understanding, from which I might learn and grow as a political subject.

  Throughout its educative practices and political work, a major focus at the BRU was on issues to do with authority, with limited scope for middle-class white males such as myself to impose institutional or interpersonal order within the organization. During my involvement at the BRU, I learned that my university background did not afford me any special privileges in that setting. Although I was bursting with energy to "make a difference" in the BRU's mass work, based upon my readings about the implosion of social movements in the 1970s and prior involvement in place-based struggles in both Australia and the United States, I understood that the involvement of white males had to be treated with caution (Kelly, 1997; Whittier, 2002; Robnett, 2002). Aware of this incipient threat, the BRU was self-conscious about issues of social composition and waged internal struggles to challenge and transform the attitudes and behaviors displayed by its activists, especially those from more privileged backgrounds.

  Although it would be wrong to say I did not wish to be at the vanguard of a militant social movement organization, I did not resent this treatment as it provided me with the opportunity to examine my white skin privilege and how I might organize my antiracist work. On the one hand, my involvement at the BRU confirmed the central importance of class as an organizing category for challenging the capitalist state, which through its actions created this community of "dissidents" that self-consciously identified as "working class" through both personal and collective forms of address and political engagement (Meyer, 2002: 13). On the other, although the class-bound material interests of members shaped the BRU as a political force, as a civil rights organization it refused to ignore forms of oppression that arise out of the exploitative nature of capitalism such as racism, sexism, and homophobia. Indeed, drawing upon the legacy of the civil rights Movement and the Women's Liberation Movement, it used the legal system, in Bernans (2002) words, "to exploit the contradiction between an economic system that legitimates itself, in part, through its claim to be politically and culturally neutral, and the existence of racist and (hetero)sexist oppression" (63). At the same time, scornful of the state, the BRU also refused to allow its political activities to be circumscribed by a narrow liberal definition of democracy.

From my experience at the BRU, it is now clear to me that reaching out and contributing to the material struggles of "broad-based coalitions" requires more than a narrow and introspective form of "abstract" theorizing (Scatamburlo-D'Annibale and McLaren, 2003). As Giroux (2002) argues, it requires overcoming our privileged role as intellectual "specialists" in the university through sustained and collective action, which includes mass work that is sometimes less than glamorous. As I see it, part of the problem is that academics are not used to doing "low-status" community organizing "housework" that receives little or no credit, especially within the reward structures of the university (Hubbard, 1996). Used to relying upon an invisible "background" army of faceless workers, "a great many of whom are female," such as secretaries and cleaners whose acts of concrete labor make their high-status "abstract intellectual work" possible, academics are prone to opportunism and often join social movements such as the BRU with a short time horizon (Hubbard, 1996).

Recognizing this tendency, I attempted to support the organization's mass work through acts of manual labor such as putting up campaign posters, entering data on the BRU's computer system, and setting up chairs and tables at meetings, which provoked unsolicited but favorable responses to my insider/outsider status in the organization. What matters here is that although I might not have been at the forefront of the struggle, I was ready to put my body on the line if requested and to call out, for example, racist statements and behaviors no matter the context (Allman, 2001). Given the history of state terror and repression in U.S. urban communities, I believe that such a personal commitment is central to building mutual respect, trust, and cooperation in social movement organizations that are actively challenging the very foundation of capitalism, which is rooted in patriarchy and white supremacy in the United States.

My conscious effort to rework my identity and subjectivity in the university brings me to a related conclusion that evolved out of my practical work at the BRU. When I first approached the BRU, the political leadership stressed the importance of organizational affiliation. Given that the organization's mass work overlapped with my potential research interests, I joined the BRU as an "active" dues-paying member through my personal involvement in a number of collective actions. The catch, however, came at an academic conference, when I discovered how this affiliation impacted my writing, which was not considered "objective" or "value-free." Indeed, acknowledging the concept of reflexivity and the "activist" politics of research that underwrites this narrative, I recognize that I was colluding with the BRU's knowledge interests, and perhaps to a lesser extent disagreeing with the views of a few of its members (Cox, 1998). To do this had certain limitations as well as a specific form of value for the way I attempted to reconstruct my identity as an academic/activist in resistance to the highly essentialising and homogenizing effects of the modern corporate university. What I gained as a form of therapy through a revolutionary critical pedagogy (Ecclestone, 2004), which ties "individual human development" to forms of collective engagement aimed at social transformation, was a greater sense of agency, hope, and solidarity (Allman, 2001: 180). Despite the relationship between production and the production of embodied subjectivity in the modern corporate university, I found that my organizational affiliation provided me with a material base from which I could act, speak, and creatively write as I came together in a voluntary unity of exchange with BRU members and MTA ridership, who ought to have PhDs based upon what they have taught me and themselves.

If Not Now…

As a socialist, I feel privileged to have inherited a rich body of theory that appears to grow only more relevant at a time capitalism has matured into a parasitic, decadent, and unsustainable system. Making no apologies for straddling the academic/activist and theory/praxis divide, I also felt the need to accept responsibility for my own privileged location by taking the time to learn from social movements in order to flesh out that theory. Scaling all sorts of geographical and social barriers, my intention was to shed light on the practical, how-to-do-it methods—which might provide educators with the sought after tools to reconstruct critical pedagogy as material force capable of "building the bridge to the future" (Support Every Outbreak, 1980: 1). After years of precarious employment in the academy as a result of his lengthy "no contest" rap sheet of political agitation and labor union activism, Dave Hill (1997, 2005) notes that these are not easy bridges to make (or traverse) within individualizing state apparatuses such as the university. Here, in the best of cases, creating for ourselves new forms of human sociability, not yet imagined or realized, requires finding and widening the cracks and openings within the firewalls of the university through which we can channel our irrepressible energies to take us outside of these strategies of confinement, control, and exploitation.

Animated by an impetus toward rebellion, Marxist theory, and a Freirian methodology, I allowed myself to be swallowed whole at the BRU and when

I was finally belched back onto the earth rather than emerge from the whale pretending to be a prophet, I wanted to honor the contributions and sacrifices of BRU staff and membership to the struggle ahead through their collective actions on the buses and on the streets. What I learned through this embodied form of political engagement is that as part of the inner struggle between living and dead labor in the social factory of life, the challenge is not to give in to blind anger, pessimism, and despair but rather to recognize that revolutionary politics are in effect, *something made*, through critical praxis.

Although we have not reached the "end of history," experiments in critical community building such as the one I engaged in at the BRU do not exist in isolation. Without pretending to have presented the entire picture or to have all the answers, I contend that the starting point for reconfiguring our identities is to abandon our "natural" impulse to work productively and efficiently as *particular* normalized subjects in our jobs. As management attempts to squeeze every last drop of surplus-value out of our labor power, we need to actively resist the imposition of the commodity form in the workplace by avoiding research activities that reflect and animate the values and priorities of the capitalist class. The reason I raise this point is because many scholars feel torn between the alienation of producing research as a commodity and directing their "left over" energies toward more creative and community-based forms of activism. There are no easy solutions here. This points to the specific need for an increasing amount of empirical research into "how, why, and in what ways" individuals can resist their interpellation as particular subjects within the "total productive processes" they experience, interpret, and negotiate in their everyday lives (Rikowski, 2002: 27). After all, if new antagonistic subjectivities bent on "raising hell" can be developed, the domination of capital can be smashed from within its inside and outside. Most crucially, for Marxist academics, the task remains of uniting the radical politics of the university with forms of grassroots political organizing. What matters here is that revolutionary critical pedagogy increases our individual and collective capacity to act, theoretically and practically, when it is tied to forms of political engagement that are based in the material interests and class conflicts of the multiracial, multilingual working class.

References

Allman, P. (1999) *Revolutionary Social Transformation: Democratic Hopes, Political Possibilities and Critical Education* (Westport, CT: Bergin & Garvey).

———. (2001) *Critical Education against Global Capitalism: Karl Marx and Revolutionary Critical Education* (Westport, CT: Bergin & Garvey).

Bernans, D. (2002) Merely Economic? Surplus Extraction and Misrecognition. *Rethinking Marxism*, 14 (1): 49–666.

Bus Riders Union. (2002a) *Bus Riders Union Asks the U.S. and Israeli Governments: Let the Palestinian People Go!* (Los Angeles, CA).

———. (2002b) *Fight Transit Racism, Billions for Buses and Environmental Justice Campaigns.* (Los Angeles, CA).

Callinicos, A. (1999) *Social Theory: A Historical Introduction* (New York: New York University Press).
Cole, M., D. Hill, P. McLaren, and G. Rikowski. (2001) *Red Chalk: On Schooling, Capitalism and Politics* (Brighton: Institute for Education Policy Studies).
Cox, L. (1998) *Gramsci, Movements and Method: The Politics of Activist Research*. Available at http://www.iol.ie/~mazzoldi/toolsforchange/afpp/afpp4.html (accessed 4 December 2002).
Cruz, C. and P. McLaren (2002) Queer Bodies and Configurations: Toward a Pedagogy of the Body. In S. Shapiro and S. Shapiro (eds.) *Body Movements: Pedagogy, Politics, and Social Change*, 187–207 (Creskill, NJ: Hampton Press).
Defilippis, J. (1998) Our Resistance Must Be as Local as Capitalism: Place, Scale and the Anti-globalization Protest Movement. Paper presented on COMM-ORG: The Online Conference on Community Organizing and Development. Available at http://comm-org.utoledo.edu/pages.htm (accessed 4 December 2002).
Dunayevskaya, R. (2001) Marx's Concept of Praxis. Available at http://www.newsandletters.org/Issues/2001/Nov/fta_nov2001.htm (accessed 28 September 2003).
Ebert, T. (1997) Quango-ing the University. *The Alternative Orange*, 5 (2): 5–47.
Ecclestone, K. (2004) *Therapeutic Stories in Adult Education: The Demoralization of Critical Pedagogy*. Available at http://www.leeds.ac.uk/educol/documents/00003586.htm. (accessed 5 October 2004).
England, K. (1994) Getting Personal: Reflexivity, Positionality, and Feminist Research. *Professional Geographer*, 46 (1): 80–89.
Feagin, J. and H. Vera. (2001) *Liberation Sociology* (Boulder, CO: Westview Press).
Freire, P. (1970/1993) *Pedagogy of the Oppressed* (New York: Continuum).
Fuller, D. and R. Kitchin. (2004) Introduction. *Radical Theory/Critical Praxis: Academic Geography beyond the Academy?* 1–20 (University of British Columbia in Kelowna, British Columbia, Canada: Praxis[e]Press). Available at http://www.praxis-epress.org (accessed 1 October 2004).
Ghosh, P. (2003) *Responses. New Correspondence: A Bulletin of Socialist Ideas and Practices*, No.–0002: 11–17, May.
———. (2004) In Search of Class. *New Correspondence: A Bulletin of Socialist Ideas and Practices*, 1 (1): 1–6.
Giroux, H. (2002) *Public Intellectual and the Politics of Education*. Avaialble at http://lihini.sjp.ac.lk/careers/edreform/n_amer/giroux/giroux_praxis_1.htm (accessed 2 August 2004).
———. (2003) Public Pedagogy and the Politics of Resistance: Notes on a Critical Theory of Educational Struggle. *Educational Philosophy*, 31 (1): 3–16.
Hill, D. (1997) Brief Autobiography of a Bolshie Dismissed. *General Educator*, 44: 15–17.
———. (2004a) Books, Banks and Bullets: Controlling Our Minds—The Global Project of Imperialistic and Militaristic Neo-liberalism and Its Effect on Education Policy. *Policy Futures*, 2 (3). Available at http://www.triangle.co.uk/pfie.
———. (2004b) Educational Perversion and Global Neo-liberalism: A Marxist Critique. *Cultural Logic: An Electronic Journal of Marxist Theory and Practice*. Available at http://eserver.org/clogic/2004/2004.html (accessed 2 February 2005).
———. (2005) Critical Education for Economic and Social Justice: A Marxist Analysis and Manifesto. In M. Pruyn and L. Huerta-Charles (eds.) *Teaching Peter McLaren: Paths of Dissent*, 146–185 (New York: Peter Lang).
Houston, D. and L. Pulido. (2002) The Work of Performativity: Staging Social Justice at the University. *Society and Space*, 20 (4): 401–424.

Hubbard, A. (1996) The Activist Academic and Stigma of "Community Housework." Available at http://commorg.utoledo.edu/si/hubbard.htm (accessed 1 March 2004).

Hudis, P. (2003) *The Future of Dialectical Marxism: Toward a New Relation of Theory and Practice*. Paper presented at Rethinking Marxism Conference.

Hunter, L. (2004) Bourdieu and the Social Space of the PE Class: Reproduction of Doxa through Practice. *Sport, Education and Society*, 9 (2): 175–192.

Hurst-Mann, L. (1998) Publishing on the Terrain of the Bourgeois Culture Industry and Crisis in the Socialist Project. *AhoraNow*, 5: 12–15.

Kelly, R. (1997) The Abolition of Whiteness and Black Freedom Movement. Available at http://www.postfun.com/racetraitor/features/rkelley.html (accessed 3 June 2005).

Kelsh, D. (1998) Desire and Class: The Knowledge Industry in the Wake of Poststructuralism. Available at http://clogic.eserver.org/1-2/kelsh.html (accessed 9 January 2001).

Landsteicher, W. (2004–2005) Autonomous Self-organization and Anarchist Intervention. *Anarchy: A Journal of Armed Desire*, 22 (2): 51–62.

Lapavitsas, C. (2003) Money as Money and Money as Capital in a Capitalist Economy. In A. Saad-Filho (ed.) *Anti-capitalism: A Marxist Introduction*, 59–72 (London: Pluto Press).

Lefebvre, H. (1974/1984) *The Production of Space* (Cambridge, MA: Blackwell).

Lenin, V. I. (1967) *The State and Revolution* (Peking: Foreign Languages Press).

Mann, E. (1998) Workers of the World Unite: The Struggle against Imperialism Is the Key to Marxism's Reconstruction. *AhoraNow*, 5: 5, 1–7.

———. (1999) Class, Community and Empire: Toward an Anti-imperialist Strategy for Labor. In E. Meiksins, P. Meiksins, and M. Yates (eds.) *Rising from the Ashes? Labor in the Age of "Global" Capitalism*, 100–109 (New York: Monthly Review Press).

———. (2001) A Race Struggle, a Class Struggle, a Women's Struggle All at Once: Organizing on the Buses of L.A. Available at http://www.thestrategycenter.org/AhoraNow/body_socialistregister.html (accessed 19 April 2002).

———. (2002) *Dispatches from Durban: Firsthand Commentaries on the World Conference against Racism and Post-September 11 Movement Strategies* (Los Angeles, CA: Frontline Press).

Mann, E. with Planning Committee of the Bus Riders Union. (1996) *A New Vision for Urban Transportation: The Bus Riders Union Makes History at the Intersection of Mass Transit, Civil Rights, and the Environment* (Los Angeles, CA: Strategy Center Publications).

Martin, G. (2005) Afterword. In P. McLaren, (ed.) *Red Seminars: Radical Excursions into Educational Theory, Cultural Politics, and Pedagogy*, 547–550 (New Jersey: Hampton Press).

Marx, K. (1967) *Capital* (New York: International Publishers).

———. (1972/1978) The German Ideology. In R. Tucker (ed.) *The Marx-Engels Reader*, 2nd ed., 146–200 (New York: W. W. Norton & Company).

———. (1988) *Economic and Philosophic Manuscripts of 1844* (New York: Prometheus Books).

Marx, K. and F. Engels. (1967) *The Communist Manifesto* (Ringwood, Victoria: Penguin Books).

Maxey, L. (2004) Moving beyond from within: Reflective Activism and Critical Geographies. In D. Fuller and R. Kitchin (eds.) *Radical Theory/Critical*

Praxis: Making a Difference beyond the Academy, 159–171 (University of British Columbia in Kelowna, British Columbia, Canada: Praxis[e] Press). Available at http://www.praxis-epress.org (accessed 4 October 2004).

McLaren, P. (1997a) Critical Pedagogy. Available at http://www.teachingeducation.com/vol19-1/mclaren.htm (accessed 24 November 1997).

———. (1997b) *Revolutionary Multiculturalism: Pedagogies of Dissent for the New Millennium* (Boulder, CO: Westview Press).

———. (1998) Revolutionary Pedagogy in Post-revolutionary Times: Rethinking the Political Economy of Critical Education. *Educational Theory*, 48 (4): 431–462.

———. (1999) *Schooling as a Ritual Performance: Toward a Political Economy of Educational Symbols and Gestures*, 3rd ed. (Lanham, MD: Rowman & Littlefield).

———. (2000) *Che Guevara, Paulo Freire, and the Pedagogy of Revolution* (Boulder, Co: Rowman & Littlefield).

———. (2005) Critical Pedagogy in the Age of Neo-liberal Globalization. In P. McLaren (ed.) *Capitalists and Conquerors: A Critical Pedagogy against Empire*, 19–73 (Lanham, MD: Rowman & Littlefield).

McLaren, P. and G. Rikowski (2001) Pedagogy for Revolution against Education for Capital: An E-Dialogue on Education in Capitalism Today. Available at http://clogic.eserver.org/4-1/mclaren&rikowski.html (accessed 9 April 2002).

Meyer, D. (2002) Opportunities and Identities: Bridge Building in the Study of Social Movements. In D. Meyer, N. Whittier, and B. Robnett (eds.) *Social Movements: Identity, Culture, and the State*, 3–24 (Oxford: Oxford University Press).

Meyerson, G. (2004) The Good Professors of Szechuan. Available at http://www.workplace-gsc.com/features1/meyerson.html (accessed 2 February 2005).

Moss, P. (2004) A "Politics of Local Politics": Praxis in Places That Matter. In D. Fuller and R. Kitchin (eds.) *Radical Theory/Critical Praxis: Making a Difference beyond the Academy*, 103–115 (University of British Columbia in Kelowna, British Columbia, Canada: Praxis[e] Press). Available at http://www.praxis-epress.org (accessed 2 October 2004).

Pastor, M. Jr. (2001) Common Ground at Ground Zero? The New Economy and the New Organizing in Los Angeles. *Antipode*, 33 (2): 260–289.

Raduntz, H. (1998) Researching and Rebuilding a Marxian Education Theory: Back to the Drawing Board. Available at http://www.aare.edu.au/98pap/rad98259.htm (accessed 1 May 2005).

Read, J. (2003) *The Micro-politics of Capital: Marx and the Prehistory of the Present* (New York: State University of New York Press).

Rikowski, G. (1999) Education, Capital and the Transhuman. In D. Hill, P. McLaren, M. Cole, and G. Rikowski (eds.) *Marxism against Postmodernism in Educational Theory*, 111–143 (Lanham, MD: Lexington Books).

———. (2000) Messing with the Explosive Commodity: School Improvement, Educational Research, and Labour-Power in the Era of Global Capitalism. If We Aren't Pursuing Improvement, What Are We Doing? Paper presented at the British Educational Research Association Conference, Cardiff University, Wales, 7–9 September.

———. (2001a) *The Battle in Seattle: Its Significance for Education* (London: Tufnell Press).

——— (2001b) After the Manuscript Broke Off: Thoughts on Marx, Social Class and Education. Paper presented at the British Sociological Association, Education Study Group Meeting, King's College London, 23 June.

Rikowski, G. (2002) Methods for Researching the Social Production of Labour Power in Capitalism. Paper presented at Research Seminar, University College Northampton, 7 March.

Robnett, B. (2002) External Political Change, Collective Identities, and Participation in Social Movement Organizations. In D. Meyer, N. Whittier, and B. Robnett (eds.) *Social Movements: Identity, Culture, and the State*, 266–285 (Oxford: Oxford University Press).

Scatamburo-D'Annibale, V. and P. McLaren. (2003) The Strategic Centrality of Class in the Politics of "Race" and "Difference." *Cultural Studies<=>Critical Methodologies*, 3(2): 148–175.

Smith, L. (1999) *Decolonizing Methodologies: Research and Indigenous Peoples* (London; New York: Zed Books).

Whitter, N. (2002) Meaning and Structure in Social Movements. In D. Meyer, N. Whittier, and B. Robnett (eds.) *Social Movements: Identity, Culture, and the State*, 289–307 (Oxford: Oxford University Press).

Chapter 14

The Making of Humanity: The Pivotal Role of Dialectical Thinking in Humanization and the Concomitant Struggle for Self and Social Transformation

Paula Allman

Introduction

Humanity, both the word, or concept, and the reality, can be perceived statically and relegated to a transhistorical stasis; alternatively, it can be perceived as a dynamic process, an ongoing struggle to become more fully human. In his "Prison Notebooks," Antonio Gramsci encapsulated the dynamic perception when he reflected on the question of humanity as follows: "in putting the question 'what is [humanity]?' what we mean is: what can [humanity] become?…can [human beings] dominate [their] own destiny, can [they] 'make [themselves],' can [they] create [their] own life?" In response to these questions, Gramsci maintains that human beings are "a process" or more precisely "a process of [their]…active relationships." Moreover, he stresses that human beings should be conceived of as an *"ensemble* of these relations"—the *individuality* of each and every human being is an *"ensemble* of these relations," and this is also true for our collective being—our humanity (Gramsci, 1971: 351–352). If we are to be the active creators of our own individualities and of the present stage of the entirety of humanity's development, we must become conscious of these relations. It is only with a critical consciousness of the relations that have shaped and formed us that we acquire a basis from which transformation can begin—we can modify ourselves and our humanity only by modifying these relations (352). In other words, democratically and collectively we can become critically and creatively engaged in the project for humanity's progressive development—in the process of humanization.

This dynamic perception of humanity has crucial implications for the meaning of revolution, or what I prefer to call *self and socioeconomic transformation*, because far more than changing governments or socioeconomic structures is required. Authentic revolution must simultaneously involve the transformation of our selves. It is, therefore, a thoroughly educational endeavor—an endeavor entailing not only knowledge and critical understandings but also an active struggle to transform educational relations. Central to this struggle will be the development of dialectical thought. Gramsci's perception of humanity, a perception he shared with and probably developed from Karl Marx, is derived from dialectical thinking. Paulo Freire also based his educational philosophy on dialectical conceptualization, and drawing on Marx, Gramsci, and Freire, I have written extensively about critical education for self and social transformation elsewhere (Allman, 1999, 2001, 2007). Therefore, in this chapter I focus on why dialectical thought/conceptualization is crucial to every facet of the struggle for humanization and the concomitant struggle for self and socioeconomic transformation. It enables us to form a critical understanding of human beings and of humanity and also to recognize the myriad of ways that our humanity is thwarted within the social relations of capitalism. In addition, dialectical thought allows us to envision an alternative socioeconomic future and the type of dynamic process of humanization suggested by Gramsci. Most importantly, with the aid of dialectical thought/conceptualization, we can grasp the essence of capitalism and thus know what must be thoroughly transformed in order to initiate a global process through which we collectively can shape our future. This would be a process through which humanity critically and creatively engages in the "making" of humanity, in the struggle for humanization—a democratically organized collaborative project through which we become everything that the best of humanity has ever been and in the process continuously create new possibilities for humanity's positive development. Freire expressed one of the most inescapable reasons for this project when he said quite simply... "I can not be unless you are" (1974). The pivotal role of dialectic thought in conceptualizing humanity, in formulating an alternative vision of the future and in developing a critical understanding of capitalism constitutes the organizational framework for this chapter.

Dialectic Thought and Dialectically Conceptualizing Humanity

Before further discussion of dialectically conceptualizing this collaborative project of humanization, I describe, in general terms, the nature of dialectical concepts and dialectical conceptualization/thought. Dialectical concepts are fluid, relational concepts. In other words, these concepts derive from grasping movement, or a process of development—for better or worse—of entities/phenomena that are unified within a relation—a very specific type of relation. In common sense thinking as well as a great deal of academic or intellectual thought, to think in terms of relations means focussing on externally related entities/phenomena. Externally related entities/phenomena bring their own characteristic attributes to the relation, or interaction, and often produce a third entity/phenomenon composed of a combination of these attributes. Significantly, the attributes of the original entities/phenomena do not change, or at least are not perceived as changed. In contrast, dialectical concepts grasp internally related entities/phenomena and focus on the way in which the relation, the *internal relation*, influences the development, or the nature, of each entity's/phenomenon's attributes (Tolman, 1981). Here, for better or worse, change or movement is the rule. The overall nature of each entity/phenomenon is shaped by its existence within the internal relation; therefore neither can be transformed without totally changing or even destroying the relation. Focussing on external relations may be entirely appropriate, that is, sufficient to a comprehensive understanding, in certain circumstances, such as conceptualizing a chemical relation/reaction or understanding the terms agreed in a negotiation. However, critical understanding frequently demands conceptualizing entities/phenomena as internally related (Ollman, 1976). For example, conceptualizing the relation between labor and capital or the relation between teachers and students as external relations allows a certain level of understanding, but not the critical level necessary to engage in self and social transformation. Although I discuss in detail the exact implications of a dialectical conceptualization of these internal relations elsewhere (e.g., Allman, 2001), the aim, in this instance, is to examine how this type of conceptualization contributes to our critical understanding of humanity and the process of humanization.

A dialectical concept of human beings, or for that matter humanity, involves grasping the most significant internal relations that shape and determine people. As Gramsci noted, individuals are constituted within an ensemble of relations (1971: 351–352). Within these relations, we continuously shape and determine others and ourselves whilst simultaneously being shaped and determined by the others to, or with, whom we are internally related. I am not talking about a "here this minute—gone the next" type of interactive, external relation but rather one that internally binds us to one another such that no member of the relation would be what they are outside of that relation. Moreover, we could not have become that which we presently are outside of these relations nor can any of us change significantly, or fundamentally, so long as the relation exists, at least, exists in its present form. Conceptualizing our material reality, including our social relations, as a vast network of internal relations, historically specific internal relations, is the basis of dialectical thinking—that is, the dialectical mode of thinking developed and utilized to great critical effect by Karl Marx. Marx's critical/dialectical explanation of capitalism focuses on how the historically specific internal relations of capitalism lead to the dehumanization of the vast majority of human beings and the less than satisfactory humanity of the supposed "winners" in this socioeconomic system. Dialectical thinking also enables us to have a vision of how humanity could be developing in a much more progressive manner—a vision of human beings collaboratively engaged in the making of humanity, in the process of humanization.

An Alternative Vision Imperative to a Democratic Future

The creation of a socially and economically just and democratic world in which each and every human being, unimpeded by unmet basic needs, will be able to develop his or her full potential cannot be stage managed by the few for the many. It is not something that can be done to or for human beings but only by the creative and critical engagement of all human beings learning and working together. Of course, there will be a division of labor, but one that is democratically decided and also rotational thus, for example, enabling people to have a wide variety of work experiences and the opportunity to develop multiple skills and all of their potentials. In other words, if humanity is to achieve its full potential and in so doing continuously create new horizons of potential, humanity must be involved in its own "making," including the making of what it means to be fully human. To some readers—maybe even all of you—this may sound so utopian as to be absurd. I fully admit to and acknowledge the enormity of the tasks, many of which will be educational, that must be accomplished and the problems and hurdles that will need to be overcome—what Freire (1972) would call limit-situations and in this case a veritable glut of them. I also acknowledge the most obvious factor of all, the considerable time that it would take to accomplish such a radically democratic global socioeconomic transformation, moreover, one dependent upon the simultaneous transformation of our selves. Nevertheless, I persist with this vision, which I consider to be a realistic or feasible utopia, because the alternatives spell disaster for humanity and dash all hope for the possibility of humanization and an authentically democratic future for the citizens of the world. If we do not envision the possibility of all of humanity taking control of its destiny, then that destiny must be decided and engineered by someone else or some cadre of the supposedly more capable members of the human species. Most democrats would rail at the notion of an elite, elected or otherwise, charting the course for the future of humanity, but is this not what our democracies have delivered? Whether we are talking about powerful politicians or multinational corporations, the results tend to be the same. The assumption that most people are not up to governing themselves and their destinies is well ingrained—so much so that the only vision of democracy that most people entertain is one wherein we actually abrogate our powers and civic responsibilities to others yet somehow continue to think we are in control because we have the right to decide at a designated time whether or not to reelect them.

The idea that somebody else must do whatever it is for the vast majority, who are incapable of taking responsibility, resounds in everyday discourse.

Statements such as the following are indicative: "why doesn't the government do something about the health service, rising crime rates, or the homeless sleeping rough in our cities" and "if there were a god, surely there would never have been the mass genocide of World War II, nor the daily reality of millions of children starving, or recurring incidents of children killed and mutilated by serial murderers and perverts." These are just a few examples that reveal this type of thinking. Although the recipient of the blame varies, the statements bare a striking resemblance. They all stem from thinking that some being or group of beings, somewhere—someone rather than ourselves together with other human beings—should intervene and prevent these problems and terrible things from happening. I think that this thinking misses the point about what humanity is supposed to be striving to be. A god who intervened to thwart every evil deed in human affairs would be much the same as parents who never allow their children to grow up and to make often painful but necessary mistakes in the process. And if a government were able to cure all the ills of its society without the active participation of its citizens, it would be left with a nation of sheep to govern. A far easier way to govern and/or control, perhaps, but would we have any right or reason, in the first case, to respect the wisdom of such a god and in the second case, to call such a government a democracy? We only relinquish these responsibilities to either the government or god (in my opinion, mistaking the purpose and/or intention of both), when we have no vision of what humanity should be and with effort can become—when we have no hope for the whole of humanity becoming more fully human, human beings.

Capitalism and Consciousness

While dialectic thinking/conceptualization is pivotal to our perception of humanity and to formulating an alternative vision of humanity's future, understanding the true nature of capitalism, and thus what needs to be transformed, is impossible without it. Unfortunately, living within capitalism's historically specific social relations militates against the human project and our hope for humanity. Capitalism's functionaries need not do anything intentionally to make this happen (which is not to say that they do not). Humanity is denied by capitalism's inner-related network of social relations, or what we normally refer to as its structure, but this is far from obvious because capitalism's dialectical structure actually promotes an undialectical way of experiencing and understanding it that is impossible to escape unless we engage in critical/revolutionary praxis and the dialectical thinking that is essential to this form of praxis. Before elaborating on this point, a short explanatory detour is required. To explain how capitalism substantially influences our consciousness, even to the extent of penetrating our subjectivities and creating a subjective response such as hopelessness, I need to discuss Marx's dialectical/materialist theory of consciousness, or what I refer to as his revolutionary theory of praxis.

Marx began his scholarly work as a philosopher. In the 1840s there were two major philosophical paradigms that Marx was seeking to challenge. One of these paradigms, idealism, was expressed in Hegel's writings (Marx, 1977a); the other paradigm was the undialectical materialism of Feuerbach (Marx, 1977b). These seemingly opposed paradigms both result from the dichotomization of consciousness (the ideal) and the material world. With idealism, consciousness is prioritized and designated as the cause of the material world. Conversely, with undialectical materialism, consciousness is subordinate, simply registering that which exists in reality. In contrast to both idealism and mechanical, undialectical materialism, Marx developed a theory of consciousness that posited an internal relation between people's real, sensuous activity and their consciousness—a dialectical unity of, or internal relation between, the material world, including the social relations in which people are engaged, and human consciousness. Moreover, he stressed that our consciousness was most indicative of the social relations within which we produced our material world (Marx and Engels, 1976). This does not mean that other social relations and other aspects of our material world are not important but simply that our productive relations, or rather our engagement in them and in a world largely determined by them, are the most important factors that influence the overall character, or framework, of our consciousness that then reciprocally serves to maintain the existing relations and conditions of the material world. Since the productive relations of capitalism are historically specific to this mode of socioeconomic organization, so too is the type of consciousness that is internally related to these relations. In other words this is the type of consciousness that prevails so long as people engage in uncritical/reproductive praxis—the type of praxis, that is, unity of sensuous activity and consciousness, that simply reproduces the social relations, or dialectical structure, of capitalism. The positive alternative is implied here. We can become aware of how we are limited and determined by these relations and engage in a transformational struggle, that is, critical/revolutionary praxis, which enables us to critically understand our world and to transform it. As the word implies, this alternative form of praxis is not a matter of thought alone, but of transforming our thought within transformed social relations, even if initially this takes place only in a specific context and for a limited duration, that is, as an abbreviated experience of transformation (Allman, 1999, 2001, 2007).

The Dialectical Nature of Capitalism

Now that the detour has been taken, I can return to focussing directly on capitalism. In the three volumes of *Capital*, Marx exposed the dialectical nature of capitalism and demonstrated that it could be understood only by means of dialectical conceptualization. However, he also suggested that it was our actual experience of capitalism's dialectical nature that makes it extremely difficult to grasp the dialectical nature of our reality. As mentioned previously, that which we normally refer to as the structure of capitalism, or alternatively the capitalist system, is actually a vast network of internal relations, or dialectical contradictions. A dialectic contradiction, the type of contradiction that Marx found to be inherent in capitalist reality, is actually an internally related "unity of opposites" (Allman, 2001). Two of these relations, which are part of the essence of capitalism, are the premise or precondition for all of the others. Yet, due to what Marx called the "noisy sphere"—the multifarious, surface manifestations of these contradictions in our daily experience of capitalism—we find it difficult to grasp the essence of capital and capitalism (Marx, 1976: 279). We may be well aware of the opposites that comprise these relations, or dialectical contradictions, but because we actually *experience them separately in both time and space*, it is almost impossible, without dialectical thinking, to understand that they are internally related and also responsible for the complex reality of capitalism's various articulations in our daily experience. The first of these dialectical contradictions is the internal relation between labor and capital. Capital cannot exist unless it is internally related to productive labor, or labor that produces the basis of capitalist profit, namely surplus-value—value constituted by labor-time that is in excess of the labor-time it takes to produce the wage or salary of productive workers. And, labor, or the proletariat—the working class—would not exist as a separate and exploited class if it were not bound into an internal relation with capital. Moreover, as capitalism develops, eventually into a global and universal system (the relentless process of globalization that we are now experiencing), the vast majority of human beings will not be able to exist unless they enter into this exploitative relation with capital. This is because competition between capitalists results in the need to increase productivity, which usually involves a reduction in the labor-time needed for the production of a commodity. As productivity increases, the value and surplus-value contained in each commodity are also reduced. Therefore, to increase or simply maintain the amount of surplus-value produced in an economy, more and more areas of work, including professional work, must be reconstituted within the labor-capital relation so that capitalism can survive. Capitalists do not need to have any understanding of how capitalism actually works or of how their behavior influences the overall functioning of the system. Competition and acting logically to assure their own survival within this system also guarantees the preservation of the system.

Not only do we tend to experience labor and capital at different times and often in different spaces, thus making it difficult to grasp their unity and inseparability, we also have a fragmented or dichotomized experience of the opposites in the other major contradiction of capital's essence—namely the internal relation between production and circulation/exchange. The separation in time and space of production and circulation/exchange creates difficulties for much more than our ability to grasp their necessary unity; it is also in this separation that the possibility of crisis perpetually plagues the system (Marx, 1976). I am talking about an extremely complex system—one that is fraught with both logical and dialectical contradictions. For example, as mentioned above, competition between capitalists leads to an unceasing drive to increase productivity. While this pursuit enables human beings to increase their productive capacity and thus, hypothetically, eventually to be able to meet the needs of all human beings, the result *within capitalist relations* actually denies this promise. Since capitalist wealth is based on the exchange-value of commodities rather than their use-values, increased productivity far too often creates the overproduction of use-values, that is, useful objects and services, that are unavailable to many of those who need them but who cannot afford them at a price that would return a profit to the capitalists. These surplus commodities are destroyed or taken off the market so that their existence does not drive down prices, and we end up with the ludicrous and immoral situation of, for example, unavailable food surpluses in a world wherein millions upon millions of people are starving and dying of starvation. The only alternative, within capitalist relations, is to "dump" the surpluses for a cheap price somewhere in the underdeveloped world, which simply serves to drive local producers out of business thus increasing the country's dependence on the developed world. Although a comprehensive explanation of the processes involved can be given, it is too lengthy for this chapter; however, the full explanation is offered elsewhere (especially Allman, 2001, 2007). Having said that, I would like to offer another example, a prime example, of just how logically contradictory capitalism is.

I must preface this example with a more detailed explanation of surplus-value, which as mentioned previously is the basis of capitalist profit. According to Marx, and even some of the leading political economists who preceded him (e.g., Adam Smith and David Ricardo), labor is the source of capitalist wealth. Marx is more precise than his predecessors in explaining how this works. The famous "Law of Value" formulated by Marx states that the "magnitude of value" of any commodity is determined by the "labor-time socially necessary for its production" (Marx, 1976: 129). Marx is talking about the value that allows commodities to be equitably exchanged. If one tailor takes eight hours to produce a shirt and another takes one hour, the value that determines the price of a shirt will be an average of these two very different labor-times. At a higher level of capitalist development, that is, fully developed capitalist societies operating in the global market, the Law of Value, in particular "socially necessary labor-time," produces amazing anomalies and unbelievably contradictory results. But please judge this for yourselves, at least with reference to the following example. I must emphasize that it was only by means of dialectic thought that Marx was able to discover the complicated processes described in this example.

Despite claims to the contrary, communism, or more precisely a redistributive form of socioeconomic system, is actually alive and very well within the capitalist system. Of course, I say this in jest because this redistribution is not a conscious or planned process and because the redistribution takes place only amongst capitalists. This is how it works. In a fully developed capitalist economy, socially necessary labor-time determines the market-value of each commodity. Like socially necessary labor-time, therefore, this market-value is an average based on the various levels of productivity that exist in all the firms that produce a particular commodity. Thus the firm with the worse productivity serves to keep the market-value above what it would be if it were based on the firm with the best productivity. The latter firm—the one with the best productivity—can sell its commodities for less than the market-value. And until all the other firms catch up to this level of productivity, thus lowering the market-value, this firm will reap super profits. Moreover, this scenario would be true even if the most productive firm were fully automated and as a consequence contributed absolutely no surplus-value to the economy. In summary, due to the competition for capital investment as well as other factors, the capitalists whose firms produce the most surplus-value (i.e., the newly created value in a commodity that is in excess of the labor-time that constitutes the wages of productive labor) are those who fare the worse. Their surplus-value, in effect, is redistributed to the capitalists who produce the least surplus-value—all thanks to market competition. Competition, therefore, is absolutely essential to the proper functioning of capitalism (Marx, 1976, 1981).

Marxists and socialists who have not managed to read the third volume of *Capital* or who do not understand it, will not be any more aware of this redistribution of surplus-value than capitalists are; and as a consequence, they easily buy into notions such as The Law of Value no longer holding in mature capitalism or the equally ridiculous idea of the demise of the working class. The example I just gave explains how the capitalist firm with no labor manages to get more than its fair share of surplus-value. Furthermore, the same example, together with other points I have made, implies that class is not a thing or a socioeconomic category but instead a relation (Thompson, 1974), while also implying that rather than shrinking, the working class, that is, the class of people who produce surplus-value, the people Marx called productive labor, is an ever expanding class of people whose work has been or is being reconstituted within an exploitative internal relation with capital. This process of reconstitution that takes place within nations as well as internationally, is neither smooth nor uncontested. One final example of capitalism's insidious and contradictory nature (actually an extension of the previous example) will demonstrate how this process of reconstitution has been in the past and will continue to be a blight on the human condition.

The survival of capitalism depends upon a global market comprised of firms operating at different levels of productivity and thus nations at varying stages of development ranging from underdeveloped to developed. Firms in underdeveloped countries will tend to be labor intensive and able to compete with highly automated firms in developed countries only by keeping the wages they pay extremely low. Historically, the same sort of productivity differentials once pertained within nations, but for social and political reasons these disparities will have a relatively limited lifespan within nations. The need to maintain, or continuously recreate, differentials in productivity is one of the factors that pushes capital into the global market, creating the supposedly "natural" and "inevitable" process of globalization. Without differing levels of productivity, the redistribution of surplus-value could not take place. As a consequence, what, in a very limited sense, could be seen as the progressive side of capitalism, that is, the unplanned drive to increase the overall productivity of human beings, would cease functioning, and the system would stagnate. When looking at capitalism from the side of production, we must never forget that capitalism needs consumers. We must grasp production and the total process of circulation and exchange as an internally related unity of opposites, that is, a dialectical contradiction. Therefore, because capitalism will always need productive labor as well as people employed who can afford to consume, two processes have to exist simultaneously. In the underdeveloped world, more and more people who are, or who have been recently, working in precapitalist relations of production, for example, self-sufficient farmers, must be drawn into the labour-capital relation. And in developed countries, workers who previously produced no surplus-value—who were unproductive in Marx's terms—must also be drawn into the labour-capital relation, and very often this involves types of work considered to be professional, for example, teaching, legal services and various aspects of health care. Once they are drawn into this relation, they too can become the producers of surplus-value and, therefore, productive labor. Sometimes these areas of work are first deskilled or deintellectualized so that untrained, unqualified, or semiskilled workers can be used to weaken professional or skilled power and resistance (CCCS, 1981). In other cases, it is simply a matter of an area of work being taken over by a capitalist firm or of being converted into an institution whose survival, at least in part, depends upon making a profit out of the service or commodity it offers. The implication of what I am explaining is dire. At present, many well-intentioned politicians are crusading for an end to world poverty and thus a substantial reduction in the global division between the rich and the poor. Within capitalist relations, it cannot be done—temporary, mild palliatives, perhaps, but the gaps will never close and will only widen over time. However, the processes just described and the relation between them are far from obvious. They can only be comprehended by dialectically conceptualizing a reality that is, itself, dialectical in nature.

Conclusion

Further explanation is beyond the scope of this chapter; nevertheless, you have been offered a glimpse at just how complex and illogical the system really is. A dialectical understanding of how capitalism actually works reveals that rather than being the ingenious and invincible system its propagandists and pundits claim it to be, it is, in fact, a blatantly and pathetically ridiculous mode of socioeconomic organization. Here, however, I simply want to stress once more that it is capitalism's dialectical composition that actually makes capitalism difficult to understand—in fact, impossible to fully understand without Marx's mode of dialectical conceptualization. This is why I have argued, over and over again, that education that aims to prepare people to take collective charge of their destinies—to engage in revolutionary social transformation—must be in and of itself a form of critical/revolutionary praxis. Only by engaging in albeit an abbreviated experience of self and social transformation, which involves transformed relations between teachers and students, that is, ontological relations, as well as transformed relations to knowledge, that is, epistemological relations (Allman, 1999, 2001, 2007), can we enable ourselves and others to dialectically conceptualize the reality that we are seeking to transform. Herein lays the only possibility for initiating an authentic abolition of capitalism—of all the absurdity and abominations that plague humanity, thwarting our collective engagement in the making of humanity—our *potential for humanization*. The transformations required are far from easy and therefore are predicated on the constant vigilance and commitment of all members of the learning group—a group no longer comprised of some people who learn and another person who educates but of people who have reabsorbed the dual functions of learning and educating within themselves and who share the cointention of developing the ability to critically and dialectically understand the world they are hopeful of transforming in the future and the learning context they are seeking to transform in the present.

References

Allman, P. (1999) *Revolutionary Social Transformation: Democratic Hopes, Political Possibilities and Critical Education* (London: Bergin & Garvey).
———. (2001) *Critical Education against Global Capitalism: Karl Marx and Revolutionary Critical Education* (London: Bergin & Garvey).
———. (2007) *On Marx: An Introduction to the Revolutionary Intellect of Karl Marx* (Rotterdam: Sense Publishers).
CCCS. (1981) *Unpopular Education*. Education Group, Centre for Contemporary Cultural Studies, University of Birmingham (London: Hutchinson).
Freire, P. (1972) *Pedagogy of the Oppressed* (Harmondsworth: Penguin).
———. (1974) *Authority versus Authoritarianism*. Audiotape. "Thinking with Paulo Freire" (Sydney, Australia: Australian Council of Churches).
Gramsci, A. (1971) *Selections from the Prison Notebooks of Antonio Gramsci*, ed. and trans. Quinton Hoare and Geoffrey Nowell Smith (London: Lawrence and Wishart).
Marx, K. (1976) *Capital*, Vol. 1 (Harmondsworth: Penguin).
———. (1977a) Critique of Hegel's "Philosophy of Right." In D. McLellan (ed.) *Karl Marx Selected Writings*, 26–35 (Oxford: Oxford University Press).
———. (1977b) Theses on Feuerbach. In D. McLellan (ed.) *Karl Marx Selected Writings*, 114–123 (Oxford: Oxford University Press).
———. (1981) *Capital*, Vol. 3 (Harmondsworth: Penguin).
Marx, K. and F. Engels. (1976) *The German Ideology* (Moscow: Progress Publishers).

Ollman, B. (1976) *Alienation: Marx's Conception of Man in Capitalist Society*, 2nd ed. (Cambridge: Cambridge University Press).
Thompson, E. P. (1974) *The Making of the English Working Class* (Harmondsworth: Penguin).
Tolman, C. (1981) The Metaphysics of Relations in Klaus Reigel's Dialectics of Development. *Human Development*, 24: 33–51.

Contributors

Paula Allman, now retired, was formerly Senior Lecturer and Honorary Research Fellow in the School of Continuing Education at the University of Nottingham. While at Nottingham, she was the Course Co-Tutor for the Diploma in Adult Education and the M.Ed. in Continuing Education both of which were aimed at the preparation of critical educators. Her publications focus on the analysis and application of Freire's, Gramsci's, and Marx's ideas within radical/critical education, and she is the author of *Revolutionary Social Transformation: Democratic Hopes, Political Possibilities and Critical Education* (Begin & Garvey, 1999); *Critical Education against Global Capitalism: Karl Marx and Revolutionay Critical Education* (Bergin & Garvey, 2001) and *On Marx: An Introduction to the Revolutionary Intellect of Karl Marx* (Sense Publications, 2007).

Dr. Elizabeth Atkinson is a Reader in Social and Educational Inquiry in the School of Education and Lifelong Learning at the University of Sunderland. She has a long-standing research interest in the application of the radical uncertainties of postmodernism to the excessive certainties of contemporary educational research, policy, and practice. She has engaged in an extended debate with the Marxist educational research community over the relative merits of poststructuralist and Marxist approaches to social justice and social change: a debate that is illustrated in her contribution to this book.

Pat Brady lectured in social and economic studies at London Guildhall University before it merged with London Metropolitan University. He also was a visiting lecturer in the School of Education at the University of Northampton. While studying at the London School of Economics he was a professional footballer with Millwall and Queens Park Rangers. Patrick taught in schools and colleges in the UK and in both parts of a divided Germany. At present he is working on a PhD on the labor process in education at the University of Glasgow. He is deputy chair of the Council for Academic Freedom and Academic Standards (CAFAS).

Mike Cole is Research Professor in Education and Equality, and Head of Research at Bishop Grosseteste University College Lincoln, UK. He is the author of *Marxism and Educational Theory: Origins and Issues* (Routledge,

2007), and the editor of *Education, Equality and Human Rights: Issues of Gender, "race," Sexuality, and Social Class*, 2nd edition (Routledge, 2006), and *Professional Attributes and Practice for Student Teachers and Teachers: Meeting the QTS Standards*, 4th edition (Routledge, 2007).

Helen Colley is Senior Research Fellow in the Education and Social Research Institute at Manchester Metropolitan University. Her central research interests focus on social inequalities of gender and class in postcompulsory and lifelong learning, particularly the construction of emotion as labour power. Publications include her book *Mentoring for Social Inclusion* (2003, Routledge Falmer); *Informality and Formality in Learning*, coauthored with Phil Hodkinson and Janice Malcolm (2003, Learning and Skills Development Agency); and an edited collection, *Young People and Social Inclusion* (2007, Council of Europe).

Rachel Gorman is a lecturer at the Women and Gender Studies Institute of the University of Toronto, where she completed a postdoctoral fellowship funded by the Social Sciences and Humanities Research Council of Canada. Her current research uses arts-based research methods to explore disability, gender, and political activism in a transnational context. Rachel created the first two courses for a new Disability Studies stream of the Equity Studies program at the University of Toronto: the first is titled "Disability Culture and Social Change" and the second "Theoretical Approaches to Disability and Work." Rachel also teaches courses on violence against women, gender and disability, and feminist research methods, and supervises upper-year thesis students. As an activist, she has at various times worked on disability-rights, antipsychiatry, trade unionist, antiwar, and antiviolence campaigns. Rachel is a choreographer and a member of the Canadian Alliance of Dance Artists.

Anthony Green teaches at the Department of Educational Foundations and Policy Studies, Institute of Education, University of London. He has been associated with Marxist and critical discourses in education for many years and, along with Glenn Rikowski established Marxism and Education: Renewing Dialogues (MERD) a forum for Marxist analysis and critique. His particular concerns focus around ideology critique of education policy and practices as well as methodological features, problems, and issues of historical materialist, critical realist analysis and empirical research.

David Harvie sells his labor power to the University of Leicester, where he is a lecturer in finance and political economy. He has written on globalization and social movements, as well as value theory and the political economy of education. He is coeditor (with Keir Milburn, Ben Trott, and David Watts) of *Shut Them Down!: The G8, Gleneagles 2005 and the Movement of Movements* (Leeds: Dissent!; Brooklyn, NY: Autonomedia, 2005).

Dave Hill is Professor of Education Policy at Northampton University was a regional political/trade union leader, and cofounded the Hillcole Group.

He Chief Edits the *Journal for Critical Education Policy Studies* (www.jceps.com) and is Routledge Series Editor for *Studies in Education* and *Neoliberalism* and *Education and Marxism*.

Gregory Martin is a Lecturer in the School of Education and Professional Studies at Griffith University, Gold Coast Campus. His research interests include Marxist theory, critical pedagogy, and participatory activist research. He is currently a member of Australia's National Tertiary Education Union and the Gold Coast branch of Socialist Alliance.

Jane Mulderrig is a Research Fellow in the University of Edinburgh's Centre for Research in Education Inclusion and Diversity. She recently completed her doctorate on the governance of education, developing a method of combining corpus linguistics with critical discourse analysis. Her MSc (Edinburgh) and BA (York) degrees were also both in Linguistics. She has taught English language and linguistics in schools and universities in both Japan (Hokkaido) and the UK (Sussex, Lancaster). Her linguistic interdisciplinary research interests include language-based analyses of political rhetoric, forms of popular culture, and problem-focused analyses of social change. She is currently examining the education of children from the armed forces, and international policies and perspectives on equality and human rights.

Mark Olssen is Professor of Political Theory and Education Policy in the Department of Political, International and Policy Studies, University of Surrey. His most recent book is *Michel Foucault: Materialism and Education*, Paradigm Press, Boulder and London, published in May 2006. Also published in 2004, a book with John Codd and Anne-Marie O'Neill of Massey University in New Zealand titled *Education Policy: Globalisation, Citizenship, Democracy*, (Sage); an edited volume *Culture and Learning: Access and Opportunity in the Classroom* (IAP Press); with Michael Peters and Colin Lankshear, *Critical Theory and the Human Condition: Founders and Praxis*, and from Rowman and Littlefield, New York, *Futures of Critical Theory: Dreams of Difference*, also with Michael Peters and Colin Lankshear. He has published extensively in leading academic journals in Britain, America, and in Australasia.

Michael A. Peters is professor of education at the University of Illinois at Urbana-Champaign. He is the executive editor of Educational Philosophy and Theory and editor of two ejournals, *Policy Futures in Education* and *E-Learning*. His interests are in education, philosophy, and social policy and he has written over 200 artciles and chapters and some 30 books, including most recently *Why Foucault? New Directions in Educational Research* (Peter Lang, 2007), *Building Knowledge Cultures: Educational and Development in the Age of Knowledge Capitalism* (Rowman & Littlefield, 2006), both with Tina (A. C.) Besley, and *Knowledge Economy, Development and the Future of the University* (Sense, 2007).

Helen Raduntz is an Adjunct Research Fellow with the Centre of Research in Education, Equity and Work, University of South Australia, whose career has involved working in industry and secondary education; education union activism; and academic teaching and research. Subsequent to her doctoral research she has continued her interests in the development of a Marxian critique for contemporary capitalism; in the continuing impact of marketization on education and education for social change; and in mounting a critique on the subject of intellectual property and the work of information professionals. Among her publications is a chapter entitled "The Marketisation of Education within the Global Capitalist Economy" in the book *Globalising Public Education: Policies, Pedagogy and Politics*, edited by Michael W. Apple, Jane Kenway, and Michael Singh (Peter Lang, 2005).

Glenn Rikowski is a Senior Lecturer in Education Studies in the School of Education at the University of Northampton, UK. He is author of *The Battle in Seattle: Its Significance for Education* (Tufnell Press, 2001), and he coedited (with Dave Hill, Mike Cole, and Peter McLaren) *Marxism against Postmodernism in Educational Theory* (Lexington Books, 2002).

Geraldine Thorpe lectured in social and economic studies at London Guildhall University and, following merger, at London Metropolitan University. She has taught in the UK, New Zealand, and Australia. She is working on a PhD on the academic labor process at the Centre for the Study of Socialist Theory and Movements, University of Glasgow, and is a committee member of the Council for Academic Freedom and Academic Standards.

Paul Warmington is a Senior Lecturer in Education at the University of Birmingham. He has taught and researched in both further and higher education. His interests include Marxist educational theory, widening access, work-related learning, and race equality.

Name Index

Page numbers in bold refer to the main chapters of the contributors.

Adorno, T M   24, 29
Ali, T   95, 96
Althusser, L   74, 79, 96
Allen, M   57, 69
Allen, R   15, 29, 279
Allman, P   8, 73, 75, 96, 121, 125, 128, 132, 250, 255, 258, 261, **267–78**, 272–4, 277
Althusser, L   74, 79, 96
Apple, M   18, 29, 72, 74, 77–8, 93, 96
Ariès, P   59, 69
Atkinson, E   **119–33**, 279
Athene, as Mentor   202 ff., 212
AUT (Association of University Teachers)   88

Ball, S   137, 140
Bakhurst, D   20, 29
Bataille, G 169
Baxter, J   120, 122, 132
Beckett, F   96
Bellamy, J   80, 96
Benn, C   69, 95, 96
Bennett, J   170
Bernstein, B   18, 29
Beveridge, W   48
Bhaskar, R   38
Blair, A   50, 79, 92
Bourdieu, P   18, 23, 29, 79
Bowen, J   60, 69
Bowles, S   18, 29, 74, 79, 156
Brady, P   4, **35–56**, 279
Braverman, H   63, 69
Burchell, G   165, 166
Bush, G, W   77, 84
Butler, J   124, 132, 133

Caffentzis, G   233, 240–1, 242 [n 10]
Calinicos, A   129, 153
Callaghan, J   50, 83
Capita   51
Chitty, C   69
Cleaver, H   232, 233, 242 [n 10]
Clinton, W J   79
Cohen, S   49
Cole, M   5, 18, 29, 57, 69, 73, 75, 76, 94, 97, 100, **103–16**, **119–33**, 181, 279–280
Colley, H   7, **201–14**, 280
Cohen, N   224

Daniels, H   20, 29
Darwin, C   154
Davidson-Harden, A   89
De Angelis, M   232, 234 [n 2]
Dehal, I   76, 97
Derrida, J   120, 133
Dewey, J   20, 29, 169
Dinerstein, A   19, 29
Dumenil and Levy   72, 97

Ebert, T   78, 97
Edgley, R   54
Edwards   18, 29
Eichoff, W   55
Engestrom, et al   20, 29
Engels, F   28, 29, 35, 78
with Marx   272, 277
England, K   258

Fahey, J   23, 29
Fairclough, T   137, 138, 141, 142

NAME INDEX

Farahmandpur, R 57, 70, 77–9, 93, 97, 100
Federici, S 240–1
Feuerbach, L 272
Fitzclarence, L 69, 70
Flax, J 125, 133
Foley, G 14, 29, 184, 192
Foucault, M 156, 157–9, 168–72
Fortunati, L 237–8
Freire, P 8, 14, 15, 256, 259, 261 268, 270, 277, 256, 259, 261
Furedi, F 20, 29

Galloway, G 95
Ganss, G 58, 69
Geras, N 121, 133
Giddens, A 152
Gillborn, D 76, 88, 97
Gimenez, M 75, 97
Gindin, J 91
Gintis, H 18, 29, 74, 79, 156
Giroux, H 18, 30, 254, 260
Goff, T 20, 30
Gordon, C 160, 162
Gorman, R 6, **183–99**, 280
Gramsci, A 8, 12, 14, 15, 77, 79, 267–9, 277
Green, A **3–9, 11–31**, 280

Habermas, J 15, 30, 129
Hall, D 91, 98
Halsey, A H 52
Hardt and Negri 74, 98, 240
Harman, C 74, 80–1, 95
Healey, D 50, *55*
Hegel, G W F 153, 156–7, 272
Heilbroner, R 44
Hilferding, R 46, 48
Hobsbawm, E H, 46
Harris, K 231
Harvey, D 24, 30, 57, 62, 64, 65, 69, 72, 80, 98
Harvie, D 8, **231–47**, 244 [n 1], [n 2], [n 10], 280
Hatcher, R 69, 188, 189, 190, 192–3, 195–6
Hegel, G W F 207, 272
Heller, A 211

Hill, D 5, 23, 57, 69, 71, **71–102**, 75, 79, 85, 88, 91, 93, 98–9, 100, 120, 133, 261, 280
Hirtt, N 69, 88–9, 99
Hochschild, A 209
Horkheimer, M 24, 29
Hursh, D 84, 99

IFC (International Finance Corporation) 89, 99
International Marxist Group 95
International Socialist Group 95

Jessop, R 137, 138

Kellaway, L 42, 55
Kelsh, D 75, 93, 99
Kennedy, P 35–6, 38, 48–50, *54*
Kenway, J 23, 29, 69, 70
Keynes, J M 47–9
Korten, D 93, 99

Laclau, E 26, 30
Lather, P 120, 122, 133
Lebowitz, M 21, 30
Leontyev, A N 30
Lloyd George, D 50
Livingstone, D 20, 30, 194, 196
Luxemburg, R 13, 83, 99

Mandel, E 35–47, 54, 55, 57, 61, 66, 67, 68, 70
Marginson, S 57, 60, 70
Martinez and García 181
Martin, G 8, **249–66**, 281
Marx, G 28, 30
Marx, K 8, 11, 35–47, 54, 55, 57, 61, 62, 63, 66, 69, 70, 78, 82, 80–1, 99–100, 133, 153–8, 193, 197, 207, 268–9, 271–7
Marx's critique of capital, 61
  Capital, vol. 1 232, 239, 245 [n 11]
  Communist Manifesto 78, 82
  Economic and Philosophical Manuscripts 232, 234
  "Results of the Immediate Process of Production" 238–9
  Theories of Surplus Value 231
  with Engels, F 272, 277

NAME INDEX

Mayo, P 14, 30
McLaren, P 16, 18, 21, 30, 57, 69, 74, 75, 76, 88, 96, 97, 98, 99, 100, 253, 255–6, 260
Mead, G H 20
Mészáros, I 40, 58, 70
Milberg, W 44
Miliband, R 83
Mirza, H 76, 97
Mohun, S 231, 237 [n 7], 238
Moisio, O-P 23
Molnar, A 85, 88, 100
Murdoch, R 83
Moss, P 254
Mouffe 26, 30
Moore, R 18, 30
Mulderrig, J 5, **135–50**, 281
Musgrave, P W 59, 70

NATFHE (National Association of Teachers in Further and Higher Education) 88
Neary, M 19, 29
Ngugi Wa Thiong'o 77
Nicolaus, M 38, 41, 54
Noble, D F 68, 70
Novak, G 207 ff

Ofsted 234
Ollman, B 194, 196, 269, 278
Olssen, M 6, **151–79**, 281
Owen, R 45
Ozga, J 60, 70

Pamuk, O 22, 30
Passeron, J-P 20, 29
Perlman, F 43
Peters, M 6, **151–79**, 281
Petty, W 42
Postone, M 19, 30, 225
Powell, C 77

Raduntz, H 4, **57–70**, 254, 281
Reimers, E 89, 100
Ricardo, D 42
Rice, C 77
Robbins, L 48
Rubin, I I 43, 274

Rikowski, G **3–9**, 8, 13, 19, 30, 54, 56, 58, 69, 70, 73, 75, 76, 79, 84–5, 89, 93, 99, 100, 101, 193–5, 233, 235, 238 [n 8], 252–3, 257, 262, 282
Rikowski, R 58, 68, 70
Rowthorn, B 82, 101
Rubin, I I 57, 63, 70

Samuelson, P 42
Sarup, M 54
Sawchuck, P 20, 30
Schugurensky, D 89, 101
Schweitzer J B 38
Sharp, R 18, 30, 54
Sharp and Green 20, 30
Shattock, M 45–6
Shor, I 67, 70
Sieber, N 43
Sikka, P 50
Simon, E 45
Slaughter and Rhodes 14, 30
Small, R 18, 20, 31
Smith, A 42, 274
Smith, David G 184–5
Smith, Dorothy 186, 192–3
Smith, M 12, 31
Smyth, J 42
Spinoza, B 170
Suoranta, J 23
Sypnowich, C 20, 29

Taaffe, P 95, 101
Thatcher, M 82, 83
Thompson, E P 275, 278
Ticktin, H H 35–6, 40, 46–7, 49, 54
Thorpe, G 4, **35–56**, 282
Tolman, C 269, 278
Toynbee, P 81, 101

UCU (University and Colleges Union) 88, 102
UNESCO 92, 101
UNICEF 92, 101
Usher, R 18, 29

Van Leeuwen, T 142, 150
Vygotsky 20

Wallace, M 218, 220
Warmington, P 7, **215–28**, 282
Weber, M 94, 166
Wenders, W 209
Wersch, J 20, 31
West, M 44
Wheen, F 26, 31
Whitty, G 85, 102
Williams, R 155
Willis, P 22, 156

Willmott, H 37, 42
Willmott, R 13, 31
Wolff, J 26, 31
Wolf, M 50
Wright, E O 74, 102
WTO (World Trade Organization) 113[n 2], 183–8, 190

Youdell, D 88, 97

Subject Index

academic activism   250, 253, 254, 261–2
academic capitalism   28
academic entrepreneurialism   251
academic labor   8, 71
academies   85, 95, 96, 191–2, 196
activism/activist   8, 119, 127, 129
Activity Theory   20
AERA (American Educational Research Association)   26
alienation   193–4
A-Levels   see Qualifications
Anti-Academies Alliance   96
appearance and essence   194, 207 ff
Atkinson/Cole debate   5, 119–33

base-superstructure   6, 154–8
"best practice"   235, 237
Biopolitics   163
Bus Riders Union (BRU)   8, 249–51, 257–61

capitalism   36, 38, 40–4, 46–53, 55, 64, 65, 66, 67, 68, 69, 268–9, 271–7
   academic   28
   alienating influence   58
   caring   22
   circulation process   62
   dialectical thought and   272–4, 276–7
   dynamics of   57, 58, 59, 61, 63, 65, 68, 69, 78
   finance capital   46–53
   globalization of   273–6
   intellectual labor power demand   68
   Keynesian management policies   60
   labor process   61, 62, 63, 68
   labor productivity   59, 61, 62, 63, 69
   market as regulator   64
   market value   58, 61, 62, 63, 64, 65
   See also value, market
   mobilisation of technology   63
   private property, capitalist form   61, 64, 66
   See also mode of production
Capitalist Agenda in/for Education   114 [n 4], 181, 189–9, 193–6
Capitalist economic crises   57, 59, 61, 64, 65, 67, 68
   causes and effects   65, 66
   crisis management and strategies   57, 64, 65, 66, 67, 68, 69
   crisis resolution   57, 58, 64, 66, 69
   dual function of   65
   effects on education   57
   expansion as crisis resolution   64–5
   expansion imperative   57, 64, 67
   industrial   47–8, 51–3
   role of education   57–70
   social class barrier to crisis resolution   66
   sources   62
   See also patriarchy
Caring work   201, 202, 203, 204, 209, 210, 211, 212
   Also see work
Casualisation   87, 95
categories   35–43, 52–3, 54
Chicago School   162, 164, 165
citizenship   151
city academies   85
   See also city academies

civil society 35, 42
class 58, 65, 66, 258, 260, 267–9, 271–4, 275, 276
  analysis 17, 25
  consciousness 82, 83
  reconstitution of 273, 275
  reductioinism 75
  as relation 273, 275
  struggle 17, 20–6
  war from above 72
  working 19
  See also capitalist economy, emotion
Commercialization/commercialism 85
commensuration 231, 235–7
commodification 6, 7, 14, 36, 38, 57, 58, 78, 193, 194
  of labour power 194
  of learning and teaching 135, 137
  of qualifications 215, 216–17
  of skills 193
  See also decommodification; quasi-commodity
commodity 37–41, 43, 53, 54
  See also quasi-commodity; decommodification
commodity production 20, 37–41, 43, 53, 273–6
community organizing housework 260
competition 273–5
competitiveness 138, 139, 140
  discourse of 143–4
consciousness 193, 195, 196, 267, 271–2
  trajectory of 186, 192
Correspondence theory 156
crisis resolution 4
critical pedagogy 16, 18, 195, 255
Critical Discourse Analysis 5, 135, 136, 137, 144
Critical Race Theory 13, 22
Critical theories, *see* emotion
Corpus Linguistic Methods 136
cultural capital 216, 218, 225
  qualifications as cultural capital 216, 218, 225

decentred unity 79
decline 35, 37, 40–1, 43, 46–7, 49–51, 53

decommodification 49
dehumanization 269
deintellectualization/deskilling 276
democracy 79, 113–14 [n 3], 181, 183–5, 268, 270–1
democratic centralism 79
development 273–6
  socio-economic 273–4, 276
  varying stages of 274–6
dialectical concept/conceptualization/thought 8, 12, 29, 35–6, 38, 40–1, 52–3, 268–74, 276–7
dialectical nature of capitalist reality, 269, 272–3, 276–7
  See also humanity/humanization; internal relation
dialectical contradiction/unity of opposites 272–3, 276
  labor-capital relation 272–3, 276
  production-circulation and exchange 273, 276
  separation in time and space 272–3
  See also internal relation/ external relation; Marx; Marxism
Dialectical Materialism 207
dialogue 11–22
discourse
  analysis 135, 136, 137, 144
  of competitiveness 143–4
  of enabling 142–5
  of globalization 139, 144–5
  of meritocracy 143
  See also Critical Discourse Analysis
disabled artists/activists, case study of 191–5
division of labor 209
domestic violence 189

economic determination 79
education 57, 58, 59, 60, 61, 65, 67, 68, 69, 167
  business model for 68
  capitalist agenda in/for 114 [n 4], 181, 189–9, 193–6
  capitalist mode of 58
  critical education 195

commodification 57, 58
　economic crises effect on 67
　See also learning
Education Act, 1944 48
Education Reform Act, 1988 45, 52, 55
education systems 59, 60, 68
　corporatisation 59
　development of 58, 59, 61, 68
　infrastructure 59
　Prussian model 59
　segmentation 59
　state monopoly 59
educational change
　direction for social change or for capital 67–8, 69
　historical development 58
　nature of 61
employability 210
emotion 204, 209, 210
　critical theories of 211, 212
　and gender, class 211
　and means of production 211
　and mode of production 211–12
　and reproduction of labor power 211
emotional intelligence 209
emotional labor 209 ff
engagement mentoring, definition of 201–2
equivalence theory 74
essentialism 24
　See also appearance and essence
examinations/exams 7, 215–16
　boards 216, 219–20, 222, 225
　results 215–17, 219–21, 225–6
　pass rates 216, 218–24, 226 [n 2]
　and standards agenda 216, 219, 222, 225, 226
exploitation 194, 195
　as crisis management strategy 57
　integration into capitalist economy 58, 59, 67, 68
　management organization 84
　marketization 57, 58, 67
　as productive force 67
　for social change 57, 60

feminism
　Marxist 6–7, 188

Feminist theory 206–7
Foucault 6, 7, 156–72
　ethics 168–72
　and Nietzsche 157–8
　political economy 6, 156
　view of Marxism 157–9
Foundation Schools 191–2, 194–5
Frankfurt School Critical Theory 24
Freiberg School 162, 163

GATS (General Agreement on Trade in Services) 181, 183–6, 196
Gender 210–11
　See also emotion
General Certificate of Secondary Education (GCSE) 7
global democracy 6
globalization 138–9, 141, 210
　discourse of 139, 144–5
Governmentality 160–2
　See also; Foucault

Hatcher/Rikowski debate 5
Hegelian dialectic 153
　Monistic Conception of Society/ Totalism 156–7, 169
　See also Hegel, G W F Idealism
Hermeneutics of the Self 167
Higher education 216–25
　massification of 217–20, 223
historical materialism 6, 57
　See also; Marx, Marxism
historical specificity 269, 271–2
History 125, 126
Homer's Odyssey 7
human capital theory, critique of 194, 196
humanity/humanization 8–9, 268–71, 277
　dialectical concept of 267–71
　making of 267–8, 270–1, 277

ICT 24–5
idealism, 272
　See also Hegel, GWF
identity 126–31
ideology 18
　news media 220–22
　of practice 212

IFC (International Finance
    Corporation) 89, 99
informal learning 6
    appearance and essence of 194
    critique of gender, race and ability
        assumptions 192
    critique of individualized theories
        of 190–1, 192
    as expansion of exploitation 193,
        194
    as a regressive concept 193,
        196
    literature review of 183–4, 192
internal relation 269, 272–3,
    275–6
    See also dialectical contradiction/
        unity of opposites
IFC (International Finance
    Corporation) 89, 99
International Marxist Group 95
International Socialist Group 95

knowledge capitalism 6
knowledge economy 58, 59, 60, 61,
    67, 68, 135, 136, 140
    information and communication
        technology 67, 68
    role of discourse in
        constructing 137–8, 141
    See also capitalist economy
Kurdish
    diaspora, politics of 187–90
    political parties, sexism of 190
    resistance to state violence 189–90
    women, case study of 6, 185–91
Kurdistan, geopolitics of 186–7

Labor 36–43, 45, 47–54
    academic 8
    affective labor 239
    alienated labor 232–5
    division of 209
    domestic labor 237–8
    "efficiency" 235, 236, 241
    emotional 209
    exploitation 61, 66, 67
    immaterial 239
    intellectual 68
    productive 36–43, 45, 47–54,
        231–47

power 38, 40, 42, 43, 49, 52, 54,
    208, 253, 262
process 38, 40, 42, 49, 52, 212,
    253, 262
reproduction 211
labor-time 273–5
    socially necessary 274–5
    See also value
laborism 36, 48–9
Labour Party 48, 51, 85, 92, 95
    See also New Labour; Third Way
law 35–6, 39–50, 52–3
learning
    credentializing of 187, 193, 195, 196
    feminist theory of 188
    impact of trauma on 187, 188–9
    resistance as 188, 189, 190, 195
    through collective struggle 188,
        189, 190, 195
    survival as 188, 189
Learning Society 135, 136, 137,
    140–1, 147
Left/Right, the 151, 152
Left-Nietzscheans 152
Leninism 13
liberal sociology 22, 23

managerial governance 143–6
markets 151, 166
Marketisation 84
Marx, G 28
Marx, K 8, 11
Marx's critique of capital 61
    Capital, vol. 1 232, 239, 245 [n 11]
    Economic and Philosophical
        Manuscripts 232, 234
    and education 16–20, 156
    Historical Materialism 155–6,
        158
    incomplete 17, 21
    Political Economy 153–5
    "Results of the Immediate Process of
        Production" 238–9
    Theories of Surplus Value 231
    of his time 20
Marxism and education 156
Marxism 3–10
    as commodity 28–9
    embodied 3
    and hybridity 17

SUBJECT INDEX

Marxism and Education Renewing Dialogues seminar, *see* MERD
Marxist feminism 6–7, 207, 212
materialism 271–2
   delectable 211
   dialectical 207
measure 235–7, 239–40
mediation 35, 37, 40–1, 43–4, 48, 51–3
methodology
   feminist research group 188
   individual vs. relational approach 187–8
   and the Left 153
   standpoint method 186, 192–3, 195
   testimony as research method 185–6
metrics *see* measure
mentoring 7, 201–14
   See also engagement mentoring
MERD 3–10, 11–22
   origins of 13
meritocracy
   discourse of 143
Miners strike 1984–1985 72, 83
mode of production 35–7, 43, 47, 50
   Also see capitalism; value; emotion; Marx
Myths, uses of 202–4

NATFHE (National Association of Teachers in Further and Higher Education) 88
National Union of Teachers (NUT), *see* trade unions
Naturalism 154–5
New Labour 5, 85, 92
   Five Year Strategy 5, 49, 50, 85, 92, 181, 191–3, 196
neoliberalism 5, 152, 163, 164, 165, 181–4, 187–8, 190, 196
   education workers 5
   higher education 251–2
   individualism 9
   policy discourse 5
new public managerialism 82
News media 7, 215–26
   headlines 216–17, 220, 221–2, 225

   use of icon terms 220, 224
   and ideology 220–2
   panics 215, 218, 220–1, 222, 225–6
   templates 220–2, 225
No Child Left Behind 84

oppression 196
Ordoliberalen 162, 163, 164

Patriarchy, patriarchal capitalism 209–10
pedagogy 19–20
Pedagogy of critique 14–18, 24
performance related pay 86
personalization
   in governance 142–5
   as legitimation strategy 144–5
political economy 6, 35–6, 41–3, 45, 53–4
   See also Marx
postmodernism 5, 7, 119–33
poststructuralism/poststructural 119–33
poverty 181, 187, 196
power 158, 160
praxis 271–2, 277
   abbreviated experience of 272, 277
   collaborative nature of 268, 270, 277
   critical/revolutionary 271–2, 277
   with respect to epistemological and ontological relations 277
   uncritical/reproductive 272
private property, capitalist form of 61
   barrier to crisis resolution 66
   intellectual property privatisation 68
   See also capitalist economy
press 7
productive labor 273, 275–6
   See also labor; value; surplus-value
productivity 273–6
   differences in 275–6
profit virus 84
proletarianisation 82, 88, 95
   education workers 68
   teachers work 60
   See also capitalist economy
public-private partnerships 68

public-private partnerships—*continued*
  role in capital crisis
    resolution 57–70
Pupils 146–7

Qualifications 215–26
  A-Levels 216–26, 226n
  commodification of 215, 216–17
  credential inflation 216–26
  cultural capital 216, 218, 225
  degrees 216, 217, 220, 223, 225
  exchange/use value of 216, 222–5
  falling standards claims 216–26
  value of 215, 217, 219–26
Quality Assurance Agency for Higher Education (QAA) 233, 234, 236
quasi-commodity 37
  See also commodification; commodity

racism/antiracism 126, 128, 128, 130
radical pedagogy 8
reality/material world 123–6, 269, 272–3, 276–7
  See also dialectical nature of capitalist reality, 269, 272–3, 276–7
Reformism 16
Regulation Theory 166–7
relative autonomy 77
reproduction theory 78
Research Assessment Exercise (RAE) see research selectivity
research selectivity 233, 236, 244 [n 1]
rich-poor divide/gap 276

self 20, 167
  abbreviated experience of 272, 277
  educational nature of 27
  Hermeneutics 167
  revolution as 268
  See also critical/revolutionary praxis and socio-economic transformation 267–70, 272
social exclusion 196
social relations 196
social identity
  government 142–5
  pupils 146–7
  teachers 145–6

social/socio-economic
  relations 267–9, 271–4, 276
social movements 73
social structure as vast network of inner-related social relations 269, 271–3
socially necessary labor time 235–7
  See also measure
social wage 86
Social Watch 91
Socialist Party 95
Socialist Teachers' Alliance 96, 101
Socialist Workers Party 95
Soviet Union 14
specialist schools 85, 191, 194–5
  utilitarianism in 60
  value to economy 68
  See also; educational change; capital economic crises learning
Stalinism 49, 53
Standard Assessment Tests (Sats) 240
state apparatuses 78
State of the World's Children Report (UNICEF) 92
Structural Adjustment Programmes (SAPs) 240–1
Subject Benchmarks 233
substitutability 88

teachers 8, 145–6
  See also proletarianization
temporary contracts 87
Thatcherism 50
Third Way 151, 162
  See also New Labour
TINA (there is no alternative) 16, 21, 24
Tomlinson Report 219–20, 225, 226 [n3]
tough love 20
trade unions 24, 240
  appropriation of skills by management 193
  gender politics of 191
  research on learning practices in 191, 193
transition 35–7, 43, 47, 50–1, 53–4
Transparency Review 236
transitional period 4

university (corporate) 8
University and College Union (UCU)
    88, 102
  See also trade unions
universities tests 45
UNESCO 92, 101
UNICEF 92, 101

value 35–53, 54, 58–61, 273–6
  as capitalist form of wealth
    274
  labor value 63
  Law of Value 58, 61,
    274–5
  market-value 274–5
  materialisation 62
  as revenue 65
  source and dialectic 62
  surplus value 61, 62
  use value 208

use/exchange 7, 58, 61, 208, 274:
  qualifications as 216, 222–5
  See also education; Marx
surplus value 61, 62, 273–6: as
  basis of profit 273–4;
  redistribution of 275–6;
  See also Marx

Wapping printworkers' strike 83
war of position 79
winter of discontent 83
work
  imposition of 232, 234, 236 [n 5],
    240, 243
  Also see caring work; labor
World Bank 87
World Social Forum 74
WTO (World Trade
  Organization) 113 [n 2], 151,
  183–8, 190

GPSR Compliance

The European Union's (EU) General Product Safety Regulation (GPSR) is a set of rules that requires consumer products to be safe and our obligations to ensure this.

If you have any concerns about our products, you can contact us on

ProductSafety@springernature.com

In case Publisher is established outside the EU, the EU authorized representative is:

Springer Nature Customer Service Center GmbH
Europaplatz 3
69115 Heidelberg, Germany

www.ingramcontent.com/pod-product-compliance
Lightning Source LLC
LaVergne TN
LVHW011802060526
838200LV00053B/3658